THE HIDDEN LIFE
BASAL GANGLIA

THE HIDDEN LIFE OF THE BASAL GANGLIA

AT THE BASE OF BRAIN AND MIND

HAGAI BERGMAN

THE MIT PRESS   CAMBRIDGE, MASSACHUSETTS   LONDON, ENGLAND

The MIT Press would like to thank the anonymous peer reviewers who provided comments on drafts of this book. The generous work of academic experts is essential for establishing the authority and quality of our publications. We acknowledge with gratitude the contributions of these otherwise uncredited readers.

This book was set in Stone Serif by Westchester Publishing Services. Printed and bound in the United States of America.

Library of Congress Cataloging-in-Publication Data

Names: Bergman, Hagai, author.
Title: The hidden life of the basal ganglia : at the base of brain and mind
  / Hagai Bergman, The MIT Press.
Description: Cambridge, Massachusetts : The MIT Press, [2021] | Includes
  bibliographical references and index.
Identifiers: LCCN 2021000493 | ISBN 9780262543118 (paperback)
Subjects: LCSH: Basal ganglia—Physiology. | Basal ganglia—Diseases.
Classification: LCC QP383.3 .B47 2021 | DDC 616.8/3—dc23
LC record available at https://lccn.loc.gov/2021000493

10   9   8   7   6   5   4   3   2   1

CONTENTS

PREFACE

The title of this book, *The Hidden Life of the Basal Ganglia: At the Base of Brain and Mind*, has a double meaning. In the first versions, I used the word "basement." A basement is the hidden part of a house, most frequently holding the most important facilities for the proper functioning of the house. However, the basement also has a negative connotation as the place where you store broken and discarded items. Deciding to study the basal ganglia meant living in the basement of the neuroscience community. But, the more I lived in this basement, the more I became convinced that it serves as the base of the brain and mind. I hope that the reader of this book will eventually agree with me regarding this position of the basal ganglia. For me, it has been a great place to live in. I hope it will continue to be this way forever.

First, I would like to explain the narrative style of this book. The flow of the description of experimental data in this book will not be strictly linear, as in a text-book. For example, I present the basic electrophysiological findings in the first chapters of the two main parts of this book that deal with the basal ganglia in health and disease. Then, I report again on physiological studies in the subsequent chapters that focus on computational models of the basal ganglia and therapies for basal ganglia–related disorders. Education experts may call it the spiral method of learning. I tried to be less conservative when submitting this book for publication. I decided to do so because my (and other) more conservative publications are already in the public domain as peer-reviewed research papers and because the goal of a monograph is to share "out-of-the box" ideas. I will refer to books that I like and recommend as "additional reading" at the end of each chapter.

Like my research life, this book is biased toward studies of nonhuman primates and human basal ganglia. By contrast, most modern academic research is conducted on rodents. The exponential growth of molecular biology and optogenetic tools is a good reason for this shift. I will try to avoid the mistake of rewriting basal ganglia history and will give credit to the significant contribution of rodent studies to our understanding of the computational physiology of the basal ganglia.

This book focuses on system physiology and in vivo extracellular recording techniques. These techniques enable prolonged (tens of minutes to hours) recording in the behaving animal, while intracellular recording is better suited for in vitro studies and is very limited in time (less than ten minutes) in behaving animals. I am inclined toward methods that make possible the reliable recording of spiking activity rather than those that, in the best case, measure the activity of populations of neurons, such as functional magnetic resonance imaging (fMRI) and EEG. Similarly, I will only briefly discuss in vitro and molecular/cellular studies.

Finally, this book is biased toward the studies of our research group. I am human and thus more empathetic to and trusting of our own local work. Additionally, like most of us I am much better at telling my students they are not reading enough than at being a good example. So this bias reflects my relative ignorance of the accomplishments of other devoted students of the basal ganglia and brain. I sincerely apologize for this. Please share your feelings with me regarding any shortcomings in this book, and if I am lucky enough to realize a second edition, I will do my best to improve it accordingly.

The book consists of four parts:

1. Introduction and Background: This part or some of its chapter content can be skipped by the expert reader.
2. Computational Physiology of the Healthy Basal Ganglia: Here we will discuss four generations of computational models of the basal ganglia and relate them to the brain and behavior.
3. Computational Physiology of Basal Ganglia Disorders and Their Therapy: In these chapters we will examine the pathophysiology of Parkinson's disease and its pharmacological and surgical therapies. I will extend the discussion from neurology and movement disorders to psychiatric diseases.
4. From the Basal Ganglia and Brain to the Mind: I will use this part to (naively) consider the implications of our findings so far and their impact on the hard questions of free will and human responsibility.

Frida Kahlo (1907–1954) wrote, "I used to think I was the strangest person in the world but then I thought there are so many people in the world, there must be someone just like me who feels bizarre and flawed in the same ways I do. I would imagine her, and imagine that she must be out there thinking of me, too. Well, I hope that if you are out there and read this and know that, yes, it's true I'm here, and I'm just as strange as you." I hope you will enjoy this book as I enjoyed writing it.

ACKNOWLEDGMENTS

So many thanks should be given.

Mentors: I did my graduate (MSc and DSc) studies with Rami Rahamimoff Z"L and Yoram Palti. These studies were devoted to the peripheral nervous system—the neuromuscular junction and the myelinated nerve. Both Rami and Yoram were very supportive of my early exploratory pathways at the expense of high-level publications product. I hope that I have learned a little from their generous teaching and mentoring life.

During my medical internship at the Rambam Medical Center, Silvia Honigman Z"L and Judith Aharon generously introduced me to the fascinating world of neurology and good medicine.

I completed my first postdoctoral fellowship with Moshe Abeles. Moshe taught me the basics of practical multiple-electrode recording and neural network analysis. Life was never easy with Moshe, but without his mentorship, I doubt if this book would exist. My second postdoctoral fellowship was with Mahlon DeLong. Mahlon introduced me to the (single-electrode) physiology of the basal ganglia and the MPTP nonhuman primate model of Parkinson's disease. Mahlon has been a dear friend from the beginning. I hope he will like the "crazy" Middle Eastern unconventional treatment of the basal ganglia given below.

In the thirty years following these postdoctoral fellowships, I have been using multiple-electrode recording techniques in the basal ganglia. This book represents the outcome of these studies.

Coworkers: Thomas Wichmann. We met in 1988 as two young postdoc fellows in the lab of Mahlon Delong at Johns Hopkins hospital. I have learned from Thomas the meaning of solid science and of humble and true friendship. Thomas, you are my best friend forever. Ikuma Hamada was also at Johns Hopkins at that time. Ikuma taught me the art of extracellular recording in the basal ganglia. Whenever I see nice spikes of the globus pallidus, I recall Ikuma telling me, "Maybe, it will be a good one."

Back in Israel, I was lucky to collaborate with Eilon Vaadia (who taught me Assembler programming in 1976 and how to train, record, and analyze monkey data), Ya'acov Ritov (my RADAR engineer), Tali Tishbi (who taught me the deep details of information theory), Zvi Israel (who shared his DBS and clinical insights with me), Renana Eitan (who introduced me to the neglected world of psychiatry), Reem and Imad Younis (of Alpha Omega, who showed me the meaning of good industry and proved that Israelis and Palestinians can collaborate and make a better world), Suzanne Haber (who taught me anatomy and connectivity in the basal ganglia), and Uri Werner-Reiss (who has taken care of all the details of lab life over the last few years).

My life was enriched by meeting, discussing, and learning from many colleagues in the basal ganglia, Parkinson's, and neuroscience fields. I will mention only a few of them: Jérôme Yelnik, Paul Krack, Marwan Hariz, Jose Obeso, Paul Bolam, Peter Brown, Tim Denison, Pete Magill, Andy Sharott, Mingsha Zhang, Harith Akram, Carine Karachi, Stephan Charbardes, Andres Lozano, Yasin Temel, Chiung Chu (Zoe) Chen, Jerry Vitek, Abdelhamid (Hamid) Benazzouz, Jonathan Dostrovsky, Christian Moll, Andrea Kuhn, Jens Volkmann, Günther Deuschl, Boran Urfali, Gili Silberberg, Andy Horn, Ziv Williams, Idan Segev, Rony Paz, Opher Donchin, Yifat Prut, Sharon Hasin, Nir Giladi, Iddo Straus, and Idit Tamir.

Students: This book is the outcome of a wonderful life with so many students who have taught and inspired me (figure A.1). I will mention them here according to their time in the lab and in the text of this book alongside their publications: Hamutal

A.1   A typical scene in the lab recording setup.

Slovin, Asaph Nini, Ariela Feingold, Aeyal Raz, Moshe Rav-Acha, Gali Havetzelt-Heimer, Izhar Bar-Gad, Josh Goldberg, Genela Morris, Shlomo Elias, David Arkadir, Alon Nevet, Michal Rivlin, Thomas Boroud, Michael Levy, Mati Joshua, Adam Zaidel, Anna Castrioto, Yoram Ben-Shaul, Shay Moshel, Ruby Shamir, Naama Parush, Avital Adler, Boris Rosin, Maya Slovik, Imelda Palsey, Eitan Schechtman, John Winestone, Odeya Marmor, Dan Valsky, Inna Vainer Pinkas, Shiran Katabi, Marc Deffains, Alex Kaplan, Aviv Mizrahi-Kliger, Omer Dauber Z"L, Nir Asch, Hodaya Abadi, Oren Peles, Levi Sherman, Pnina Rapel, Lily Iskhakova, Noa Rahamim, Daniel Sand, Omer Linovski, Shai Heiman Grosberg, Orilia Ben-Yishay Nizri, Halen Baker, Shimon Firman, Xiaowei Liu, and Jing Guang.

There are many qualities that are critical for a graduate student in a physiology lab. Probably the most important ones are resilience and endurance (Charney and Southwick 2018). A good example of a test of endurance is the 1915 South Pole expedition of Ernest Shackleton and colleagues (Lansing 2000). Figure A.2 depicts an artist's view of Shackleton's famous advertisement for team members. Unfortunately, it is too good to be true, as the original source has never been located (http://www.antarctic-circle.org/advert.htm). That's how I would like to present my way of thanking all of you students at the Hebrew University Basal Ganglia (HuBG) lab: "Men and women, you have joined a hazardous journey, ignoring the low wages, cramped conditions and long hours of hard work. Safe return to the academy was doubtful, and you have no real hope of honor and recognition in the event of success." This book is an echo of your striving and a way of thanking you.

Monkeys, veterinarians, and the Israel Primate Rehabilitation Park: This book is based on monkey experiments. The decision to do experiments on monkeys was and is a hard decision. It is driven by my belief that solutions to human suffering due to brain disorders can be better achieved by experiments rather than by theoretical thinking or models. Moreover, these experiments should be done on animals that are close to us. Needless to say, these experiments should be done under the 3Rs of ethical animal research (Replacement, Reduction, and Refinement, or replace

MEN WANTED
FOR HAZARDOUS JOURNEY. LOW WAGES, BITTER COLD, LONG HOURS OF COMPLETE DARKNESS. SAFE RETURN DOUBTFUL. HONOUR AND RECOGNITION IN EVENT OF SUCCESS.
ERNEST SHACKLETON. 4 BURLINGTON ST.

A.2 The mythical advertisement of Ernest Shackleton.
Source: Adapted from Paul Ward, CoolAntarctica.com.

species, reduce the number of animals, and reduce suffering). Making the related ethical decisions does not make daily life easy. I was very lucky when twenty years ago I met Dr. Tamar Ferdman of the Israel Primate Sanctuary (https://www.ipsf.org.il /en/). Since then, we haven't been "sacrificing" our monkeys. Instead, at the end of the electrophysiological experiments (and for the last few years even after the MPTP experiments), we have been removing their recording chambers and caps, waiting for their recovery, and sending them to the sanctuary park for rehabilitation. I am really thankful to Tamar, the veterinarians, and all the workers at our animal facility for making the lives of our monkeys better. I hope that in the future, we will find cures for human disease and suffering without the need for monkey experiments. Until then, good and brave scientists will keep searching for better treatments for human suffering by studying animal models. Thank you all for enabling us to do it in the most humanitarian way.

Patients: Since 2002 and the start of the deep brain stimulation (DBS) program by Zvi Israel at the Hadassah medical center, I have been participating in DBS procedures and performing the physiological navigation to the target. I am very thankful to all our patients. They have always been very tolerant of my research into their basal ganglia and brain. They gave me more than I gave them. If any are reading this book, they should know how grateful I am to them.

Supporting agencies and foundations: The Hebrew University of Jerusalem has been my second family. The Hebrew University, the Department of Physiology (now Neurobiology) of the Faculty of Medicine, and the Interdisciplinary Center for Neural Computation (ICNC, now the Edmond and Lily Safra Center for Brain Research; ELSC) have provided me a warm nest, providing lab space, administration and technology support, and, most important, a social net.

My postdoctoral fellowships and first three years in a tenure-track position at the Hebrew University were sponsored by the Levi-Eshkol, Chaim Weismann, and Charles Smith foundations. My research lab was generously supported by grants from the Israel Science Foundation (ISF), the United States-Israel Binational Science Foundation (BSF), the German-Israeli Foundation for Scientific Research and Development (GIF), the Israel Institute for Psychobiology, the European Research Council (ERC), the Netherlands and Canadian Friends of the Hebrew University, the Israel Authority for Renovations, the Rosetrees Foundation, the Adelis Foundation, the Vorst family foundation and many anonymous donations. "If you have two pennies, spend one on bread and the other on a flower. The bread will sustain life. The flower will give you a reason to live." Thank you all for giving us the bread; this book is the flower we give back.

Industry: I have been a friend of Reem and Imad Younis for many years and have worked with Alpha-Omega, Nazareth, Israel, since its start. Recently, I have been appointed a consultant to Alpha-Omega. Over the years I have received travel honoraria to scientific and clinical meetings from Alpha-Omega, Medtronic, and Boston Scientific. I believe that my thinking and writing here are not influenced by these highly valued connections with these companies.

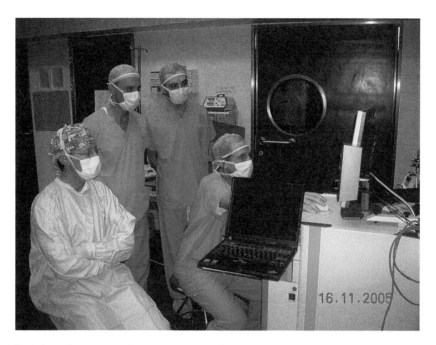

A.3 Physiological navigation during a DBS procedure.

Editors and referees: I thank Robert (Bob) Prior and Anne-Marie Bono of the MIT Press, Melody Negron of Westchester Publishing Services, and the anonymous referees of the first draft of this book for their support and many efforts on behalf of this book.

Family: First, my parents, David and Ahuva Bergman Z"L. I have accumulated more years of formal education than both of them together, and still I am the arrow that their bow released to the sky. Without their continuous, solid, and loving support, I could never have done it. My father, David Z"L, is a Holocaust survivor who has never spoken of his experience. I never got to meet his parents and five brothers and sisters who were killed in the Holocaust. I learned the names of my grandparents Dereizl and Yom-Tov Lipman Bergman only very late in my life. Indeed, silence speaks louder than a scream. ABBA and IMMA, may these words, written on the strongest material in the world—a book—be my late message of love to you.

The first edit of this book was done by Eugene (Gene) Bergman. Gene is a cousin of my father and, like him, a survivor of the Holocaust (see his book *Survival Artist: A Memoir of the Holocaust*, 2009, and DVD *Memories of the Warsaw Ghetto*, 2016). Post-Holocaust Jewish families have created bridges over remote family ties and distances. So, when I, Revital, Lottem, and Marva arrived for my postdoctoral fellowship in the US, we were "adopted" by Gene, Claire Z"L, and Sabrina. Thank you, Gene, for being my big cousin.

Revi; Lottem, Mori, Gili, and Razi; Marva, Yannai, Carmeli, Yuvali, and Aloni; Re'em; Tzufit, Tsuf, and Tevel—you are the better part and the beacons of my life. I love you.

TODA and SHUKRAN (thanks in Hebrew and Arabic) to all of you.

I

INTRODUCTION AND BACKGROUND

1

THE BASAL GANGLIA: WHY AND WHERE?

The basal ganglia were probably first named as a set, or a network, of subcortical structures by Sir David Ferrier in his 1876 book, *The Functions of the Brain*. The epithet "basal" correctly reflects the location of this set of neuronal masses at the base of the forebrain. However, the term "ganglia" is misleading. It is used today for clusters of neurons located in the peripheral nervous system. Thus, the basal ganglia would be more accurately called the "basal nuclei." However, it serves no use to cry over spilled milk, so we will use the term "basal ganglia" in this book.

WHY SHOULD WE STUDY THE BASAL GANGLIA?

The ancient Greek aphorism "know thyself" is the first of three maxims inscribed in the forecourt of the Temple of Apollo at Delphi, according to the Greek writer Pausanias. To know ourselves is to know our brain. No big differences exist between our genes, liver, and heart and those of a mouse or an elephant. But our brain and mind are very different. Thus, the biggest mystery of humanity is the brain and how it generates the mind and self-awareness.

The human brain is a huge structure, with an average of eighty-six billion ($\sim100 * 10^9 = 10^{11}$) densely connected neurons. Why do I suggest you study the basal ganglia rather than other brain structures such as the cortex, cerebellum, hippocampus, or hypothalamus? George Herbert Leigh Mallory (1886–1924) replied to the question "Why did you want to climb Mount Everest?" with the retort "Because it's there." The basal ganglia are also there, in the middle of the brain. For a real understanding of the brain, we should understand all its parts. I will try to explain below why I believe that the basal ganglia network is one of the best starting points in the long journey to comprehending the brain.

Evolution, rather than careful engineering design, has shaped the human brain. Thus, the same function is performed by more than one brain system or network. Brain systems can compensate each other for their failures, and the borders between them are fuzzy. In this book I espouse the view that the main goal of the basal ganglia is to optimize and execute our automatic behaviors. Behaviors fall along a big spectrum. The most automatic behaviors are innate and may be controlled by (monosynaptic) reflexes. I claim that the basal ganglia are involved in controlling the behavior that can lead to the achievement of innate reward (food, water, and sex, for example) and to the minimization of pain and life-threatening events. Most of our behavior is automatic, and we are aware of our behavior only infrequently. Even then, such awareness is often achieved only after we have already completed our actions. Thus, the basal ganglia are a good starting point for the long journey toward understanding the brain and behavior.

The brain can be the cause of extreme human suffering. Brain disorders are often incredibly distressing to patients and their close family and friends. Because the body and the brain are connected, most body dysfunctions (e.g., in the heart and the liver) affect the brain. But at least in the early stages of a disease, one can fight it. Brain disorders affect and modify the central features of what we define as ourselves. Many neurological and psychiatric disorders are due to basal ganglia dysfunctions or can be treated by pharmacological or surgical manipulation of the basal ganglia. If you are looking for a field in which your research may lead to a reduction of human suffering, the basal ganglia provide a logical place to begin.

In the next paragraphs, I will describe the major brain nuclei that constitute the basal ganglia and related structures. In the following chapters of this part, I will describe the morphology of their cellular elements, their synaptic connectivity, and their physiological function in health and disease.

GROSS ANATOMY OF THE BASAL GANGLIA

The three-dimensional anatomy of the brain is usually depicted over three orthogonal planes: coronal (dorsoventral, or up-down), sagittal (anterior-posterior, or front-back), and axial (lateral-medial-lateral, or left-right). Figure 1.1 shows a coronal section of the human brain and depicts the main structures of the basal ganglia. These include the following:

- Striatum, divided into:

   □ Caudate nucleus (*dark green*). The caudate has a complex, C-shaped, three-dimensional structure and can be divided into a "head" (figure 1.1) that tapers to a "body" and a "tail" (cauda). In the nonhuman primate, the volume of the tail of the caudate is only 2 percent of the total volume of the striatum (Percheron et al. 1984), so I will use the term "caudate" to refer to the caudate head.

- ◻ Putamen (*light green*).
- ◻ Nucleus accumbens (NAc, the main structure of the ventral striatum [VS]. Here we will not discriminate between the NAc and the VS). The NAc is not shown in figure 1.1.
- ◻ The caudate and putamen are often combined into one unit and termed the *dorsal striatum.*

- Globus pallidus, divided into:

  - ◻ External segment of the globus pallidus (GPe, *blue*).
  - ◻ Internal segment of the globus pallidus (GPi, *red*).
  - ◻ Ventral pallidum (VP, not shown in figure 1.1 because it is located in a more anterior coronal section).

- Substantia nigra, divided into:

  - ◻ Substantia nigra pars reticulata (SNr, *orange*).
  - ◻ Substantia nigra pars compacta (SNc, *light blue*).
  - ◻ The ventral tegmental area (VTA, not shown in figure 1.1) is located medial to the SNc. We will refer to dopamine neurons in the SNc and VTA as midbrain dopaminergic neurons (the SNc and the VTA are located in the midbrain, the upper part of the brain stem).

- Subthalamic nucleus (STN, *pink*).

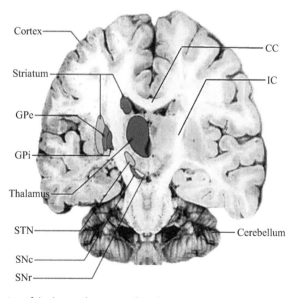

1.1 A coronal section of the human brain revealing the main structures of the basal ganglia. CC, corpus callosum; GPe, GPi, external and internal segments of the globus pallidus; IC, internal capsule; SNc, SNr, substantia nigra compacta and reticulata; STN, subthalamic nucleus.
Source: Adapted from Obeso et al. 2014.

The putamen and the globus pallidus arise at the junction of the diencephalon and telencephalon and do not exhibit a sagittal curved shape as does the caudate nucleus. Together they constitute the lenticular nucleus. The shape of the lenticular nucleus is triangular when seen on coronal sections (figure 1.1) but far more elongated, like a banana, when seen on axial sections (figure 1.2).

Two additional major structures, the cortex and the thalamus (*dark blue*), are shown in figure 1.1, and since they are strongly connected with the functional network of the basal ganglia, we should not neglect them and will further discuss them below. Most of the borders of the basal ganglia and related neuronal structures are delineated by white-matter axonal tracts. Figure 1.1 depicts the corpus callosum (CC) that connects the two hemispheres and the internal capsule (IC) that connects the cortex and subcortical structures such as the thalamus, basal ganglia, brain stem, and spinal cord.

The gross anatomy of the basal ganglia has been known for many years. See figure 1.2A for the description of the basal ganglia in an axial section by Andreas Vesalius (1514–1564), the founder of modern anatomy.[1] The figure shows the right (*A*) and left (*B*) hemispheres. The putamen is marked *C, D*, and the outline of the globus pallidus is given. The anterior and posterior limbs of the internal capsule are denoted as *E*. Medial to the internal capsule, one can see the caudate (more anterior) and the thalamus (marked as *D*). Figure 1.2B shows the same section as revealed by today's

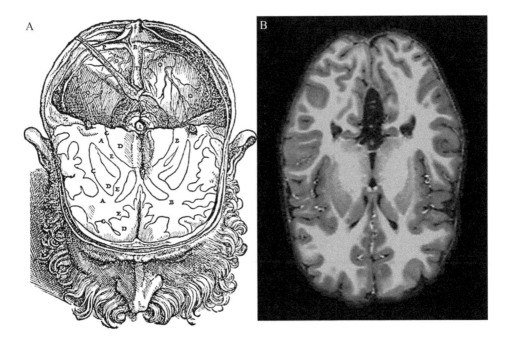

1.2 Old and new views of the basal ganglia. *A*, Vesalius's basal ganglia. *B*, an axial 7T T1 slice of normal human brain.
Sources: A. Adapted from Parent 2012. B. Courtesy of Stijn Michielse and Yasin Temel, Maastricht University.

most advanced seven-tesla MRI techniques. The MRI T1 axial plane is shown with the anterior portion of the brain pointing downward to comply with the older version. One hundred hours of scan time in seven-tesla MRI of the ex vivo human brain allows even better 100 μm resolution (Edlow et al. 2019). The gross anatomy of the human brain has probably not significantly changed over the last five hundred years; nor have descriptions of it. I hope that readers who make their way deeper into this book would agree that our understanding of the computational physiology of the basal ganglia has improved since then.

BASAL GANGLIA VOLUMES

The volumes of the different structures of the basal ganglia can be estimated from serial histological sections and from MRI imaging. Table 1.1 gives the volumes of the main structures of the basal ganglia for the rat, macaque monkey, and human, as reported by several studies. We will use these numbers in later chapters when we discuss the density and number of neurons in each structure and the reduction in volume and number of neurons along the cortex-striatum-globus pallidus axis.

The total volume of the human basal ganglia is consistent among the different studies and ranges between 15,000 to 25,000 mm³. In comparison, the volumes of the human left hippocampus, thalamus, cerebellum, and cortex are 5,000, 9,000, 95,000, and 400,000 mm³, respectively (Brain Development Cooperative Group 2012; Manera et al. 2020).

"TELL ME WHO ARE YOUR FRIENDS?": THE CORTEX AND THALAMUS

An old proverb states, "Tell me who are your friends, and I'll tell you who you are." This is probably also correct for neighbors. As shown in figure 1.1, the cerebral cortex and the thalamus are close neighbors of the basal ganglia, and these three brain networks are strongly interconnected. Here, I will provide a very short introduction to these structures.[2]

The cerebral cortex (hereafter, the cortex) is the outer layer of neural tissue of the telencephalon (the uppermost part of the central nervous system). Most of the cerebral cortex is part of the six-layer neocortex. The cortex can be divided into many subareas (e.g., 52 areas according to Korbinian Brodmann in 1909 or 107 areas according to Constantin von Economo in 1925; figure 1.3).

All cortical areas bilaterally project to the major input nucleus of the basal ganglia—the striatum. The cortical projections to the striatum are topographically arranged, but with overlaps and fuzzy borders. The STN receives inputs mainly from the ipsilateral motor and the somatosensory cortices. Finally, the output nuclei of the basal ganglia (GPi and SNr) project to neurons in the thalamus that are connected with the frontal cortex.

Table 1.1 Histological and MRI estimation of volumes of the main nuclei of the basal ganglia in the rat, macaque monkey, and human. *A*, histological, *B*, MRI.

	Rat		Monkey		Human		
A.	Oorschot 1z996	Hardman et al. 2002	Harman and Carpenter 1950	Hardman et al. 2002	Harman and Carpenter 1950	Hardman et al. 2002	Yelnik 2002
CD	10		348		5,270		4,316
PUT	10		474		6,690		5,625
STN	0.1	0.4		17		120	158
GPe	2. 7	5.2	96.4	126	1,970	1,160	808
GPi	0.2	1.1	61.6	64	982	477	478
SNr	1.0	2.2		75		380	412

	Monkey	Human			
B.	Frey et al. 2011	de Jong et al. 2012	Manera et al. 2020	Ewert et al. 2018	Pauli et al. 2018
CD	432	8,900	4,280		6,361
PUT	607	9,700	5,400		7,016
NAc		1,380	410		490
STN				140	115
GPe				929	850
GPi				316	426
SNr					244

Notes: All volumes are given as cubic millimeters (mm^3) for one hemisphere (*left*, if indicated). Oorschot (1996) gives a value of 19.9 mm^3 for the caudate and putamen. In the table this volume was divided equally between the two structures. Data from Hardman et al. (2002) are for two hemispheres and normalized here for one hemisphere. SNr volume is reported as total volume of the substantia nigra for studies that do not discriminate between the pars reticulata and compacta. Other rodent studies gave comparable values to the results of Oorschot and report a volume of 37 mm^3 and 90 mm^3 for the rodent dorsal striatum and whole striatum, respectively (Andersson et al. 2002; Mengler et al. 2014). CD, caudate; GPe, GPi, globus pallidus external and internal segments; NAc, nucleus accumbens; put, putamen; SNr, substantia nigra reticulata.
Source: Data for the California Institute of Technology (CIT) probabilistic high-resolution in vivo atlas of the human subcortical brain nuclei Cit-atlas (Pauli et al. 2018) were calculated by Andy Horn.

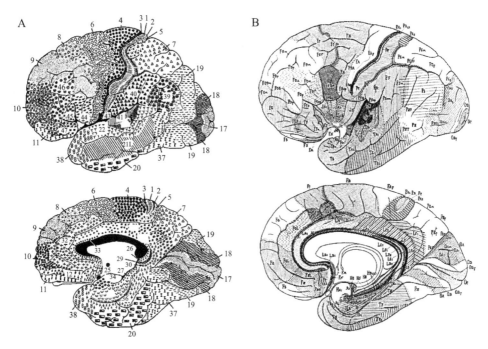

1.3 Histological and functional subdivisions of the cortex. *Upper row,* lateral surface; *lower row,* medial surface. *A,* according to Brodmann. *B,* according to von Economo.
Sources: A. Adapted from Pandya and Yeterian 1990. B. Adapted from Triarhou 2013.

The thalamus is one of the largest masses of subcortical gray matter (neurons; axons are white matter). Naively, it may be considered a relay structure of sensory peripheral information (e.g., visual, auditory, and somatosensory information through the lateral geniculate, the medial geniculate, and the ventral posterior thalamic nuclei) to the cortex. However, the thalamus is much more than a mere relay station between the periphery and the cortex. First, it is a complex structure with numerous (~15–25) subnuclei, many with no input from the periphery. Second, the connectivity between the cortex and the thalamus is reciprocal rather than unidirectional from the thalamus to the cortex. Actually, in most areas of the thalamus there are more afferent fibers coming from the cortex than efferent fibers going from the thalamus to the cortex.

The thalamus is anatomically divided by the internal medullary lamina (Y-shaped bundles of intrathalamic axons; figure 1.4) into three main anatomical domains: the medial (cognitive) domain, which is reciprocally connected with the frontal cortex; the anterior (limbic, emotional) domain, which is connected with limbic structures; and the lateral (sensorimotor) domain (figure 1.4). The nomenclature and the identities of different thalamic nuclei are passionately debated (Ilinsky and Kultas-Ilinsky 2001; Percheron et al. 1996; Mai and Majtanik 2019). The medial and the lateral thalamic lobes are also broadly divided into dorsal and ventral tiers. The ventral lateral

A

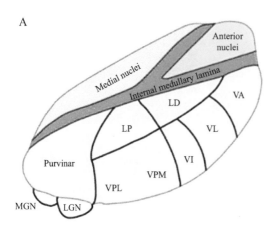

B

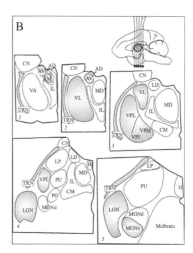

1.4 The thalamic nuclei. *A*, a schematic diagram of right thalamus. *B*, the major thalamic nuclei in the left hemisphere of a generalized nonhuman primate. The coronal sections go from rostral to caudal, are numbered *1* through *5* and are indicated in the upper-right midsagittal view of a monkey brain. The nuclei filled in yellow are first-order nuclei, and those filled in blue are higher order. Nearby nuclei are left blank. AD, anterodorsal nucleus; AM, anteromedial nucleus; AV, anteroventral nucleus; CM, centromedian nucleus; CN, caudate nucleus; H, habenular nucleus; IL, intralaminar (and midline) nuclei; LD, lateral dorsal nucleus; LGN, MGN, lateral and medial geniculate nucleus; LP, lateral posterior nucleus; MD, mediodorsal nucleus; MGN, medial geniculate nucleus; PO, posterior nucleus; PU, pulvinar; TRN, thalamic reticular nucleus; VA, ventral anterior nucleus; VI, ventral intermediate (Vim) nucleus; VL, ventral lateral nucleus; VPI, VPL, VPM, inferior, lateral, and medial parts of the ventral posterior nucleus. *Sources:* A. Adapted from Wijesinghe et al. 2015. B. Copied from Sherman 2006. Creative Commons BY-NC-SA license.

thalamic tier is the classical "relay" tier of the thalamus. From anterior to posterior, the ventrolateral tier can be divided into the ventral anterior (VA), ventral lateral (VL), ventral intermediate (Vim), and ventral posterior (or VC, ventral caudal) nuclei that receive projections from the SNr, GPi, cerebellum, and ascending spinothalamic sensory pathways and project to the frontal, motor, and somatosensory cortices, respectively.

The intralaminar thalamic nuclei participate in the control of arousal and awareness (Matsumoto et al. 2001; Minamimoto et al. 2005). In the primate the intralaminar nuclei are divided into rostral and caudal complexes. The rostral complex is composed of the central medial, paracentral, and central lateral nuclei. The caudal complex includes the centromedian and parafascicular nuclei (CM and PF). The CM is not clearly delineated in the rodent, and the lateral part of the rodent PF is considered to be the homologue of the primate CM (Smith et al. 2004). The ethmoid thalamic nucleus (Eth) is included in the rodent intralaminar nuclei (Deschênes et al. 1996). The thalamic intralaminar nuclei are the main source of thalamic innervation of the input stages of the basal ganglia—CM and part of the PF to the striatum and the PF to the STN. Schematic diagrams of the thalamus (figure 1.4*A*) may give the impression that the intralaminar nuclei are smaller in comparison with other thalamic nuclei. Less schematic schemes (figure 1.4B) and stereological counting (chapter 2) reveal that the volume and number of neurons in the thalamic intralaminar nuclei is much like the other major thalamic nuclei. Other thalamic nuclei that receive basal ganglia output like VA and VL also project to the basal ganglia input stages. About half of the glutamatergic innervation of the striatum originates in the thalamus (the other half is from the cortex).

Finally, the thalamus is encapsulated by the thalamic reticular nucleus (TRN). The TRN GABAergic neurons receive inputs from collaterals of the corticothalamic fibers and provide diffuse inhibition of the thalamocortical projection neurons. They are also innervated by the brain stem reticular formation and therefore play a major role in the orchestration of the thalamocortical networks during the sleep/wake cycle. It has been suggested that the TRN receives massive projections from the GPe (Hazrati and Parent 1991). However, other studies have not confirmed this, and it is probably incorrect.

A CORD OF THREE MOTOR STRANDS IS NOT QUICKLY BROKEN: THE CEREBELLUM

King Solomon (Ecclesiastes 4:10–12), as well as Gilgamesh of Mesopotamia (the *Epic of Gilgamesh*, tablet 5) and many others (e.g., Alexandre Dumas's *The Three Musketeers*), have pointed out the strength of the companion of three. The basal ganglia, the cerebellum, and the motor cortices have been the three musketeers of the motor system since the beginning of modern neurology. Figure 1.5 depicts Kemp and Powell's classical description showing the similarity (and variation) between the cerebellar and basal ganglia circuitry. As a whole, like the basal ganglia the cerebellum recieves information from the cortex and projects back to the cortex through the thalamus. The connectivity of the cerebellum with the spinal cord and periphery is much more prominent than that of the basal ganglia. Like the cortex and thalamus, the cerebellum deserves its own monograph, so it will be only briefly discussed below and in the following chapters.[3]

Anatomically, the primate cerebellum (little brain) has the appearance of a separate structure located at the bottom of the brain (figure 1.1). The cerebellum can be divided into three rostral to caudal (or dorsal to ventral in the human) lobes. The cerebellum is functionally divided into three domains: the medial spinocerebellum and the larger lateral cerebrocerebellum are in the anterior and posterior lobes; the flocculonodular lobe represents the vestibulocerebellum. These domains are of different evolutionary age and varied (but overlapping) functions. The cerebrocerebellum (neocerebellum) is the largest in human and nonhuman primates. It receives input from the cerebral cortex (especially the parietal lobe) via the pontine nuclei and sends output through the dentate deep cerebellar nucleus to the ventral lateral

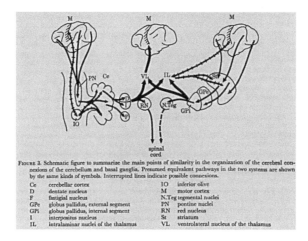

FIGURE 3. Schematic figure to summarize the main points of similarity in the organization of the cerebral connexions of the cerebellum and basal ganglia. Presumed equivalent pathways in the two systems are shown by the same kinds of symbols. Interrupted lines indicate possible connexions.

Ce	cerebellar cortex	IO	inferior olive
D	dentate nucleus	M	motor cortex
F	fastigial nucleus	N.Teg	tegmental nuclei
GPe	globus pallidus, external segment	PN	pontine nuclei
GPi	globus pallidus, internal segment	RN	red nucleus
I	interpositus nucleus	St	striatum
IL	intralaminar nuclei of the thalamus	VL	ventrolateral nucleus of the thalamus

1.5 The functional connectivity of the cerebellum, basal ganglia, and motor cortex.
Source: Copied with permission from Kemp and Powell 1971.

thalamus. Finally, the inferior olive provides the climbing fiber teaching input to the Purkinje cells of the cerebellar cortex.

EMOTION AND MOTION: THE PAPEZ LIMBIC CIRCUIT

Our brain faculties are usually divided into cognitive, emotional, and motor domains. In this book I will take the motorcentric point of view—namely, that the major function of the cognition and emotion faculties is to support optimal motor behavior. Indeed, the word "emotion" originated in the mid-sixteenth century from "motion." So emotion is a feeling and involves movement. In this section we will discuss the neuroanatomy of our emotional system.[4]

In 1937, James Papez proposed that the neural circuit connecting the mammillary bodies of the posterior hypothalamus to the limbic lobe (the cingulate and parahippocampal gyri or formation, according to Pierre Paul Broca) is the basis for emotional experiences. The classical Papez circuit is a sagittal C-shaped structure that starts and ends in the hippocampal formation and goes through the following structures and pathways: hippocampal formation → fornix → mammillary bodies → mammillothalamic tract → anterior thalamic nucleus → cingulate gyrus → cingulum → entorhinal cortex → hippocampal formation.

Paul D. MacLean elaborated upon Papez's proposal and coined the term "limbic system." MacLean's evolutionary triune brain theory proposed that the human brain is composed of the reptilian complex (the basal ganglia), the limbic system, and the neocortex. MacLean's definition of the basal ganglia as reptilian was driven by the comparative anatomy thinking of his time that the (hyper-)striatum, not the cortex, dominates the forebrain of reptiles (and birds). MacLean's reptilian basal ganglia complex is responsible for the species-typical instinctual behavior involved in aggression, dominance, territoriality, and ritual displays. MacLean redefined Papez's limbic network to include the amygdala and septum. These structures are connected by the stria terminalis. The septal nuclei provide the cholinergic innervation of the hippocampus and are essential in generating the hippocampal theta rhythm, thus closing a loop between the amygdala and the hippocampus.

The Papez circuit is not a complete list of the brain structures devoted to brain emotional processing. For example, the habenula plays a major role of bridging between the limbic system and the basal ganglia (Pelled et al. 2007; Hong and Hikosaka 2008). A very elegant series of papers by Hikosaka and colleagues has shown that the lateral habenula encodes negative prediction errors (Matsumoto and Hikosaka 2007; Bromberg-Martin and Hikosaka 2011) and modulates the activity of midbrain dopaminergic neurons (Hong et al. 2011). A recent study revealed a neural pathway from retinal melanopsin-expressing ganglion cells to the dorsal perihabenular nucleus to the nucleus accumbens. This retinal-habenula-basal ganglia circuit probably plays a role in the generation of depressive symptoms and sleep disorders in Parkinson's patients (Ortuño-Lizarán et al. 2020).

Today, the (dorsal) hippocampus and its related structures are considered part of the cognitive systems encoding spatial (and even social) maps. It seems that like the basal ganglia, even the hippocampus can participate in the neural control of more than one domain of our behavior. Coming back to the basal ganglia, both the amygdala and the hippocampus project to the nucleus accumbens (ventral striatum), the limbic domain of the striatum. Thus, the basal ganglia—especially their ventral domains, including the ventral striatum, ventral pallidum, and ventral tegmental areas—are well connected to the main structures of the limbic (emotional) system.

FORM AND FUNCTION

Unlike some of the cortical areas whose names (e.g., primary or secondary visual, auditory, and motor cortices) reflect their functions, the names of many basal ganglia structures reflect their anatomical appearance and our lack of understanding of their functions. These include, for example, the striatum, so named following Thomas Willis's (in *Cerebri Anatome*, 1664) description of the striped (striated) appearance of the dorsal striatum and of the gray-matter bridges connecting between the caudate and the putamen across the internal capsule; the globus pallidus, named by Karl Burdach as the *blasser Klampen* (the pale clump, or globe) to describe its pale appearance in fresh sections of the brain; the substantia nigra, "the black substance," because of its melanin-containing dopamine neurons; the zona incerta because of its location between the thalamus and the subthalamic area; and finally, the substantia innominate, the unnamed structure located below the globus pallidus and the dorsal striatum. I most strongly agree with Francis Crick's (1988) statement that "if you want to understand function, study structure." The description provided here is very schematic, and we should keep in mind that the anatomy of the basal ganglia, like that of all brain structures, is complex and three-dimensional.[5]

CHAPTER SUMMARY

- The main structures of the basal ganglia are the striatum, the subthalamic nucleus, the globus pallidus, and the substantia nigra.
- The striatum is divided into the dorsal striatum (caudate and putamen) and the ventral striatum.
- The globus pallidus is divided into external and internal segments and the ventral pallidum.
- The substantia nigra is divided into the substantia nigra pars reticulata and the substantia nigra pars compacta.
- The basal ganglia are connected to and work in tandem with the cortex, the thalamus, the cerebellum, and the limbic system.

2

CELLULAR ANATOMY AND BIOCHEMISTRY OF THE BASAL GANGLIA

In the previous chapter, I provided a brief introduction to the gross anatomy of the basal ganglia and their related structures. Now I will zoom down to the cellular level and examine the cellular anatomy and biochemistry of basal ganglia neurons.[1] The cellular anatomy and biochemistry of three nuclei of the striatum (caudate, putamen, and NAc) are very similar, and therefore I will discuss the striatum as one structure. Additionally, most of the neurons in the downstream structures of the basal ganglia—the GPe, GPi, and SNr—share similar morphology and biochemistry and therefore will be discussed together. The STN is positioned in between these two groups of basal ganglia neurons. I will also consider the basal ganglia input/output structures such as the cortex, thalamus, and brain stem motor nuclei.

THE CORTEX

The cortex is the starting point of most computational models of the basal ganglia (chapter 9). Even for the circular reinforcement learning and state-to-action models (chapters 10–12), the cortex plays a critical role mediating between the external world and the basal ganglia. The cellular structure of the cortex is quite homogenous, with parallel vertical pyramidal cells with a basal (near the soma) horizontal dendritic field, an ascending (upgoing) apical dendritic field, and a descending axon (figure 2.1*A*, *B*). This parallel anatomical organization of the pyramidal cells makes possible the summing of their electrical activity (especially when this activity is more synchronized) into an electric field that can be recorded from outside the scalp—the electroencephalogram (EEG). About a fifth of the cortical neurons are GABAergic interneurons, with diverse families and morphological structures (figure 2.1*C*).

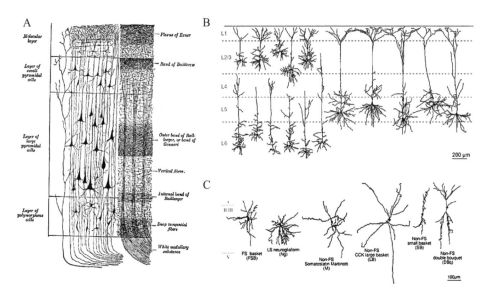

2.1   The cellular structure of the cortex. *A*, cortical layers of neurons and axons (*left and right*). *B, C*, camera lucida reconstructions of pyramidal and nonpyramidal neurons from the rat cortex. Layer location is marked on the left. Note the different scale bars for *A* and *B*.
Sources: A. Copied with permission from Henry Vandyke Carter—Henry Gray (1918) *Anatomy of the Human Body*, plate 754, public domain, https://commons.wikimedia.org/w/index.php ?curid=541599B. B. Adapted from Ledergerber and Larkum 2010. C. Adapted from Kawaguchi et al. 2006.

THE STRIATUM

The cytoarchitecture (the arrangement of cells in space) of the striatum is very different from the layered structured cytoarchitecture of the cortex. The neurons of the striatum are densely packed, with no clear vertical or horizontal layout of their soma (cell body) or axons. The majority (95 percent in the rodent, 80 percent in monkey and human) of striatal neurons are medium spiny neurons (MSNs). They are medium size (14–20 μm soma diameter), and they have a dense (mean number of dendritic tips equals 34 and 42 in monkey and human respectively; Yelnik et al. 1991) and compact spherical (three-dimensional symmetric, diameter ~0.3 mm) dendritic field (figure 2.2*A*, *B*). The dendrites are covered with spines contacted by cortical and thalamic glutamatergic excitatory axons. The MSNs are also innervated by dopaminergic axons emitted by midbrain SNc and VTA dopaminergic neurons that innervate the dorsal and ventral striatum, respectively. The MSNs have a rich system of local (lateral) axons (of about the same size as that of their dendritic field) that innervate nearby MSNs and other striatal neurons.

Like all basal ganglia structures, the striatum is a one-layer nucleus. Namely, the MSNs receive inputs from the cortex and thalamus and project to downstream basal ganglia structures (GPe, GPi, and SNr). This one-layer arrangement is very different from the layered organization of the cortex and the cerebellum, with complex local

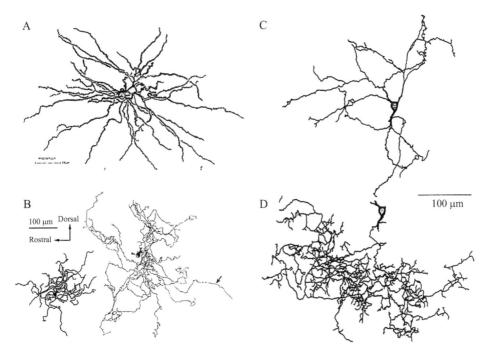

2.2   The striatal neurons. *A*, camera lucida drawing of Golgi-stained medium spiny neuron (MSN) of a baboon. Scale bar: 50 μm. *B*, biocytin-filled rodent MSN; *left*, dendritic field; *right*, axonal arborization. Arrow indicates the main axon projecting to the GPe. Scale bar: 100 μm. *C, D*, the dendritic (C) and axonal (D) fields of a striatal cholinergic aspiny interneuron.
Sources: A. Adapted from Yelnik et al. 1991. B. Adapted from Kawaguchi et al. 1990. C, D. Adapted from Aosaki and Kawaguchi 1996.

microcircuits (figure 3.2). Like most other basal ganglia structures (with the single exception of the STN), the MSNs use gamma-aminobutyric acid (GABA), an inhibitory transmitter, as their main transmitter.

The densely packed MSNs are traditionally divided into two populations, one with D1 and another with D2 dopamine receptors. The two populations intermingle in space and cannot be differentiated by classical histological staining methods. Traditionally, the D1 MSNs project to the output structures of the basal ganglia—the GPi and the SNr (the direct pathway), while the D2 MSNs project to the GPe (the indirect pathway). Both MSN populations use GABA as their main transmitter; however, they have additional cotransmitters: Substance P in the D1 MSNs and enkephalin in the D2 MSNs. The physiological roles of these cotransmitters have not been intensively studied.

The minority (5–20 percent) of striatal neurons are interneurons (i.e., they do not project outside the striatum). Most striatal interneurons are GABAergic medium aspiny neurons. Several families of striatal GABAergic interneurons exist, and their physiology and connectivity is rapidly elucidated with modern optogenetic methods

(Tepper et al. 2004; Burke et al. 2017; Hjorth et al. 2020). Today, extracellular recording techniques enable us to discriminate only the fast-spiking, parvalbumin-expressing interneurons (FSI; figure 5.5), which probably provide a massive network of lateral inhibition in the striatum.

A distinct population of striatal interneurons consists of the giant aspiny cholinergic interneurons (figure 2.2*C*, *D*). They constitute 1 percent of the total population of striatal neurons. Their dendritic field is less dense than that of the MSNs; however, it spreads over a larger distance, to about 1 mm from the soma. The axons of the cholinergic interneurons are long and highly branched and enable the widespread release of acetylcholine in the striatum. Indeed, the density of cholinergic biomarkers in the striatum is highest for the overall central nervous system and resembles the distribution of dopaminergic markers in the striatum (figure 11.3).

Ann Graybiel pioneered the neurochemical analysis of the striatum. Graybiel's biochemical studies have established that the neurochemical organization of the striatum is not uniform. Rather, the striatum can be differentiated into neurochemical discrete compartments known as striosomes (or patches) and matrix. These compartments differ in their expression of neurochemical markers, as well as in their afferent and efferent connectivity. The striosomes occupy 10–15 percent of the striatal volume and exhibit high-level dopamine D1 receptors and opioid receptors. The matrix is enriched in D2 dopamine receptors and cholinergic markers. The two classes of MSNs, the D1 and D2 expressing, can reside in either striosome or matrix compartments. The dendritic and axonal fields of some MSNs respect the striosome-matrix boundary, but other MSNs do not respect this boundary. Striatal interneurons (e.g., the cholinergic interneurons) tend to lie at the border. The functional roles of the two compartments are still debated.

THE SUBTHALAMIC NUCLEUS

The cytoarchitecture of the subthalamic nucleus (STN) lies in between the striatum and the pallidum (Marani et al. 2008). As for the striatum, the STN neurons are densely packed. They are medium to large cells (25–40 μm soma diameter). Each cell gives rise to five to eight dendritic stems that branch less extensively than the striatal MSNs to give rise to a mean number of twenty-seven dendritic ends. The secondary dendrites are thin and have few spines on their distal portions (figure 2.3). The dendritic field is symmetric (spheroidal) but can be disk-shaped—that is, two-dimensional in structure (Yelnik and Percheron 1979; Kita et al. 1983; Koshimizu et al. 2013; Emmi et al. 2020), resembling the pallidal neurons. The subthalamic long dendrites can extend as far as 600 μm from their parent cell bodies (Yelnik and Percheron 1979). The extent of the dendritic domain of an individual STN neuron can cover about half, one-fifth, and one-tenth of the STN in the cat, monkey, and human, respectively (Yelnik and Percheron 1979). In all cases this is a significant fraction of the volume compared to the

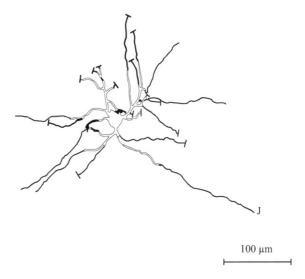

100 μm

2.3   Camera lucida drawing of a Golgi-stained subthalamic neuron from a human brain. A large number of dendrites are cut. Small bars indicate the points of section.
Source: Adapted from Yelnik and Percheron 1979.

dendritic field of striatal cells, suggesting that the corticosubthalamic system is more functionally convergent than the corticostriatal system.

The STN dendrites are innervated by glutamatergic inputs from the (frontal) cortex and thalamus and GABAergic inputs from the GPe. As for other structures of the basal ganglia, the vast majority of STN neurons are projection neurons (again, a one-layer network). These projection neurons provide glutamatergic (excitatory) innervation to the downstream structures of the basal ganglia—the GPe, the GPi, and the SNr (Parent and Parent 2007). Among the core structures of the main axis of the basal ganglia, only the STN is glutamatergic.

GPe, GPi, AND SNr

The last, but not least, family of basal ganglia neurons to be discussed here are the projection neurons of the downstream structures of the basal ganglia—the GPe, the GPi, and the SNr. In these three structures, as well as in the ventral pallidum (VP), most of the neurons are projection neurons (i.e., one-layer network). The GPe neurons innervate all basal ganglia structures: the striatum, STN, GPi, and SNr. Consequently, the GPi and the SNr innervate the ventrolateral motor tier of the thalamus, as well as brain stem motor nuclei such as the pedunculopontine nucleus (PPN) and the superior colliculus (SC).

The cellular morphology of the pallidal and SNr neurons is very unique (figure 2.4). The neurons are large (20–60 μm soma). They have a very sparse dendritic field, with few dendrites (four dendritic stems, one to two branches, leading to an average number of twelve to fourteen dendritic ends). The long and sparsely ramified dendrites

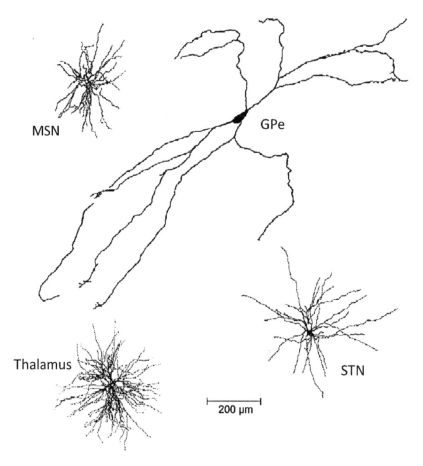

2.4 The dendritic structure of pallidal neurons in comparison with common basal ganglia neurons. MSN, striatum (medium spiny neuron); GPe, globus pallidus external segment; STN, subthalamic nucleus; thalamus, thalamic relay neuron.
Source: Adapted from Yelnik 2002.

of pallidal neurons and SNr neurons differ from the short but densely ramified dendrites of the striatal medium spiny and thalamocortical relay neurons (figure 2.2 and 2.6*A*). In the core of the two segments of the globus pallidus, the sparse and aspiny dendritic field is large and flat—that is, disk shaped. Typically, the disk has a diameter of 1–1.5 mm and a depth of 0.25 mm. It is orthogonal to the incoming striatal axons (figure 2.5), enabling sparse sampling of information from remote striatal domains. Exceptions exist; SNr neurons and pallidal neurons located on the border of the nuclei do not have a discoidal dendritic arborization.

Most (90–95 percent) of the synaptic input to the GPe, GPi, and SNr is GABAergic and from the striatum. The striatal axons take most of the volume of these structures, leading to large (0.2–0.5 mm) distances between the cell bodies of these nuclei and to their pale appearance. The dendrites of the GPe, GPi, and SNr neurons are covered by synapses of striatal axons, and the soma is innervated by GPe axons (figure 3.1).

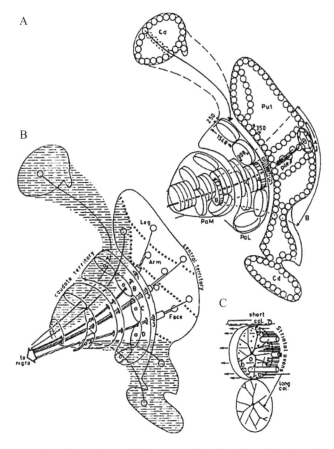

2.5 Spatial organization of the striatopallidal system. The dendritic disks of pallidal neurons (A, C) are orthogonal to the incoming striatal axons (B).
Source: Adapted from Percheron et al. 1984.

Of the synapses located on the dendrites of pallidal and SNr cells, 5–10 percent are glutamatergic, excitatory, and of STN origin.

THE THALAMUS

The thalamus is strongly connected with the cortex. There is also a rich connectivity between the basal ganglia and the thalamus. Most important are the projections of the output structures of the basal ganglia—the GPi and the SNr to the ventrolateral and ventroanterior nuclei of the thalamus and the projections from the centromedian and parafascicular intralaminar thalamic nuclei to the input structures of the basal ganglia—the striatum and the STN.

In human and nonhuman primates, the ratio of thalamic projection (glutamatergic, relay) cells to interneurons is similar to the ratio found in the cortex between pyramidal neurons and interneurons (four to one). Rodents are different and interneurons are essentially missing from most thalamic nuclei. The neurons of the

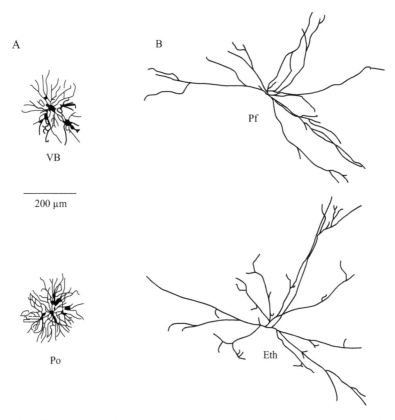

A

B

Pf

VB

200 μm

Po

Eth

2.6 The dendritic field of thalamic neurons. *A*, the bushy structure of a typical thalamic relay neuron. *B*, the sparse dendritic field of intralaminar thalamic neurons. VB, ventrobasal; Po, posterior; Pf, parafascicular; Eth, ethmoid nucleus.
Source: Adapted from Deschênes et al. 1996.

thalamic reticular nucleus are GABAergic and inhibit the relay cells of the main thalamic nuclei. Most thalamic relay neurons have a small three-dimensional symmetric and a dense dendritic field (figure 2.6*A* and figure 2.4, *bottom left*) resembling the medium spiny neurons of the striatum. The dendritic field of the intralaminar thalamic nucleus differs (figure 2.6*B*) and looks like the sparse and large dendritic field of the pallidal neurons.

THE BASAL GANGLIA IN NUMBERS

Several research groups have measured or estimated the number of neurons in the various structures of the basal ganglia in the rat (Oorschot 1996), macaque monkey (Percheron et al. 1984, 1994), and human (Lange et al. 1976; Kalanithi et al. 2005; Salvesen et al. 2015). Glenda Margaret Halliday (Hardman et al. 2002), working on five species (including marmosets and baboons), reported a logarithmic relationship between the number of neurons in the basal ganglia and the brain mass. The major exception is the relative smaller number of neurons in the rodent

Table 2.1 Number of neurons in the basal ganglia nuclei of one hemisphere of rat, macaque monkey, and human. Numbers are based on different studies.

Source	Rat		Monkey		Human			
	Oorschot 1996	Hardman et al. 2002	Percheron 1984, 1994	Hardman et al. 2002	Lange et al. 1976	Kalanithi et al. 2005	Salvesen et al. 2015	Hardman et al. 2002
Dorsal striatum	$2.79 * 10^6$		$31 * 10^6$				$162.5 * 10^6$	
STN	$13.6 * 10^3$	$11.3 * 10^3$		$76 * 10^3$	$870 * 10^3$		$450 * 10^3$	$280 * 10^3$
GPe	$46 * 10^3$	$42.0 * 10^3$	$166 * 10^3$	$250 * 10^3$	$1,004 * 10^3$	$1,695 * 10^3$	$1650 * 10^3$	$725 * 10^3$
GPi	$3.2 * 10^3$	$6.3 * 10^3$	$63 * 10^3$	$78 * 10^3$	$314 * 10^3$	$540 * 10^3$		$176 * 10^3$
SNr	$26.3 * 10^3$	$23.8 * 10^3$	$54 * 10^6$	$65 * 10^3$				$144 * 10^3$

Note: GPe, GPi, external and internal segments of the globus pallidus; SNr, substantia nigra reticulata; STN, subthalamic nucleus.
Sources: Hardman et al. 2002 and Lange et al. 1976—adjusted from original numbers given for both hemispheres; Kalanithi et al. 2005 and Salvesen et al. 2015—numbers reported for GP without division for the internal and external segment.

entopeduncular nucleus (GPi homologue) and the relative higher number of neurons in the SNr. Table 2.1 below provides the number of neurons in the main axis of the basal ganglia.

The numbers are quite consistent between the studies and reveal 1) a drastic reduction (of two or three orders of magnitude) of the number of neurons from the striatum to the basal ganglia downstream structures (GPe, GPi, and SNr); 2) that the number of neurons in the STN is of the same order of magnitude or slightly smaller than in each of the downstream structures of the basal ganglia; 3) that the number of neurons in the GPe is higher (two to three times more in the nonhuman primate and human) than the number of neurons in the GPi and the SNr.

Dorothy Oorschot and Glenda Margaret Halliday (Oorschot 1996; Hardman et al. 2002) report also the number of putative (tyrosine hydroxylase-positive) neurons in the substantia nigra pars compacta: $7.2 * 10^3$ and $12.7 * 10^3$ in the rat, $203 * 10^3$ in the macaque, and $382 * 10^3$ in the human.

NUMBER OF NEURONS OUTSIDE THE HARD CORE OF THE BASAL GANGLIA

The six structures discussed so far (dorsal striatum, STN, GPe, GPi, SNr, and SNc) are indeed the hard-core structures of basal ganglia physiology. However, our understanding of the basal ganglia network cannot be completed without examining the following other structures.

The ventral basal ganglia   The first is the network of the limbic (ventral) basal ganglia. The network includes the ventral striatum (nucleus accumbens), the ventral pallidum, and the ventral tegmental area. The dorsal and lateral borders of the putamen and the caudate nucleus are clearly defined. However, the ventral border with the ventral striatum/nucleus accumbens is fuzzy. Some atlases (e.g., Bowden and Martin 2000) draw a clearly artificial straight line as the border. The rostroposterior extent of the ventral striatum is still debated. Functional domains are fuzzy and may be different than anatomy. For example, the anterior putamen might be part of the associative rather than the motor territory of the striatum. Additionally, there might be species variations; the limbic ventral striatum of the rodent probably occupies a bigger fraction of the striatum in comparison with nonhuman primates (Karachi et al. 2002; Jan et al. 2003). Many papers on the nucleus accumbens in the rodent (but not in the nonhuman primate) literature emphasize the importance of the accumbens' core and the shell divisions and their different roles in normal and pathological behaviors (Zahm and Brog 1992; Mannella et al. 2013; Berridge and Kringelbach 2015). Table 1.1B indicates that the nucleus accumbens occupies 3–7 percent of the total volume of the striatum of nonhuman primates and humans. On the other hand, histological and MRI quantitative data (Wong et al. 2016; Welniak-Kaminska et al. 2019) indicate that the nucleus accumbens occupies 16–45 percent of the total volume of the rodent striatum.

The division of the STN into three functional territories is even more debated. Based on projections from different cortical areas, the STN is divided into three nearly equal and overlapping territories (Haynes and Haber 2013; Ewert et al. 2018). However, the normalization algorithm that Ewert and colleagues used is biased toward the equal division of territories (Andreas Horn, personal communication). Other studies report a much smaller limbic territory in the nonhuman primate and human (Mathai and Smith 2011; Horn et al. 2017) in line with the relatively small limbic domain in the striatum in these species.

Thalamus   The number of neurons in the intralaminar (CM, PF) and main (VA, VL) thalamic nuclei is about $0.5–1 * 10^6$ and $1–5 * 10^6$ per nuclei in the nonhuman primate and the human, respectively (Brooks and Halliday 2009; Villalba et al. 2014). The mediodorsal nucleus of the nonhuman primate has a larger volume than the intralaminar nuclei while still having about the same number of neurons as the CM or PF (Villalba et al. 2014). In line with the expansion of the prefrontal cortex in humans, there are more neurons in the human dorsomedial nucleus ($5–7 * 10^6$; Dorph-Petersen et al. 2004; Abitz et al. 2007; Karlsen et al. 2014).

Brain stem motor nuclei   The input/output relationships of the basal ganglia and the brain stem are not well delineated. The number of neurons in the superior colliculus and pedunculopontine nucleus that are innervated by GPi and SNr efferent projections is still an open question.

Cortex   The cortex is not a monolayer structure and has a complex local microcircuitry, with a small fraction of input and output neurons (figure 3.2). Charles

(Charlie) Wilson (Zheng and Wilson 2002) used axon-tracing sample as representative of all corticostriatal axons and estimated the total number of rat corticostriatal neurons to be seventeen million, about ten times the number of striatal projection neurons. This suggests a continuous reduction in the number of neurons along the cortex-striatum-GPe and GPi/SNr axis (Bar-Gad et al. 2003). The number of neurons projecting to the STN and the number of cortical neurons innervated by the projection (relay) neurons of the basal ganglia receiving areas in the thalamus can only be estimated at the time of writing this book.

The whole brain The total number of neurons in the brain of the rat, macaque monkey, and human is estimated as $200 * 10^6$, $6,000 * 10^6$ and $86,000 * 10^6$ neurons, respectively (Herculano-Houzel 2009). Table 2.1 gives the number of basal ganglia neurons in one hemisphere. Even with the numbers doubled, the basal ganglia neurons represent only a very small fraction (~0.1–1.5 percent) of the total number of neurons in the brain. The cerebral cortex and the cerebellum hold about 20 percent and 60 percent, respectively, of all brain neurons (yes, there are many more neurons in the cerebellum than in the cerebral cortex). The volume and number of neurons do not imply importance. The volume of your computer CPU (here used as an abbreviation for central processing unit, rather than for the caudate and putamen) is small in comparison with the size of your printer. The author hopes that by the end of this book, the reader will be convinced that the basal ganglia, with their tiny fraction of all brain neurons, play a critical role in shaping our behavior in health and disease.

THE OTHER HALF: THE GLIA

The yin and yang of the central nervous system are the neurons and the glial cells. Rudolf Virchow discovered the glial cells in 1856 in his search for a connective tissue in the brain. The term "glia" derives from the Greek "glue" and suggests Virchow's original impression that the glial cells are the glue of the nervous system. Even today, most neuroscientists see the main job of the glial cells as housekeepers for the main players—the neurons. Unlike the neurons, the glial cells do not generate action potentials and therefore are believed to provide morphological, metabolic, and tissue protection. However, there is increasing evidence that glia do exert certain physiological effects, such as the modulation of neurotransmission.

Glial cells are very diverse in their morphology, function, and distribution in different brain areas. Astrocytes account for 20–40 percent of brain glia, and they interlink neurons and the brain blood supply. They regulate the external chemical environment of neurons by removing excess potassium ions and by recycling neurotransmitters. For example, striatal astrocytes may act as a reservoir of L-dopa, the gold standard of Parkinson's therapy. Oligodendrocytes are the most common (40–70 percent) glial cells and, like the Schwann cells in the peripheral nervous system, provide the myelin sheet for axons in the central nervous system. The microglia account for about 10 percent of all glial cells. They are mobile and provide reactive protection for brain damage.

Glia were thought to outnumber neurons by a ratio of 10:1;[2] however, recent studies (von Bartheld et al. 2016) provide evidence that overall, the brain contains about an equal number of glia and neurons. The glia-to-neuron ratio is highly variable among different brain regions; it equals only 0.2 in the cerebellum and 1.5–3.5 in the cortex. The glia-to-neuron ratio is also highly variable within the basal ganglia. In the human striatum, the glia-to-neuron ratio is similar to that in the cortex and equals 3.7, while the ratio reaches a record of near 160 in the globus pallidus (Kalanithi et al. 2005). The huge number of glial cells in the pallidum is not surprising given its low density of neurons and high density of striatal axons. Indeed, oligodendrocytes probably compose 60–80 percent of the glia in the pallidum (Salvesen et al. 2015), providing the myelin cover for incoming striatal axons. Still, the pallidal high astrocyte-to-neuron ratio might also reflect the need for high-frequency-discharge pallidal neurons for more substantive metabolic support. This book is more focused on extracellular recording techniques and studies of neuronal spiking activity. The relative effects of glia subthreshold activity on local field potentials are still unknown. The other half of the brain—its glia—is still waiting for its equal rights amendment.

CHAPTER SUMMARY

- Most neurons in the basal ganglia are projection neurons. Each basal ganglia structure is therefore a one-layer network in which the projection neurons receive information from their afferent structures and output the processed information to the next network structure.
- Most neurons in the main axis of the basal ganglia (with the exception of the STN neurons) use GABA, an inhibitory transmitter, as their main transmitter. The STN neurons use glutamate, an excitatory transmitter.
- The spiny dendritic fields of striatal projection neurons are dense, three-dimensional, symmetric, and of a short radius of 0.2–0.3 mm. In addition to their projections to downstream basal ganglia structures, they have extensive local axonal fields. This anatomical structure suggests that lateral interactions play a major role in the function of the striatum. Anatomical lateral connectivity also exists in other structures of the basal ganglia.
- The aspiny dendritic fields of pallidal and SNr neurons are sparse, wide, and disk-shaped with their large axis orthogonal to the incoming striatal axons.

3

THE SYNAPTIC CONNECTIONS OF THE BASAL GANGLIA NETWORK

In the previous chapters, we described the gross and cellular anatomy of the basal ganglia. Here, we will discuss the connections inside and outside the basal ganglia core structures. Connectivity is the name of the game in our central nervous system (CNS). The CNS is usually divided into gray matter—that is, areas with cell bodies, such as the cortex, thalamus, and basal ganglia, and white matter areas—areas with tracts of nerve fibers (axons), such as the corpus callosum and internal capsule (figure 1.1). However, in the gray matter most of the volume of the structure comes from the neuropil (axons and dendrites) that connects between the neurons. For example, the fraction of the mouse cortical volume occupied by the neuropil is 84 percent, while the cell bodies of neurons and glial cells amount to 12 percent and the blood vessels to 4 percent (Schüz and Palm 1989; Abeles 1991). Integrating the gross anatomy, cellular anatomy, and connectivity knowledge in quantitative terms is of critical importance.[1]

I will use this chapter to examine the qualitative and quantitative aspects of basal ganglia connectivity. We will take advantage of the feedforward structure of the basal ganglia network and discuss the number of synapses emitted by the neurons in the source structure and their convergence versus divergence properties in the recipient structure (Nauta and Mehler 1966; Haber 2016; Petersen et al. 2019). In a way, I am following the trails of the giants like John Eccles, Masso Ito, David Marr, and James Albus in the cerebellum (Eccles et al. 1967; Marr 1969; Albus 1971; Ito 1984).

Detailed mapping of the synaptic microcircuitry, mesocircuitry, and macrocircuitry of the mammalian brain (a brain connectome) is still a dream.[2] Mapping the connectome at the micrometer resolution—namely, at the synapse scale—is a huge technological (high-throughput serial electron microscopes, machine pattern recognition) and scientific challenge (Denk and Horstmann 2004). Our approach will therefore be statistical. We will use the notion that all nuclei in the main axis of

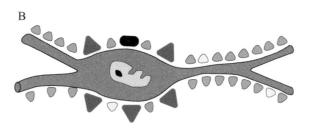

3.1 A scheme of the pallidal synaptic map. *A*, GPe neuron. *B*, GPi neuron. *Blue*, striatal; *red*, GPe; *yellow*, subthalamic; *black*, unknown afferent synapses.
Source: Adapted from Shink and Smith 1995.

the basal ganglia can be considered as one homogenous layer of projection neurons (Deister et al. 2013). We will further assume that all synapses are equal. We will keep the notion of excitatory and inhibitory synapses but will analyze each group as a single group and neglect critical issues like excitation/inhibition balance (van Vreeswijk and Sompolinsky 1996; Yizhar et al. 2011).

The assumption of a homogenous synaptic effect is more extreme than the homogeneity of the neurons (Sizemore et al. 2016). There is a big a difference between synapses on the soma versus synapses located on remote dendrites. For example, an old major paradox of the basal ganglia is the robust effects of STN over pallidal neurons despite providing only a tiny fraction of the synaptic input of the pallidal neurons. A possible solution, according to Parent and Hazrati (1993), is the unique location of the subthalamic-pallidal synapses on the soma of the pallidal neurons. They suggested that the subthalamic-pallidal synapses closely surround the soma of pallidal neurons. A later study (Shink and Smith 1995) revealed that the large synaptic boutons around the soma of pallidal neurons are of GABAergic and of GPe origin (figure 3.1). This finding hints at the importance of the lateral connectivity with the GPe and the GPe to GPi connectivity. However, the subthalamic paradox is back.

THE SYNAPTIC MAP OF THE FEEDFORWARD CONNECTIVITY OF THE BASAL GANGLIA

We are looking for a quantitative description of the number of neurons in the recipient structure that are innervated by one neuron in the source structure (the divergence ratio). On the other hand, we will ask what number of source neurons are targeting one neuron in the recipient area (convergence ratio). Quantitative data of basal ganglia synaptic connectivity are still scant. I will provide an estimate whenever I failed to find a reference. Table 3.1 will be used to check the consistency of the number of synapses from the source and the recipient point of view for the different parts of the basal ganglia networks.

Cortex to striatum connectivity   Corticostriatal projections typically arise from small- to medium-sized layer III and V pyramidal neurons from all cortical areas. In rodents, corticostriatal neurons are categorized into two main types: the intratelencephalic (IT) and the pyramidal tract (PT) neurons. The IT neurons send axonal projections to both the ipsilateral and contralateral cortex and striatum. They are mainly located in layer III and upper layer V of the rat cortex. In contrast, PT neurons are located in the lower layer V and send axonal projections to the brain stem and spinal cord, from which originate thin axon collaterals that innervate the ipsilateral striatum. The anatomical and physiological significance of the striatal projection of the PT neurons in nonhuman primates is still debated (Turner and DeLong 2000; Parent and Parent 2006; Mathai and Smith 2011).

Zheng and Wilson (2002) have shown that the total number of boutons formed by individual corticostriatal axons in the rodent is highly variable (ranging from 22 to 2,900), with an average of 879 boutons/neuron. The striatal medium spiny neurons receive glutamatergic synapses from the cortex and the thalamus, dopaminergic and cholinergic synapses from the substantia nigra pars compacta and striatal cholinergic interneurons, and GABAergic synapses from collaterals of other MSN neurons and striatal GABAergic interneurons. The number of dendritic spines, or glutamatergic synapses, targeting a single medium spiny neuron is between six thousand and fifteen thousand, with an average of ten thousand (Kincaid et al. 1998). Half of these synapses are from cortical areas.

The devil is in the small details. For me the striatum's devil is the small fraction of interneurons (5 percent in the rodent) that statistically can be neglected. Nevertheless, they may have a profound effect in striatal microcircuitry. A recent study revealed that the synaptic efficacy of cortical and thalamic inputs to the striatum is determined by the identity of the source and target cells. Ipsilateral corticostriatal projections provided stronger excitation to fast-spiking interneurons (FSIs) than to MSNs and only weak excitation to cholinergic interneurons. Projections from contralateral cortex evoked the strongest responses in low-threshold spiking interneurons, whereas the thalamus provided the strongest excitation to cholinergic interneurons (Johansson and Silberberg 2020).

Cortex to subthalamic nucleus connectivity   The corticosubthalamic projections differ from the corticostriatal projections. They originate from the ipsilateral somatomotor and frontal cortical areas (Monakow et al. 1978; Haynes and Haber 2013; Emmi et al. 2020). The corticosubthalamic projections originate primarily from cortical deep layer V neurons and were suggested to be collaterals of the thick and fast-conducting axons of the pyramidal corticospinal tract (Iwahori 1978). However, recent studies have shown that the cortico-STN projection is formed mostly by collaterals of a small population of small-to-medium-sized pyramidal neurons. The corticosubthalamic axons also emit collaterals that innervate multiple other brain sites, including the superior colliculus and the pedunculopontine nucleus and the spinal cord (Kita and Kita 2012; Coudé et al. 2018).

The small size of the STN combined with the large extent of the dendritic tree of a single subthalamic neuron suggests synaptic convergence of different cortical inputs onto single subthalamic neurons. The exact degree of convergence and divergence of the cortex-subthalamic projections remains to be established. Mark Bevan and Paul Bolam reported that it is likely that individual cortical neurons innervate many subthalamic neurons over a large extent of the nucleus, and it is also likely that many cortical neurons innervate individual subthalamic neurons (Bevan et al. 1995). The number of cortical axon varicosities per subthalamic neuron ranges between 3 and 299 in the nonhuman primate and between 1 and 94 in the rat STN (Kita and Kita 2012; Coudé et al. 2018). I failed to find an estimate for the number of cortical neurons innervating the STN. In the human, there are $20 * 10^6$ axons in the pyramidal tract, and $1 * 10^6$ are corticospinal axons (Saliani et al. 2017). Since the STN innervation is given by a small population of pyramidal neurons (Kita and Kita 2012), I suggest using $1 * 10^6$ as an estimate for the number of corticosubthalamic axons.

Thalamus to striatum and subthalamic nucleus connectivity The synaptic connectivity between the thalamic nuclei and the main axis of the basal ganglia is still under investigation. Here I will discuss the connections from the intralaminar thalamic nuclei to the input nuclei of the basal ganglia—the striatum and the STN (Kita et al. 2016).

Tracing studies of rodent thalamic projections to the basal ganglia reported no or a small fraction (20 percent) of intralaminar thalamic neurons that innervate both the striatum and the STN. Virtually all striatal innervation is carried by the same thalamic neurons that innervate the cortex (Féger et al. 1994; Deschênes et al. 1996; Lanciego et al. 2004). Axon tracing in nonhuman primates (Sadikot et al. 1992; Parent and Parent 2005) reveals that only a subgroup of the intralaminar neurons innervate both the striatum and the cortex. Other thalamic neurons innervate the cortex or the striatum only. The axonal arborization in the striatum was focal and dense, while the cortical innervation was more diffuse. I failed to find quantitative information regarding the number of boutons emitted by a single thalamic neuron.

Striatum to GPe and GPi/SNr connectivity The axonal tree of a single rat medium spiny neuron contains 50–250 boutons in each of its downstream recipient structures (Kawaguchi et al. 1990; Wu et al. 2000). About a third of the rodent striatal neurons projects only to GPe, while the others project to GPe and to one or two of the basal ganglia output structures (GPi and SNr). In the monkey, most neurons project to GPe, GPi, and SNr (Lévesque and Parent 2005). It is estimated that in the nonhuman primate each striatal neuron has 250 synapses in its pallidal target (Percheron et al. 1994; Yelnik et al. 1996).

Traditionally, it was believed that striatopallidal projections are arranged in a fashion favoring multiple synaptic contacts between one striatal axon and an individual pallidal neuron. The striatal myelinated axons enter the pallidum perpendicular to the disklike dendritic field of pallidal neurons (figure 2.5) but then give off orthogonal unmyelinated collaterals that run in parallel with the pallidal dendritic disk. It was suggested that striatal axons closely entwine with the pallidal dendrites, forming a typical "woolly" fiber arrangement and repeated synaptic contacts with a single pallidal dendrite (Hazrati and Parent 1992). However, the striatopallidal connectivity reveals a high degree of anatomical specificity, and axons of striatal neurons from two small adjacent locations do not converge upon the same pallidal neurons; rather, they project to different sets of pallidal neurons. Quantitative analysis reveals that striatal axons had either a single or few (less than ten) successive varicosities on a given pallidal dendrite. Even with large biocytin injections to the striatum, no dendrite completely covered by biocytin-filled varicosities has been observed. Thus, one pallidal neuron is receiving information from many and probably remote striatal neurons (Yelnik et al. 1996). A reanalysis of electron microscopy data (Fox et al. 1974; Difiglia, et al. 1982; Cano et al. 1989) showed that the long dendrites of pallidal neurons are entirely covered by synaptic boutons with no glial interposition (Yelnik 1996; Percheron et al. 1994). Measurements of the average surface of a striatopallidal synapse and the total length of pallidal dendrites give an average of thirty thousand to forty thousand striatal synapses per pallidal neuron. Thus, only a tiny fraction (1/3000–1/4000) of the striatal synapses of a given pallidal neuron are emitted by a single striatal neuron.

Subthalamic nucleus to GPe and GPi/SNr connectivity   STN neurons send branched projections to GPe, GPi, and SNr, as well as to other targets in the rodent (Van der Kooy and Hattori 1980) and the nonhuman primate (Sato et al. 2000). The total number of boutons of a single subthalamic neuron equals 1,269, with an average of 500 boutons in the GPe (Koshimizu et al. 2013). On the recipient side, we will use the estimate of 30,000 synapses contacting the dendrites of single pallidal neurons, of which only approximately 5 percent are glutamatergic and of subthalamic origin, to set an estimated number of 1,000 subthalamic synapses per GPe neuron.

GPi/SNr to thalamus   The output of the basal ganglia is directed toward the thalamus, as well as to the lateral habenula and brain stem (Parent et al. 2001). The thalamic route is better described, but the information is still patchy and even less robust than the information regarding the inter–basal ganglia connectivity.

The GPi and the SNr provide GABAergic projections to the thalamic ventrolateral tier. Additional projections are sent to the intralaminar thalamic nuclei and to the dorsomedial nucleus. The terminal tree of the pallidal axons is probably different for the different thalamic nuclei. In the thalamic ventrolateral region, axons divided several times before ending in different parts of the territory in a characteristic dense terminal arborization (bunch) that covers a group of about twenty thalamic neurons. The number of boutons per bunch is probably between thirty and fifty (Arecchi-Bouchhioua et al. 1996, 1997; Parent et al. 2001), yielding a total number of two hundred to three hundred synaptic boutons emitted by a single pallidal neuron. A similar number of axonal varicosities (in the range of a few hundred) are reported in the rodent (Kha et al. 2000, 2001). The pallidal varicosities are aligned along the soma and the proximal dendrites of the thalamic projection neurons, providing highly effective inhibition to each individual cell (Ilinsky et al. 1997).[3]

I failed to find stereological data regarding the number of thalamic neurons inner-vated by a single neuron of the basal ganglia output structures. Using the informa-tion given above, I will assume that each GPi/SNr neuron innervates approximately two hundred thalamic neurons with minimal overlap between thalamic targets of pallidal neurons. Thus, the sequential reduction in the number of neurons from the cortex to the striatum, the GPe, and the GPi/SNr is reversed at the pallidothalamic junction and the network expands.

Thalamus to cortex connectivity   The main relay neurons of VA and VL project to the small pyramidal (stellate) neurons of layer IV of the cortex. Early reports maintain that intralaminar thalamic neurons project to layer I (Royce and Mourey 1985); how-ever, single-axon tracing in primates reveals that they project to the deep layers (e.g., layers V and VI) of the cortex (Parent and Parent 2005). The connectivity between the thalamus and cortex is a complex one and includes the reciprocal connectivity between the thalamic relay neurons, the inhibitory neurons of the thalamic reticular nucleus, and the microcircuitry of the cortex (Sherman and Guillery 2005). For a very schematic model of this connectivity, see figure 3.2.

As for the projections from the GPi/SNr to the thalamus, I do not have solid data for the number of cortical neurons innervated by the thalamic relay cells. Glenda Halliday and colleagues (2005) counted the number of neurons of patients with Parkinson's disease and progressive supranuclear palsy (PSP) in the three nuclei of the ventral lateral tier and in their corresponding motor cortical areas. The number of neurons in the cortex was forty to one hundred times the number of neurons in the thalamus. In the macaque monkey, there are $1.4 * 10^6$ cells in the lateral geniculate nucleus (Williams and Rakic 1988) and $276 * 10^6$ neurons in the primary visual cortex (area 17) of one hemisphere (O'Kusky and Colonnier 1982)—namely, a thalamocortical divergence ratio of 1:200. If there are about 25 thalamic nuclei in each hemisphere, with $1–2 * 10^6$ neurons/nuclei, the total number of neurons in the thalamus of a human is around $50 * 10^6$. The total number of neurons in the human cortex is $10–20 * 10^9$ (20 percent of the number of neurons in the whole brain)—that is, a thalamocortical divergence ratio of the order of 1:200. I will therefore assume another expansion from the thalamus to the cortex.

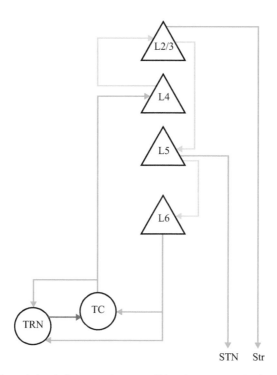

3.2 The microcircuitry of the thalamus and cortex. *Triangles*, neurons in the different layers (L) of the cortex; *circles*, thalamic neurons. *Red arrow*, inhibitory connection; *green arrows*, excitatory; *light green arrows*, connections within the cortex. STN, subthalamic nucleus; Str, striatum; TC, thalamocortical relay neuron; TRN, thalamic reticular nucleus neuron.

Table 3.1 provides a summary and sanity check of the quantitative anatomy of the feedforward synaptic connectivity in the basal ganglia.

LATERAL CONNECTIVITY IN THE BASAL GANGLIA

The rich lateral dendritic and axonal fields of medium spiny neurons (figure 2.2*A*, *B*) led to the suggestion that lateral inhibition plays a major role in the striatal network. Careful analysis of the density of synapses in the striatum (Groves et al. 1994; Ingham et al. 1998) suggests a total of about fifteen thousand synapses per striatal cell (ten thousand glutamatergic, four thousand GABAergic, and one thousand dopaminergic). Here we will emphasize the significant number of GABAergic synapses targeting a single MSN. Although a significant fraction of them may be the efferents of striatal GABAergic interneurons, or GPe back projections to the striatum, the majority are probably the efferent of neighboring MSNs.

Early physiological studies with sharp microelectrodes failed to reveal the functional effects of this lateral inhibition (Jaeger et al. 1994; note the careful title acknowledging the limit of negative results in biology: "Surround inhibition among projection neurons is weak or nonexistent in the rat neostriatum"). New technology

Table 3.1 Sanity check of the synaptic map of the main axis of the basal ganglia. The numbers are given for a general nonhuman primate. The total number of synapses (*middle column*) should be achieved by multiplying the number of source/recipient neurons and the number of synapses emitted/received by a source/recipient neuron given on the right and left sides.

	Number of source neurons	Number of synapses emitted by a source neuron	Total number of synapses	Number of synapses on recipient neuron	Number of recipient neurons
Ctx-2-Str	$300 * 10^6$	500	$150 * 10^9$	$5 * 10^3$	$30 * 10^6$
Ctx-2-STN	$1 * 10^6$	60	$60 * 10^6$	$1 * 10^3$	$60 * 10^3$
Str-2-GPe,GPi,SNr	$30 * 10^3$	300	$9 * 10^9$	$30 * 10^3$	$300 * 10^3$
Str-2-GPe,GPi,SNr	$60 * 10^3$	500	$30 * 10^6$	$1 * 10^3$	$300 * 10^3$
GPi,SNr-2-Thal	$100 * 10^3$	250	$25 * 10^6$	10	$2.5 * 10^6$

Note: Ctx-2-Str, cortex to striatum; Ctx-2-STN, cortex to subthalamic nucleus; Str-2-GPe, GPi, SNr, STN-2-GPe, GPi, SNr, striatum, subthalamic nucleus to globus pallidus external and internal segments and to substantia nigra reticulata; GPi, SNr-2-Thal, globus pallidus internal segment and substantia nigra reticulata to thalamus.

(near-infrared microscopy, averaging of hundreds of postsynaptic potentials, organo-typic cultures, transgenic mice) enabled the recording of very close MSNs (Czubayko and Plenz 2002; Koós et al. 2004; Tunstall et al. 2002; Taverna et al. 2008; Planert et al. 2010) to reveal that the collateral inhibition between MSNs is indeed weak (in comparison with the inhibitory effects of FSIs) but existent. Only 30 percent of paired MSN recordings produced a synaptic response, and these connections were always unidirectional. D2 MSNs are more likely to have collateral effects than D1 NSNs. These new techniques are leading to a complex map of striatal microcircuitry that includes the lateral inhibition between MSNs, as well as the lateral effects of striatal interneurons (Tepper et al. 2004; Burke et al. 2017; Hjorth et al. 2020).

Lateral connectivity is not limited to the striatum. Although less massive, one can observe axon collaterals in most basal ganglia structures. Thus, all GPe cells have local axon collaterals that probably innervate the soma of their neighbors (figure 3.1). The GPe neurons could be divided into two populations with a mean of 250 and 600 local axonal boutons. The local axon collaterals give rise to arborization close to, or within, the parent dendritic field (Sadek et al. 2007). For a comprehensive review of lateral connectivity in the basal ganglia, see Parent et al. (2000).

FEEDBACK AND RECIPROCAL CONNECTIVITY IN THE MAIN AXIS OF THE BASAL GANGLIA

Never say never in biology. Although the basal ganglia can be characterized as feed-forward networks, exceptions exist. Here, we will discuss the back projections from the GPe to the striatum and the STN.

Single-axon tracing of GPe efferents have frequently reported projections to the striatum (16 percent of GPe neurons in the nonhuman primate; Sato et al. 2000). The pallidostriatal pathway was ignored by most basal ganglia students up until proto-typical and arkytypical GPe neurons were described (Mallet et al. 2012; Mastro et al. 2014; Abdi et al. 2015; Hernández et al. 2015). The prototypic GPe neurons project to the STN, GPi, and SNr, and the arkypallidal neurons exclusively innervate the striatum, targeting both striatal projection neurons and interneurons (Glajch et al. 2016). The number of striatal axon varicosities from a single arkypallidal neuron was 736 (Fujiyama et al. 2016). If indeed all GPe striatal projections are given by arky-typical neurons (but see Fujiyama et al. 2016) and the arkytypical neurons account for a quarter of all GPe neurons, the number of GPe boutons in the striatum of the idealized nonhuman primate is $0.25 * 160,000 * 1,000 = 40 * 10^6$, a small fraction in comparison with the estimated number of $12 * 10^9$ MSN-MSN synapses. Still, GPe-striatum connectivity might have a major role in the physiology of the striatum, either from a more strategic location of the synapses on the MSN soma or by affect-ing the highly efficient network of FSIs that provide a general inhibition blanket to striatal MSNs.

Old students of the basal ganglia might be surprised by the inclusion of the GPe to the STN connectivity in the section on feedback connections in the basal ganglia network. The common model of the basal ganglia network is the D1/D2 direct/indi-rect box-and-arrow model. In this model the indirect pathway starts with D2 MSNs that inhibit the GPe that inhibits the STN (disinhibition, figures 9.1 and 9.2). How-ever, strong and reciprocal connections exist between the GPe and the STN (figure 9.4). Here, we took the view of the three-layer model of the basal ganglia (Deffains et al. 2016; figure 9.4), moving the STN to the input tier of the basal ganglia, with reciprocal connections with the GPe in the central layer of the basal ganglia net-work. The feedforward STN-GPe connectivity has been described above. GPe to STN feedback connectivity is surprisingly sparse. A single GPe neuron gives about 250 synaptic terminals in the STN. Thus, even if each GPe synapse is contacting a differ-ent STN neuron, a single GPe neuron contacts fewer than 2 percent of STN neurons ($10 * 10^3$ and $50 * 10^3$ in the rodent and nonhuman primate, respectively). The total number of GPe GABAergic terminals in the rodent STN is therefore estimated as $40,000 * 250 = 10 * 10^6$, and the average number of GPe synapses per STN neuron is $10 * 10^6/10 * 10^3 = 1,000$. A single STN neuron maximally receives input from 2 per-cent ($1,000/40 * 10^3$) of GPe neurons. However, this sparse connectivity is highly potent. A single GPe axon gives rise to multiple synapses with both the soma and the dendrites of an individual STN neuron. Dynamic clamp experiments revealed that small changes in the GPe input to the STN can inhibit and synchronize the subthalamic activity (Baufreton et al. 2009).

Quantitative data comparing the cortical-glutamatergic innervation and the GPe-GABAergic innervation of the STN in the same species are still missing. If in the nonhuman primate a single GPe cell is also giving 250 synapses at the STN, the total number of GABAergic synapses in the STN is $200 * 10^3 * 250 = 50 * 10^6$, about the same order of the number of cortical glutamatergic synapses (table 3.1). This is a very different scenario than the striatum, in which the number of excitatory afferents (~10,000 synapses/MSN) outnumbers the number of GABAergic afferents (~1000 synapses/MSN). It is in line with the report that the density of GABAergic symmetric synapses on the subthalamic dendrites is larger than the density of glutamatergic asymmetric synapses (Bevan et al. 1995). Maybe the number of cortical (and thalamic) neurons projecting to the STN is smaller than $1 * 10^6$. In any case, the two subthalamic afferent systems interact (Chu et al. 2015).

CONVERGENCE/DIVERGENCE OF BASAL GANGLIA CONNECTIVITY

Most of the connections along the main axis of the basal ganglia, starting with cortical projection to the striatum, to the GPe, and then to the GPi/SNr, are characterized by a high degree of numerical reduction in the number of neurons from the source to the recipient nuclei. The convergence is never complete, and in most cases neurons in the source structures innervate hundreds of neurons in the recipient areas. The STN is an exception. It is the only divergent and glutamatergic structure in the basal ganglia network. The outputs of the basal ganglia are probably divergent, with two levels of expansion from the GPi/SNr to the thalamus and then to the cortex. Data are missing regarding the basal ganglia projections to the brain stem motor nuclei.

Despite the high levels of divergence and convergence in the feedforward networks of the main axis of the basal ganglia circuitry, the average connectivity is very sparse. Individual neurons in the source structure innervate only a small proportion of the neurons in the recipient area, and a single neuron in the recipient area receives input from a limited fraction of the source neurons (Kincaid et al. 1998; Baufreton et al. 2009). Thus, recipient neurons with totally overlapping dendritic volumes have few presynaptic axons in common, and neighboring source cells with overlapping axons have few target neurons in common.

The synaptic connectivity map of the basal ganglia is very different from that of the cortex. On a first order of approximation, we can describe it as a feedforward monolayer network versus the multilayer and reciprocal connectivity in the cortex. A second major difference is that the limited number of synapses emitted by basal ganglia neurons, and therefore the divergence properties of the basal ganglia network, is much smaller than in the cortex network. The maximal divergence of basal ganglia neurons is five hundred to one thousand (assuming no multiple contacts). Cortical pyramidal neurons emit seven thousand to ten thousand synapses, with a minimal number of multiple contacts. The divergence ratio of cortico-cortical connectivity is one order higher than in the main axis of the basal ganglia.

CONVERGENCE, DIVERGENCE, AGAIN AND AGAIN

Alternation between convergence and divergence or between contraction and expansion is very common in the brain networks. Marr (1969) and Albus (1971) have inspired us by showing that huge expansion and contraction in the cerebellum circuit enable the dynamic classification of complex data. The convergent networks of the basal ganglia probably enable the bottleneck dimensionality reduction selecting the most critical features of the states to be used for optimization of the next actions. The output of the basal ganglia is expanded while routed to the frontal cortex and brain stem motor nuclei, to converge again on a small number of spinal motor neurons, leading to our motor behavior.

Deep neural networks are buzzwords today for advanced methods of information processing. The basal ganglia network is a good example of a natural network. Unlike artificial networks, we do not have to assume that the same kind of computation is performed at each level in our basal ganglia or brain networks. Switching between excitatory/inhibitory, linear/nonlinear, and convergent/divergent layers might significantly improve the speed and the accuracy of processing in the brain network (Bar-Gad et al. 2003; Jeanne and Wilson 2015). Figure 3.3 is a toy model of the

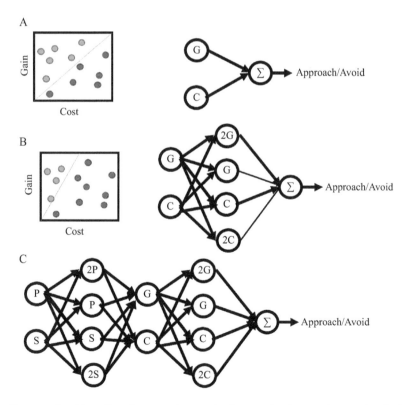

3.3 Computational benefits of repeated expansion/contraction in neural networks. *A*, simple classification problems. *B*, expansion of the network enables risk-aversive policy. *C*, repeated expansion/contraction network enables optimal approach/avoidance behavior for different objects.

advantages of repeated expansion/contraction architecture in the nervous system. Figure 3.3*A* depicts the simple classification problems of a simple behaving agent, with two neurons representing the gain (G neuron) and the cost (C neuron) of her action. The decision neuron can do the subtraction operation G – C and, if the sum is positive (gain is larger than the cost), the agent will select an approach action. If the sum is negative, she will select the avoidance action and would probably explore better alternatives. However, situations might arise in which simple subtraction is not good enough—for example, if our agent is already well fed or is tired. In this case she will be ready to approach only if the gain is much larger than the cost. Figure 3.3*B* shows that expansion of the two neuron input layers into a middle layer with neurons representing the gain, cost, and K * gain and K * cost (K=2,3, . . .) easily makes possible risk-aversive (e.g., approach only if 2G>C) or risk-seeking policy. The decision criteria do not have to be linear, and having neurons representing G^2 or the like enriches the classification repertoire. Unfortunately, objects in the world do not come with gain/cost tags. The agent can estimate their gain/cost values by their external properties—for example, their color and size (a big red apple is probably better than a small yellow one). An expansion/contraction network would help the agent make a flexible and accurate conversion from the object's external properties to a representation of the object's gain and cost and, finally, to a decision of approach or avoidance (figure 3.3*C*).

This chapter and the two previous chapters have focused on the gross, cellular, and connectivity anatomy of the basal ganglia. This is a monograph on the computational physiology of the basal ganglia and their disorders. Kemp and Powell (1971)'s seminal review of the anatomy of the basal ganglia and cerebellum (figure 1.5) cites John Eccles (1953): "Precise physiological investigation has to await the construction of reliable maps of the nerve connections." I failed to locate this citation in my copy of Eccles's book; still, I trust Kemp and Powell to read it more carefully than me. In any case, "Accept the truth from whatever source it comes" (foreword to the eight chapters of Maimonides[4]), and the truth is that anatomy creates the boundaries of the physiology space.

CHAPTER SUMMARY

- The synaptic connectivity of the basal ganglia is of a feedforward converging-diverging nature with lateral connectivity.
- There is a numerical reduction in the number of neurons from the cortex to the striatum and to the STN.
- Striatal projections are converging, while subthalamic projections are diverging.
- Pallidal projection to the thalamus and thalamic projections back to the cortex are of a diverging nature.

4

COMPUTATIONAL MODELING OF BRAIN AND BEHAVIOR

Our aim is to understand the computational principles of the basal ganglia and the brain. I do not neglect the importance of the blood and oxygen supply and metabolic processes in the brain. As our computers' abilities depend on a proper power supply, our brain processes rely on proper brain metabolic activity. However, the brain's unique feature is information processing, and we will devote this book to this attribute.

TOP-DOWN VERSUS BOTTOM-UP APPROACHES

There are two extreme approaches to the old question of what the best pathway is for understanding the brain. On one side is the bottom-up, reductionist approach.[1] In general, the reductionist approach would suggest that physics leads to chemistry, to biology, to psychology, and, finally, to sociology. In neuroscience, this approach states that one should start with detailed modeling of each neuron, taking into account the morphology and physiology of the different compartments of those neurons. The fully connected simulated brain would enable us to test stimulation-response chains and to study the effects of manipulation (simulating disease-like states) on these chains.

The ability of thermodynamics to explain and predict the macrobehavior of a gas in a container by describing the Brownian movements of the gas molecules inspired the thinking that good science should follow the reductionist bottom-up approach and go from the microlevels to the mesolevels to the macrolevels. However, chemistry was not replaced by physics, and in most cases complexity does not change as we move from the micro to the macro. This point has been nicely illustrated in P. W. Anderson's 1972 paper "More Is Different," which stresses that each level of nature is as complex as the other levels.[2] The "more is different" principle is probably correct as we move from molecular biology to neurons and to neural networks. "More is obviously different" when we move from brain activity to the complex behavior of humans in society, or during disease.

Furthermore, the brain is probably a chaotic system.[3] A chaotic system is a deterministic (not random) system in which physical laws and equations describe the evolution of the system according to its current state and inputs. Nevertheless, the system is highly sensitive to the initial conditions (e.g., current state). No matter how accurate our measurements and models of the brain might be, they will always have one more level of accuracy not measured. Thus, bottom-up models of the brain will never be able (in my opinion) to predict behavior. Unlike behavior controlled by the complex biology of the brain—for example, body and social interactions—weather is controlled by much better understood Newtonian physics. But even so, no one can predict the exact temperature and wind intensity in his yard a year from now.

On the other extreme is the top-down approach. One should start with psychology or psychiatry and set goals for the physiological investigation of the brain (Marom 2015). If you are looking for a computational model of the brain, start with a model such as the Hopfield mean field model, state its prediction, and test whether these predictions are verified or ruled out.[4]

There is no one correct way to understand the brain and the basal ganglia. The optimal way is also a function of the tools available. Nikos Logothetis (2008) and Shimon Marom (2015) cite Valentino Braitenberg in saying that "it makes no sense to read a newspaper with a microscope." The battery of tools we have is more a function of our time, education, and resources than of our scientific goals. Actually, I am afraid our tools strongly affect our goals. Therefore, only a combination of bottom-up and top-down approaches will take us closer to our goal of understanding the basal ganglia and the brain. I have therefore decided to study the spiking activity of basal ganglia neurons and to use the top-down approach toward the computational modeling of the basal ganglia.

HIERARCHICAL LEVELS OF UNDERSTANDING OF THE NERVOUS SYSTEM

David Marr is one of my heroes.[5] Marr claimed that understanding a neuronal system should be built on three levels: What is the computational goal of the system? How does the system do what it does? And how is the neuronal system physically realized? David Marr's 1969 computational model of the cerebellum is a masterpiece. However, I am not aware of many neurologists treating patients with cerebellar diseases (e.g., spinocerebellar ataxia) that know and care about David Marr's thinking and models. My hope is that we will be luckier with our basal ganglia models, and they will be used as a framework to better understand and treat patients with basal ganglia disorders.

We will therefore ask in the second part of this book the following major questions:

- What is the computational goal of the basal ganglia?
- How does the basal ganglia system do what it does?
- And how is the basal ganglia system physically realized?

Our second-order questions would consist of:

- Why does the main axis of the basal ganglia use GABA (inhibitory transmitter) as a carrier of information?
- What are the advantages of the high-frequency discharge of neurons in many basal ganglia structures?
- What are the physiological/computational roles of GPe pauses?

In the third part, we will turn our attention from the normal and healthy basal ganglia to the nonhealthy basal ganglia and ask:

- What happens when the basal ganglia are not doing their job? That is, we will try to better understand the computational pathophysiology of basal ganglia–related disorders (e.g., Parkinson's disease).
- Can we use our understanding of the physiology and pathophysiology of the basal ganglia to improve the treatment of patients with severe Parkinson's disease and other basal ganglia–related disorders (e.g., by closed-loop deep brain stimulation therapy)?

CHAPTER SUMMARY

- Computational models of the brain span the spectrum of bottom-up versus top-down approaches.
- Chaotic systems are deterministic systems that are highly sensitive to their initial conditions. My working hypothesis is that the brain is a deterministic chaotic system, and therefore predicting future animal behavior is a mission impossible.
- Applying David Marr's computational questions to the basal ganglia, we will ask: What is the computational goal of the basal ganglia? How does the basal ganglia system do what it does? And how is the basal ganglia system physically realized?

5

PRINCIPLES OF NEUROPHYSIOLOGICAL DATA ACQUISITION

The study of neurophysiological phenomena started in 1790 with Luigi Galvani demonstrating the role of electrical activity in the nervous system. Edgar Douglas Adrian provided the second major breakthrough by recording the activity of single sensory fibers in a series of experiments reported between 1926 and 1929. Adrian said, "I had arranged electrodes on the optic nerve of a toad in connection with some experiments on the retina. The room was nearly dark and I was puzzled to hear repeated noises in the loudspeaker attached to the amplifier, noises indicating that a great deal of impulse activity was going on. It was not until I compared the noises with my own movements around the room that I realised I was in the field of vision of the toad's eye and that it was signalling what I was doing." This led to the discovery of the all-or-none nature of the action potential (spike) and to the notion that stimulus intensity is encoded by the frequency of these spikes.

This book is biased toward the in vivo extracellular recording of neuronal activity or to Adrian's repeated loudspeaker spikes. This chapter will be devoted to the acquisition of neurophysiological data. More advanced insight should be sought in textbooks and peer-reviewed literature.[1]

ELECTRODES, FREQUENCY DOMAINS, AND TYPES OF NEUROPHYSIOLOGICAL DATA

The first description of human brain electrical activity was provided by Hans Berger's 1924 recordings of the electroencephalogram (EEG).[2] The EEG is measured by macroelectrodes placed on the skull that detect the synchronous activation of cortical neurons (figure 2.1A). Ed Evarts (1964) pioneered the extracellular recording by microelectrodes of action potentials (spikes) of single neurons (units) in the cortex of awake-behaving monkeys (figure 5.1). In between the EEG and the single-unit

A Awake

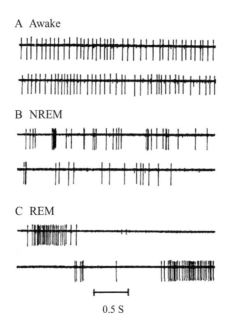

B NREM

C REM

0.5 S

5.1 Evarts's extracellular recording of pyramidal neurons in the primary motor cortex. Recordings were done during an awake state (AWAKE) and sleep (NREM, REM).
Source: Adapted from Evarts 1964.

activity, we can find the recording of local field potentials (LFP; for the origin of LFP in layered brain structures, see Buzsáki et al. 2012) and multiunit (population) spiking activity recorded by intracerebral macro- and microelectrodes.

EEG and LFP are slow electrical phenomena. LFP is sometimes called "local EEG" to denote the similarity between the two signals. In the frequency domain, EEG and LFP phenomena span the 0.1–70 Hz domain. They probably represent the neuronal subthreshold activity—for example, the synaptic input to a brain region. Nevertheless, LFP may be contaminated by spiking activity, and one should be careful with the interpretation of the high-frequency domain of LFP (Waldert et al. 2013). Single- or multiunit spiking activity is a fast phenomenon. The typical duration of an action potential is 1 ms, and therefore in the frequency domain, the maximal energy of spiking activity is around 1,000 Hz. However, there are broader action potentials (e.g., of striatal tonically active neurons and SNc dopaminergic neurons; figure 5.2*B*) and so typical filters used for recording spiking activity span the range of 300–6,000 Hz. Figure 5.2 demonstrates the "best typical" cases of our group with high signal-to-noise activity. The sharp-eyed reader may notice that in some of the recordings (e.g., the GPe low-frequency discharge neurons with bursts and the SNc) the background contains small spikes. Our working assumption holds that the background (figure 5.2*A*, *heavy horizontal black line*) is composed of the summation of many small-unit activities. Therefore, even if one cannot discriminate single spikes in his/her 300–6,000 Hz band-passed activity, the activity recorded is of the population spiking activity.

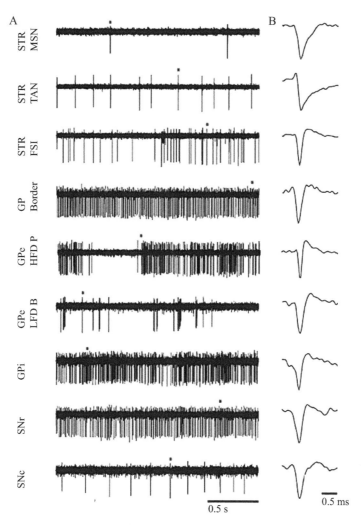

5.2 Extracellular recorded spontaneous spiking activity of different basal ganglia neurons. In (A), each line is a spike; in (B) the waveform of the spike marked with a * in (A) can be observed. STR MSN, striatum medium spiny neurons; TAN, tonically active neurons; FSI, fast-spiking neurons; GP border, pallidal border cell; GPe HFD-P and LFD-B, high-frequency discharge with pauses and low-frequency discharge with bursts globus pallidus external segment neurons; GPi, globus pallidus internal segment; SNr and SNc, substantia nigra reticulate and compacta.
Source: Adapted from Haber et al. 2011.

Extracellular recording techniques enable reliable and prolonged (tens of minutes to hours) recording in the behaving animal, while intracellular recording is better suited for in vitro studies and is very limited in time (<10 minutes) in behaving animals. More important, I do believe that the single action potential (spike) is the basic computational unit of the brain. Following Santiago Ramón y Cajal's neuron doctrine (Shepherd 2015), I would humbly suggest a "spike doctrine." To understand the functional anatomy of the brain, we should understand the structure of each neuron and how that neuron is connected to other neurons. Similarly, to understand brain information processing, we should understand the single spikes of a single neuron and how these spikes are related to other spikes of the same and other neurons or to the subthreshold LFP activity.

THE BLACK MAGIC OF NEUROPHYSIOLOGY RECORDING

When I started my journey into electrophysiological studies of the brain and the basal ganglia, my mentors and admired exemplary physiologists were technology wizards. They designed and built their own recording amplifiers and microelectrodes. They used "black magic" to change the location of the ground connections in the setup to improve the signal-to-noise ratio of the recordings. Computers and digital electronics were in their early stages of development. Therefore, electrophysiological studies of behaving animals used one microelectrode and a 300–6,000 Hz filter to eliminate the LFP and other low-frequency artifacts (movement, 50 or 60 Hz power line noise) and to ensure a flat background and a reliable detection of spikes by threshold-crossing methods. These devices emitted a 0/1 digital signal when a spike was detected, and the timing of these 0/1 signal detections, along with the timing of behavioral events, was recorded on the early computers of that day (Digital's PDP-15, with thirty-six kilowords memory, in my case). LFP were filtered out to optimize spike detection and also because hard-core physiologists presumed they (as EEG) were often confounded with low-frequency artifacts.

These days are over; today, data-acquisition technology is often based on multi-electrode arrays, fast analog-to-digital converters, and high-speed computers with practically unlimited memory. The recording equipment is provided by professional companies, and black magic to properly set the recording cables is seldom needed. Nevertheless, this technology advance may also lead to "garbage in, garbage out" (GIGO) conditions, so even today's lucky physiologists should be aware of the hardware (HW) limits of their equipment. We will briefly review these topics below.

Microelectrodes The extracellular electrical activity of the brain is usually recorded by metal (e.g., tungsten, platinum iridium) coated with insulation material (e.g., glass, parylene C) electrodes. The distal end of the electrode is tapered to enable penetration of the dura, and the last 5–25 μm are not insulated and provide the electrical connection with the brain tissue and electrical fields. The electrical model of the microelectrode is complicated. First, current is carried by electrons in the metal microelectrodes and the recording apparatus, whereas current is carried by ions (mainly sodium and chloride) in the extracellular brain environment. The long conduction shaft of the electrode is separated from the brain's conducting tissue by a thin coating of insulation material, therefore creating a capacitance-like structure between the microelectrode tip and the input to the first amplifier. In the current practice of today, we neglect these complicated issues, measure the impedance of the electrode for the main frequency of interest (1,000 Hz for spikes, 10–30 Hz for EEG,

LFP), and report the impedance of the electrode at a specific frequency, as if it was a pure resistance (ohms). I have no doubt that future technology will provide us with better insight into the biophysical model of our electrodes.

The typical value of the impedance at 1,000 Hz for extracellular recording electrodes is 0.5–1 MΩ (megaohm). You will find people like us that use 0.2–0.3 MΩ microelectrodes, with larger exposed tips to enable recording from more than one unit by a single microelectrode (figure 5.3). On the other hand, Ed Evarts used 2 MΩ microelectrodes with approximately 5 μm exposed tips that must be very close to the cell body of the neuron but then will record only the activity of a single unit (figure 5.1). The electrical field generated by a single action potential is a function of the geometry of the neuron, the distance from the neuron, and many other factors. In any case, the electrical field generated by a single action potential is tiny. The amplitude of single spikes recorded in the mammalian nervous system is 50–200 μV (if you are lucky, you may get spikes with an amplitude of 500 μV depicted in your papers as "best typical" examples). This tiny amplitude of the spikes, in combination with the high impedance of the recording microelectrodes, is why black magic is needed for good extracellular recording of spiking activity. The long path from the tip of the microelectrode to the input of the head-stage amplifier is a conductor with high impedance, and therefore any miniscule noise transmitted through the air will induce significant noise at the input of the head-stage amplifier (assuming a pure resistance path with 1 MΩ resistance, then a current of 10^{-10} amperes, will lead to a voltage swing of 10^{-4} V or 100 μV—i.e., the amplitude of a medium-size spike).

A few tricks can be used to improve the noise sensitivity of microelectrode recording systems. First, reducing the distance and the wiring to the first amplifier—for example, by mounting the first amplifier on the proximal end of the electrode and on the skull—is a very useful method. Second, lower the impedance of the electrode. The easiest way is to increase the length of the exposed tip. However, this may increase the sensitivity of the electrode to the activity of other units in the local circuit and eventually increase the background activity, thus worsening the balance between the recorded single- and multiunit activity (Zaidel et al. 2010). Increasing the roughness of the tip would enlarge the tip surface without affecting the length of the exposed tip and the balance between multi- and single-unit activity. Several materials (gold, titanium nitride, iridium oxide, and others) and methods for plating are used, but their stability in the liquid environment of the brain is still an open issue.

Amplifiers The first head-stage amplifier is the most critical. It should have high input resistance and minimal input capacitance to minimize the distortion of the recorded signal. At the same time, it should have minimal intrinsic noise. This amplifier serves mainly to reduce the setup impedance and can be used with a gain of one for this purpose. Since the output of the amplifier is of low impedance and the next wiring can be protected by ground shielding (coaxial cables), the first amplifier is the weakest link in our recording chain and will usually set the intrinsic noise level of the recording setup.

Filters Filters are defined as being low-, high-, and band-pass filters. Filters are further defined by their cutoff frequency (–3 dB, or half-power attenuation) point and by the power decline slope (filter order, number of poles, –20 dB/decade for first order, one pole, passive resistor-capacitor filter). Physiologists often use band-pass filters. The role of the high-pass filter is to minimize the big low-frequency deflections. Thus, in EEG and LFP studies the cutoff frequency is typically 0.07 (or 0.1) Hz. Old setups of spike activity that were trying to eliminate the LFP used a cutoff of 300 Hz. The low-pass cutoff frequency is set to minimize interfering noise while maximizing the information sought. In LFP studies the cutoff frequency is set to 70 Hz, or even 30 Hz in a case with high 50 or 60 Hz power line interference. For spike recording it is set around 6,000 to 9,000 Hz. In my experience you can get most of the information about spiking activity even with very conservative 500–2,000 Hz band-pass filters.

Because of the limited dynamic range of previous (12 bits) generation analog-to-digital converters and because of the one-order difference in magnitude, spike and LFP data were often collected through two parallel channels, LFP (0.1–30 Hz band-pass filter) and spike (300–6,000 Hz). Real HW filters induce different phase shifts and inaccurate relationships between the LFP and the spike data—for example, when calculating the spike-triggered average of LFP. I therefore recommend the use of one filter (e.g., 0.1–9,000 Hz) for data acquisition and separation of the data into two channels in later digital off-line processing that enables zero-phase shift filtering.

Analog-to-digital (A/D) converters A/D converters transform analog continuous data into discrete numerical representation. The numerical representation is discrete in time (set by the sampling rate) and in amplitude (set by the dynamic range and the resolution of the A/D converter).

The Nyquist-Shannon sampling theorem bridges continuous-time signals and discrete-time signals. It establishes that a sample rate greater than twice the maximal frequency of an analog signal permits a discrete sequence of samples to capture all the information from a continuous-time signal. Sampling at lower frequencies would yield aliasing error and data distortion that cannot be corrected by postfactum analysis. Because it is impossible to get an infinite filter slope to minimize the risk of aliasing error, the sampling rate of a modern electrophysiological data acquisition system is set to at least three to four times the cutoff frequency of the system low-pass filter. Thus, for four-pole (–24 dB/octave, –80 dB/decade) Butterworth low-pass filters with a cutoff frequency of 6,000 Hz, the common sampling frequencies would be 20–40 kHz.

The resolution of an A/D converter is set by the number of bits in each sample. Old A/D converters were 8–12 bit converters; namely, the range between their minimal and maximal voltage (dynamic range) was divided into $2^8 - 2^{12}$, or $256 - 4,096$ digital outputs. Today, A/D converters are 16–24 bit converters. However, not all bits are useful, and some of the last bits of these very high-resolution converters are within the level of their intrinsic noise. Practically, the amplitude resolution of the A/D converter should be weighted together with the dynamic range (maximal and minimal converted voltages) of the converter and the typical range of its input (after amplification). As always, the golden mean, or the optimal trade-off, should be found. You may have a 12-bit A/D converter with a plus/minus 5 V dynamic range and a pre-amplifier gain equal to 10. Thus, the 10 V dynamic range is divided into 4,096 discrete values, or 244 μV resolution. Most of your spikes will have 0 peak values, and some will reach 1, but none will show values of 2,047 (maximal positive voltage). To overcome this unacceptable situation, you may use an amplifier with a gain of 10,000. Your resolution will equal 0.25 μV; however, any voltage larger than 1 mV (i.e., some of the spikes of M1 layer 5 large pyramidal cells, typical LFP fluctuations, or stimulation artifacts) will exceed the dynamic range of your A/D converter. In the best case, their value will be registered as the maximal values; however, electronic devices need time to recover from saturation, and a significant portion of your data might be lost. The bottom line is that careful adjustments of the A/D resolution, the dynamic range, and the characteristics of your signals are needed to optimize neurophysiological recording (currently, we are using amplifiers with a total gain of 20- and 16-bit A/D converters with plus/minus 1.25 V range).

SPIKE DETECTION, SPIKE SORTING, AND ISOLATION QUALITY

In the early days of my mentor Mahlon DeLong and his mentor Ed Evarts, "when boats were made of wood and men were made of steel," extracellular recording techniques were optimized to near perfect (figures 5.1 and 11.8), and the signal-to-noise ratio of the recorded units was very high (it is rumored that Evarts moved his electrode up and down to ensure recording of a single isolated cell). Spikes with such high signal-to-noise ratios were detected online using a simple amplitude threshold with a minimal number of errors. However, those days are over, and we rarely see such high-quality recording today.

Our fast and efficient digital computer tools enable us to simultaneously sort the activity of several units recorded by a single electrode (figure 5.3). Many methods and types of commercial equipment make spike sorting possible (Lewicki 1998; Pedreira et al. 2012; Rey et al. 2015). Spike sorting reveals the activity of close neurons (usually less than 0.1 mm apart) and opens the door for questions regarding the synchronization of the discharge rate. Thus, it upgrades the level of our understanding from the level of a single neuron to the level of the (local) network. There is a big range of spike-sorting methods spanning the spectrum between manual adjustment of the detection and sorting parameters and completely automatic and the spectrum between online (real time) and off-line sorting methods. The choice of method (manual/automatic, online/off-line) is strongly affected by the number of electrodes used and whether they can be moved during the recording. Research groups that use

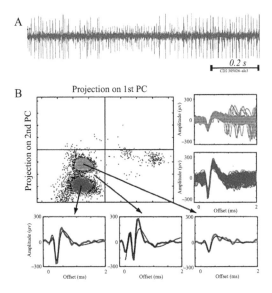

A

B   Projection on 1st PC

Projection on 2nd PC

Amplitude (μv)

Offset (ms)

Amplitude (μv)   Offset (ms)

5.3 Spike sorting of spiking activity of two neurons recorded on a single electrode. An example of off-line analysis of two cells recorded on the same electrode in the external segment of the globus pallidus is shown. *A*, 300–6,000 Hz band-passed filtered raw data. The two units can be easily distinguished by their different amplitudes. Note the continuous discharge of small neuron while the larger one pauses. *B*, *left*, principal-component spike sorting. The spike shapes of all of the spikes within the polygons are plotted (*right*). In the single-spike examples (*bottom*), the actual spike shapes are depicted by the colors of the clusters, and principal-component reconstructions are plotted in black. The unidentified signal, composed of a green spike immediately followed (and distorted) by a purple spike, is shown in blue.
Source: Copied (with permission) from Bar-Gad et al. 2003.

arrays with hundreds of contacts favor fully automatic methods (Yger et al. 2018). Single-electrode users will favor manual adjustment of the electrode position and the detection/sorting parameters. Our trade-off between quality and quantity is the use of several (four to eight electrodes) that can be moved independently. The position of the electrode and the spike detection/sorting parameters are continuously adjusted and documented during the recording.

Noisy recording due to technical and methodological problems will lead to detection and sorting errors. Nevertheless, in most physiological papers you will find the note "Only well-isolated units were included in the study," hinting at high-quality recording like Evarts's. When Mati Joshua joined my lab, he took advantage of our continuous high-frequency sampling of the electrode voltage to develop a method for quantifying isolation quality (Joshua et al. 2007). The isolation quality score quantifies from zero to one the isolation quality of each recorded neuron. Today, we use and report a threshold for the units included in our studies. There are other scales for spike isolation (Hill et al. 2011).

Can one hear the shape of a drum? (Kac 1966). It is the classical inverse problem. The shape of a drum sorts out the frequencies at which a drumhead can vibrate (the forward problem). However, multiple and even infinite numbers of different drum shapes may yield the same set of frequencies. The answer to the drum inverse question is that for many shapes, one cannot hear the shape of the drum completely. However, some information can be inferred. Spike-sorting methods also face the inverse problem, and many different neuron morphologies may yield the same shape of extracellular recorded spikes. Thus, a high level of isolation quality is not enough to ensure that only the activity of one neuron has been recorded. It could be that our electrode shifted to record the activity of another neuron, or damaged the recorded one, and still we would not detect a significant change in the spike waveform. To improve the reliability of single-neuron recording, one can measure the stability (stationarity) of the discharge rate (Gourévitch and Eggermont 2007; Valsky, Heiman-Grosberg, et al. 2020). Theoretically, other discharge parameters, like the pattern and response to behavioral events, could be tested as well. However, all stability tests will do poorly in cases in which the time constant of the natural variability of the discharge (e.g., due to different sleep stages) is short compared to the total duration of the recording. Neurophysiology is still an art, and most importantly, it does not pretend to be perfect. As Salvador Dali said, "Have no fear of perfection— you'll never reach it."

ELECTRICAL STIMULATION OF THE BRAIN

Stimulation is used in electrophysiological studies of the brain to support causality (e.g., Eduard Hitzig's and Gustav Fritsch's 1870 demonstration that electrical stimulation of the cortex motor strip evoked movements) and functional connectivity (antidromic and orthodromic stimulation).

Voltage or current sources can be used for electrical stimulation of the nervous system. Voltage sources are simpler; however, the current delivered is a function of the impedance of the electrode, the tissue, and the interface between the electrode and the tissue. Therefore, current sources should be preferred, but they have their limits. The maximal amplitude of the injected current is limited by the voltage range of the current source and the impedance of the load (usually determined by the electrode rather than by the tissue). For example, if you try to inject 100 μA through a microelectrode with impedance of 1 MΩ, the voltage of your current source needs to reach values of $100 * 10^{-6} \times 1 * 10^6 = 100$ V (Ohm's law, V = I * R). Since the impedance of the microelectrode and of the stimulated tissue is not purely ohmic, the actual load is a function of the stimulation waveform, and the amount of current delivered to the tissue might be different than expected.

Electrical stimulation of the nervous system of behaving animals is usually delivered as monopolar or bipolar stimulation. In monopolar stimulation the stimulation circuit is closed by a remote reference (ground). In bipolar stimulation the current flows between two adjacent contacts (e.g., concentric electrodes), and therefore the effect is more limited in space. Microelectrode stimulation is usually carried in a monopolar mode. Neural tissue is activated by cathodal (negative, attract cations or positively charged ions) stimulation. However, stimulation may also occur near the anode (virtual cathode).

There are safety issues to remember when stimulating the brain. First, if the current density is too high, damage to the brain tissue may occur. The upper limit for typical microelectrode stimulation is 60–100 μA. For macroelectrodes like those used for deep brain stimulation (DBS) therapy, current amplitudes at the range of several milliamperes are often used. Second, prolonged stimulation may lead to the release of metal particles into tissue and chronic damage. This can be compensated by balanced cathodal-anodal stimulation. In a clinical setting, a charge balance is achieved by asymmetric cathodic and anodic pulses to avoid hyperpolarization during the anodic pulse. The cathodic pulse is short (30–120 microseconds) and of high amplitude (1–6 mA given to a macrocontact with impedance of a few KΩ). The following anodic pulse is of long duration (3 milliseconds) and of much lower amplitude.

Microstimulation is often used to identify the primary motor cortex (e.g., Goldberg et al. 2002) or related structures (Alexander and DeLong 1985a, b). To stimulate neural tissue short trains (0.2–0.5 ms), short stimuli (30–200 μs) are given at high frequency (200–300 Hz). These stimuli evoke brisk movements of a single joint at small (<40 μA) intensity. Much more complex movements are obtained with longer (>500 ms) trains (Graziano 2009). However, longer (>10 s) trains might lead to inhibition of the stimulated area.

As for recording with analog-to-digital (A/D) converters bridging between the electrode amplifier and the lab computer, digital-to-analog (D/A) converters often bridge between the computer and the stimulator. The amplitude and the temporal resolution should be carefully adjusted to the needs of the experiment. Because short pulses are often used, a higher sampling frequency is needed (the temporal resolution of 100,000 Hz D/A is 10 μs). Sampling of the stimulation pulse (e.g., by the voltage drop on a resistor connected in series to the electrode and the tissue) should be done more quickly than the typical sampling frequency (25–40 kHz) of spiking activity.

Finally, if electrical stimulation is used for the study of brain connectivity—that is, the effects on the neural activity of the local and remote elements are studied—one should be aware of the confounding effect of the stimulus artifact. Depending on the filter used for the recording, the stimulus artifact can be of a large duration. High-pass filters (e.g., those used for removing LFP from spiking activity) are built with a capacitor in the amplifier input, leading to very long artifacts. Direct-current (DC, with no high-pass filtering) amplifiers are recommended in this case. However, no filtering of the low-frequency potentials might lead to a different trade-off of the amplification and the dynamic range of the A/D converters. A stimulus artifact big enough to saturate the amplifier and the A/D input will disallow the recovery of the original data. If the A/D dynamic range is larger than the stimulus artifact, one can do numerical off-line reduction of the stimulus artifact template. Unfortunately, the stimulus artifact is not stereotypical and may change as a function of time and even during a single train (Bar-Gad et al. 2004; Wichmann and Devergnas 2011). Most of the available techniques enable stimulation and recording by two different electrodes. Multiplexing a single electrode between stimulation and recording mode creates a new field of problems. Recording and stimulation in behaving animals is still not an easy mission.

INACTIVATION OF BRAIN ACTIVITY

In 1990 Thomas Wichmann, Mahlon DeLong, and myself published that inactivation of the STN of MPTP-treated nonhuman primates ameliorates their parkinsonian symptoms (Bergman et al. 1990). A year later, I was surprised to receive a letter from a French physician whom I have never met congratulating us for providing an explanation to Nature's previous experiment (Sellal et al. 1992). Indeed, Mother Nature pioneered brain ablation experiments before scientists. However, sometimes you need a theory to understand what you see.[3]

In the neuroscience research era, inactivation of specific brain areas was initially done by ablation with mechanical aspiration using a small-gauge sucker (Mishkin 1978) and heating by transfer of long (>20 seconds) DC electrical pulses. In the human operating room, more delicate tools such as radio-frequency electrical heating, stereotactic radiation (Gamma Knife), laser, and focused ultrasound are used. The major drawback of all ablation techniques is their lack of specificity and the damage caused to passing fibers and possibly to nearby structures (Murray et al. 2017). Lesions by axon-sparing agents like ibotenic acid (a nonselective glutamate agonist) provide better selectivity for neurons (Bergman et al. 1990; Wichmann et al. 1994). Interestingly, the chemical structure of ibotenic acid is very similar to muscimol (GABA$_A$ agonist). In mushrooms, ibotenic acid is converted to muscimol via decarboxylation (i.e., serves as a muscimol prodrug) when the mushroom is ingested or dried.

Ablation is not revisable, and compensatory processes may mask or modulate the effects of the inactivated tissue. Temporal chemical inactivation by pressure or microdialysis of lidocaine (a local anesthetic that prolongs the inactivation of fast sodium channels) or muscimol (GABA$_A$ agonist; figure 5.4) overcomes these effects. Pressure injections might affect neural activity because of their mechanical effects or by changing the chemical structure of the environment. As for ibotenic acid, muscimol effects are axon sparing, while lidocaine would affect bypassing axons as well. The kinetics of chemical inactivation are slow (minutes to hours) and depend on the amount, diffusion, and washout of injected drug.

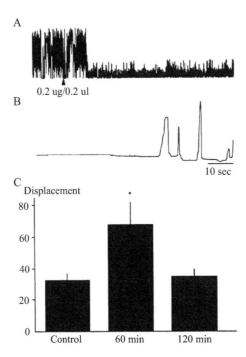

5.4 Transient inactivation of the subthalamic nucleus (STN) of an MPTP-treated monkey. *A, B*, STN neuronal activity and limb movement before and after injection of muscimol to the STN. *C*, elbow movements following torque pulse reveal a reduction in elbow rigidity sixty minutes after the STN muscimol injection.
Source: Copied (with permission) from Wichmann et al. 1994.

WHAT WAS LEFT OUT?

The physiological methods described so far have some major drawbacks. The number of neurons recorded is limited by the number of electrodes we can insert into the brain. Extracellular recording of spiking activity does not enable an accurate definition of the recorded neuron (Vigneswaran et al. 2011), and the release of transmitter from axon terminals might be independent of the spiking activity (Threlfell et al. 2012; Mohebi et al. 2019). Recording in the behaving animal teaches us about the correlation between neural activity and behavioral events but does not reveal causality. Electrical stimulation and chemical inactivation of the brain was my and previous generation's rescue path for the causality issue. However, the polarity of the effect of stimulation (excitation, inhibition) is a complex function of stimulation parameters (like frequency and duration). Additionally, which neuronal elements (neurons, bypassing fibers, afferent or efferent axons) the electrical stimulation affects is a riddle. Inactivation (temporary or ablation) faces similar problems of affecting nearby structures and tracts. Ablation effects might be overcompensated by other structures, and often it is less easy to discriminate between the core negative symptoms of the ablation and the positive compensatory symptoms.

New twenty-first-century methods may overcome the above-mentioned drawbacks—at the price of introducing other biases. Below I will discuss only briefly some methods that open new horizons for our understanding of the basal ganglia and the brain. These methods include optogenetics, measurements of tissue neurotransmitter concentrations, microelectrode arrays and tetrode recording, and, finally, intracellular recording techniques. Understanding the current advantages and disadvantages of these and other methods is the key to a good experimental strategy.

Optogenetic methods Voltage-sensitive dyes (molecules capable of emitting light in response to their electrical environment) were developed to enable broad sampling of the activity of many neurons (Grinvald 1985). However, these dyes were limited to studies of flat cortical areas and could not discriminate between different neuron types in the studied area. Optogenetic methods overcome these problems. Genetic targeting strategies such as injections of viral vectors and the creation of transgenic animals with Cre recombinase enable the delivery of light-sensitive probes to specific populations of neurons in the brain of living animals (Cui et al. 2013; Tecuapetla et al. 2016; Parker et al. 2018). Optogenetics also enables the insertion of light-sensitive ion channels (actuators) that can excite or inhibit injected genetically targeted neurons with millisecond precision (Boyden et al. 2005). Thus, optogenetics might be the dream solution for the specificity issue at the three levels of recording, stimulation, and inactivation.

Optogenetic methods have been extensively used in studies of the basal ganglia (Kravitz et al. 2012; Lüscher et al. 2015). Nevertheless, caution is advised. First, the genetic manipulation might affect other systems than those targeted and might shift a gray scale to a black-and-white picture. Second, most studies use intracellular calcium levels as proxies for spiking neural activity. The dynamics of the calcium fluorescence probes and intracellular calcium itself are slower and probably do not reflect an exact integral function of the neuronal discharge (Wei et al. 2020; Legaria et al. 2021). Recent studies suggest optogenetic monitoring of the full spectrum (including subthreshold activity) of the membrane potential (Adam et al. 2019)—however, at the expense of possible damage to the neurons following a few minutes of recording. Thus, early studies of optogenetic stimulation of the subthalamic neurons pointed toward antidromic activation of the motor cortex as the major mechanism of subthalamic DBS (Gradinaru et al. 2009). However, the kinetics of the opsin used were too slow to follow the high rates required for effective DBS. A recent study shows that using ultrafast opsin ameliorates the abnormal basal ganglia activity and the behavioral deficits of parkinsonian rats and indicates that subthalamic high-frequency stimulation plays a critical role in DBS (Yu et al. 2020). Viral expression in tissue might depend on the tissue and the species, and efforts to translate optogenetic methods to primates have been limited (De et al. 2020). Finally, optogenetic modulation of a population of neurons might shift the system into an extreme synchronized activity beyond the physiological domain (Phillips and Hasenstaub 2016). For example, Jim Teper and colleagues studied the functional connectivity between the cholinergic interneurons and the medium spiny neurons (MSNs) of the mouse striatum (English et al. 2011). They found that direct nicotinic excitation of neuropeptide Y (NPY)-expressing neurogliaform (NGF) GABAergic interneurons leads to rapid inhibition of the striatal MSNs. On the other hand, Avital Adler recorded the spiking activity of striatal MSNs, tonically active neurons (TANs, putative cholinergic interneurons), and fast-spiking interneurons (FSI; Adler, Katabi et al. 2013; figure 5.5A). Almost all TAN-MSN pairs were not correlated (figure 5.5B, *left*). In contrast, many of the MSN-FSI pairs displayed a broad and significant peak in their cross-correlation function (figure 5.5B, *right*). I suggest that optogenetic activation of the cholinergic interneurons created an extreme synchronized burst of activity and exposed sparse connectivity between two populations of neurons rarely used in the normal physiology. Admittedly, the sensitivity of in vivo cross-correlation analysis of spiking activity is limited. Low-efficacy synaptic interactions might be missed.

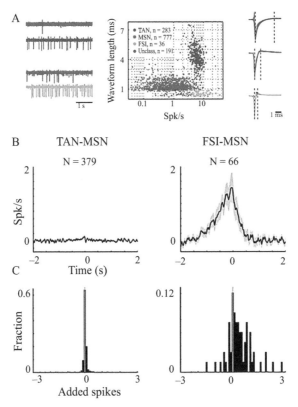

5.5 Different correlation patterns of striatal interneurons and projection neurons. *A*, identification of medium spiny projection neurons (MSNs; *red*), tonically active neurons (TANs; cholinergic interneurons, *blue*), and fast-spiking interneurons (FSI) by their discharge rate and spike waveform. *B*, average cross-correlation functions of 379 TAN-MSN pairs and 66 FSI-MSN pairs. *C*, distribution of number of spikes added to the reference spike around the timing of the trigger spike.
Source: Copied (with permission) from Adler, Katabi et al. 2013.

Measurements of tissue neurotransmitter concentrations Tissue neurotransmitter levels—for example, striatal dopamine levels—can be measured by three independent methods: microdialysis (Song et al. 2012), fast-scan cyclic voltammetry (Phillips et al. 2003; Wightman et al. 2007; Aragona et al. 2009), and the optical biosensor dLight (photometry; Patriarchi et al. 2018; Mohebi et al. 2019). The microdialysis probe is designed to mimic a blood capillary and consists of a shaft with a semipermeable membrane connected to inlet and outlet tubing. The probe is continuously perfused at a low flow rate with an aqueous solution that closely resembles the ionic composition of the extracellular environment. The solution (dialysate) is collected at certain time intervals for analysis by high pressure/performance liquid chromatography (HPLC) and accurate electrochemical detection. However, the microdialysis technique samples over a relatively large area and lacks the temporal resolution to examine phasic dopamine release. Fast-scan cyclic voltammetry or amperometry with carbon-fiber electrodes enables subsecond measurements of dopamine level, but other chemicals in the brain environment might confound the results. Photometry is an optogenetic-designed biosensor (Patriarchi et al. 2018). Photometry enables fast (subsecond resolution) and specific (submicro molar) measurement of extracellular dopamine levels; however, it does depend on the expression of virus in the monkey brain.

Microelectrode arrays and tetrode recording The activity of one cell cannot reveal the nature of local interactions in the studied structure or between structures. Therefore, many researchers (including me) have switched from one to multiple electrode recording. There is a trade-off between quantity and quality. Careful manipulation of a single electrode located as close as possible to the soma of the recorded neuron, without damaging it, leads to the exceptional recording of Ed Evarts and Mahlon DeLong (figures 5.1 and 11.8). The trade-off is still positive with eight electrodes that one or two experimenters can separately manipulate (see figures 5.2, 7.1, and 7.2). However, the yield of eight-electrode recording, even when supported by spike sorting, is the activity of ten to twenty neurons, and this yield comes with a price tag of a few hours of manipulating the position of the electrodes before recording starts.

A possible solution is the use of microelectrode planar arrays, like the Utah array invented by Richard Normann (Campbell et al. 1990) and now distributed by Blackrock. The Utah array is built of 10 * 10 sharp microelectrodes evenly spaced over an array of 4 * 4 mm. The array is inserted and secured to a flat region of the cortex and then connected to a head-fixed connector. After a period of recovery, one can bring the monkey to the lab, connect the recording setup to the head connector, and immediately see the spiking activity of 50–150 neurons. Brave research groups implant more than one array, in the same or at different cortical areas (Mitz et al. 2017). Utah array recording methods have been extensively used in nonhuman primates and even in human patients in studies of brain-machine interfaces (BMIs). Classical single-electrode/single-neuron studies are limited to stationary behavioral states. The array method allows the recording of enough units/session to achieve a reliable follow-up of learning processes. Again, one should be aware of the limits of this technique. Surgery needs a high level of training and expertise. Utah arrays are implanted above the pia after opening the dura (the dura is sutured after the implantation). Infections and other surgical problems are often observed in less-experienced groups. It is more difficult, or impossible, to change the position of the electrodes, so the experimenter cannot optimize the recording quality. The spike yield is not always as high as expected, and often the amount of recorded, well-isolated units decreases over the one to three months following surgery. Finally, Utah-like arrays do not enable the recording of the activity of neurons in deep subcortical structures. Other array methods with longer wires have been suggested; however, the cumulative experience of these methods is not enough to encourage their use by risk-aversive researchers.

A planar electrode array provides good x-y two-dimensional maps of neuronal activity. However, in many cases researchers are interested in the Z dimension—for example, the activity in different lamina of the cortex, or along the oblique axis spanning over the cortex, putamen, GPe, and GPi. Linear microelectrode arrays (LMAs) fill this gap. Typical LMAs have 16–32 contacts over a length of a few millimeters; however, a new generation of ultradense extracellular recording methods (Neuropixels, https://www.neuropixels.org/; Jun et al. 2017; Juavinett et al. 2019) is emerging. Each Neuropixels probe has 960 recording contacts, and the experimenter can select the best 384 channels to record from. A 6×9 mm probe base is fabricated with the shank and provides analog data processing and digitization, allowing less noisy digital data transmission from the probe. Some brave research groups have already started using Neuropixels probes in nonhuman primates (Trautmann et al. 2019).

In addition to the information about activity along the vertical axis of the brain, LMAs are very useful in studying the current source density (CSD) of LFP. Polarity inversion of the sensory-evoked activity may provide unique information for the local generation (source) of the evoked activity (Karmos et al. 1982; Kajikawa et al. 2017). Commercial DBS leads are a linear macroelectrode array with four to eight contacts. Polarity reversal (sometime called phase reversal) of the simultaneously recorded field potentials from DBS leads indicates that they are locally generated in the DBS targets and are minimally affected by volume conductance (Brown et al. 2001; Kühn et al. 2004; López-Azcárate et al. 2010). On the negative side, as for the horizontal array, LMAs can't easily penetrate the dura, are more suited for chronic experiments, are less easily moved, and often yield much less than expected.

Both horizontal and vertical arrays provide sparse sampling of the neural space (e.g., at a distance of 0.2–0.4 mm). However, anatomical lateral connectivity and functional interaction are probably a function of distance, and there are more interactions between close than between remote neurons (Smith and Kohn 2008; Rosenbaum et al. 2017). Spike-sorting methods enable the recording of the activity of few neurons by a single electrode; however, because of the difficulties in sorting time-overlapping spikes, they have limits in very dense neuronal areas. Tetrodes are built of four glued thirty-to-fifty micron insulated wires that record the activity of the same neuronal local space. This adds spatial information to the tools of the spike sorting because the spikes of one neuron will have different shapes on the different tetrode contacts. Tetrodes have been extensively used in rodent hippocampus recording; however, I am less aware of their use in the basal ganglia. I wonder if technologies like the Neuropixels, providing us with the benefits of both tetrode and LMA recording, will change the future of neurophysiology?

Intracellular recordings Physiological intracellular recording techniques have a long history. In the early to mid-twentieth century, researchers started to record from large cells or axons. Renshaw, Forbes, and Morrison in 1940 used glass microelectrodes to record the discharge of pyramidal cells in the hippocampus. Kenneth Cole and Howard Curtis in 1939 used the axial wire technique in the giant axon of the squid to show that membrane conductance increased during the action potential. Alan Hodgkin and Andrew Huxley in 1952 used the same technique to develop the voltage-clamp method and to describe the ion-channel kinetics of the action potential. Bernard Katz used intracellular recording to reveal the quantal mechanism of synaptic transmission in the neuromuscular synapse. The microelectrode method was expanded by the invention of the patch clamp by Erwin Neher and Bert Sakmann (1976) and the shift from sharp microelectrode recording to cell-attached methods.

Intracellular electrodes enable the recording of subthreshold activity. Additionally, tracers like biocytin can be injected into the cell to allow the reconstruction of the cell dendritic and axonal fields. Great progress has been made in our understanding of the cellular physiology of the basal ganglia and other brain structures using intracellular recording methods. The Memphis band, led by Steve Kitai, Hitoshi Kita, Charlie Wilson, and Jim Surmeier, is one example. But again, caution is needed. Glass microelectrodes have a tip size of approximately 1 μm with a resistance of 10–50 MΩ. The small tips make it easy to penetrate the cell membrane; however, they can injure the recorded cells. Cell-attached techniques may minimize the damage to the membrane. However, both sharp and cell-attached micropipettes are filled with an ionic solution, usually potassium chloride (KCl), and the electrode solution can wash out or buffer changes in the intracellular fluid composition. Perforated attached-cell techniques may minimize this effect on the intracellular fluid balance. Intracellular recording is much better suited for in vitro (brain slices) than in vivo recording, where small movements of the electrode might damage the recorded cell and significantly limit the recording duration. Finally, intracellular recording is usually carried out in the cell soma, and dendritic phenomena might be attenuated and ignored. The question of the effects of the synaptic inhibition of pallidal dendrites at a distance of 1 mm from the cell body remains open.

CHAPTER SUMMARY

- Neural activity is recorded by diverse methods, including metal microelectrodes and macroelectrodes used for extracellular recording and KCl-filled glass sharp and patch microelectrodes for intracellular recording.
- In the frequency domain, we discriminate between spiking activity (300–6,000 Hz) and local field potentials (0.1–70 Hz). The spiking activity represents the output of the neural structure studied, and the local field potentials represent the subthreshold activity, including the synaptic input to the cells.
- The quality of extracellular recording of spiking activity is a function of the equipment (microelectrode impedance, amplifier, filter, A/D converter, and sampling rate), the software used for spike detection and sorting, and the expertise and motivation of the experimenter.
- A short train of cathodic electrical stimulation of the brain can be used for excitation of neuronal tissue. However, the effects of prolonged high-frequency stimulation are still debated.
- Temporary and permanent (ablation) of neural activity can be achieved by several pharmacological compounds.
- Modern methods of optogenetic recording and activation of specific neuronal groups for in vivo measurement of the concentration of brain neurotransmitter and for array recording of the activity of thousands of neurons are in the pipeline. They are not perfect, but they will shed new light on the physiological functions of the basal ganglia and other brain structures.

6

PRINCIPLES OF NEUROPHYSIOLOGICAL ANALYSIS

Sir William Thomson, 1st Baron Kelvin, once said, "In physical science a first essential step in the direction of learning any subject is to find principles of numerical reckoning and practicable methods for measuring some quality connected with it. I often say that when you can measure what you are speaking about and express it in numbers you know something about it" (1891). I agree. We should try to express in numbers our knowledge of the basal ganglia and behavior. Below I will provide the basic analytical tools to be used. More advanced insight should be sought in textbooks and peer-reviewed literature.[1]

TEMPORAL AND FREQUENCY DOMAINS AND NEUROPHYSIOLOGICAL DATA ANALYSIS

The spontaneous discharge rate of neurons in the cortex is very low (median of one spike/second), and therefore spontaneous activity is often considered as noise and neglected. Figure 5.2 shows that some of the neurons of the basal ganglia have high spontaneous discharge rate (fifty to eighty spikes/second; see also figures 7.1, 7.2, and 11.8). We will therefore devote a major portion of this book to the analysis of spontaneous activity.

Spontaneous activity is usually characterized by the amplitude (discharge rate and power for spiking and analog activity, respectively), pattern (e.g., periodic, Poisson (random), or bursting), and synchronization (of units or local field potential [LFP] sites). Evoked activity is characterized by the physical stimuli that are efficient or less efficient in the activated region of interest, the polarity (increase or decrease), the amplitude, and the duration of the response. We can also test the similarity of responses of pairs of cells.

Classical analysis of neural activity has been conducted in the time domain (e.g., autocorrelation and cross-correlation functions). However, in cases with periodic activity we will often use the frequency domain and describe the power spectral density (PSD) of single sites and the coherence of multiple site recordings. Most of the methods described here can be equally applied to digital spike data and to analog continuous data such as EEG and LFP.

Finally, the correlation function, in the time and the frequency domains, can be calculated between different modality signals. Thus, correlation of the LFP and the spiking activity recorded by the same electrode may indicate the input-output relationships of the studied areas. The number of correlation pairs when one is recording EEG, LFP, and spiking activity with four electrodes in the cortex and four electrodes in the basal ganglia is very high and offers us a good glimpse into the functional connectivity of the cortex-basal ganglia networks.

TIME DOMAIN ANALYSIS OF RATE AND PATTERN OF SPONTANEOUS ACTIVITY

Discharge rate   The spontaneous activity can be recorded during a period when the subject is not engaged in behavioral tasks (e.g., quiet awake or during different stages of sleep) or in the last epoch of the intertrial interval when the subject is engaged in a behavioral task. The discharge rate may fluctuate, and therefore averaging is necessary for a reliable estimation of it.

Discharge pattern   Several functions can be used to estimate the discharge pattern of a neuron. As with the blind men and the elephant, each of these descriptions captures one feature of the whole story, and the "absolute truth" is the sum of all these functions. Below I will discuss the autocorrelation function, the histogram of the interspike intervals, the coefficient of variation of the interspike intervals, and specialized methods for the detection of bursts and pauses during spontaneous discharge.

The autocorrelation function, sometime called the autocorrelogram, describes the discharge rate as a function of time since a spike that occurs at time zero. The autocorrelation function will take into account all spikes occurring between time zero and the maximal delay of the autocorrelation function (e.g., 0.5–1 second). The autocorrelation function of single units will start with zero or very low discharge rates during the absolute and relative refractory periods of a neuron. This feature can be used to assess the quality of the isolation of the unit. If the spikes of more than one neuron are recorded as one unit, we should not expect to find a consolidated refractory period. After the refractory period, the autocorrelation function may have different profiles, reflecting different discharge patterns. The autocorrelation function may stabilize on the discharge rate of the neuron, might have a short peak and stabilize after, or might depict periodic peaks. The first case is associated with random (Poisson-like, with the exception of the refractory period) firing. A peak in the autocorrelation is associated with the tendency of neurons to discharge in bursts (Abeles 1982a). However, in neurons with a high discharge rate and a long refractory period, one might observe a peak in the autocorrelation function that is not due to a bursting discharge pattern (Bar-Gad et al. 2001a). Periodic peaks in the autocorrelation function will be observed in cells with periodic (single spike or burst) discharge.

Another method to estimate the discharge pattern of a cell is the time interval histogram of the interspike intervals (TIH of ISIs). The (first order) TIH of ISIs takes into account only the first spike after the trigger. The TIH of ISIs is very useful in detection of periodic activity. Thus, if a neuron regularly fires a (single) spike every 100 ms, the TIH of ISIs will have a single peak at 100 ms. If the neuron is missing one or few spikes, we will see a peak at 100 ms and other peaks at 200 ms, 300 ms, and so on. The TIH of ISIs of neurons with periodic bursting activity (e.g., a neuron that emits a burst of spikes every 100 ms) will show bimodal distribution of the ISIs, a short-term peak for the intraburst intervals, and a second peak for interburst (in-between) intervals. This second peak will be more clear in TIH of ISIs with a logarithmic time (x-axis) scale.

The firing pattern (being Poisson-like, bursty, or periodic) can be synopsized by the coefficient of variation (standard deviation [SD]/mean) of the ISIs (CV ISI). The CV ISI is equal to 1 for a Poisson process, for which the SD of the ISIs is equal to their mean. As the neuron becomes more periodic—for example, fires a single spike every 100 ms—the SD and the CV ISI approach zero. On the other hand, as the cell changes its firing mode toward a bursty mode, the CV will be greater than 1 since the SD of the ISI histogram will be larger than the mean. In the case of periodic bursting, as is common in the basal ganglia of animal models and patients with Parkinson's disease, the CV ISI may be smaller or larger than 1. Thus, the CV ISI is a less useful tool in the case of periodic bursting.

A characteristic feature of the high-frequency discharge pattern of GPe neurons is their pausing activity (figures 5.2, 7.1, 7.2, and 11.8). GPe neurons fire continuously at fifty to seventy spikes per second and on average stop every 5–10 seconds for a complete pause of 300–900 milliseconds of their discharge. Pauses of discharge, as well as bursts, can be detected by their features (many more or fewer spikes in a time epoch compared to the average discharge of the cell). Assuming a discharge pattern (e.g., Poisson), one can calculate the surprise for finding a burst or a pause and set a threshold for such detection (Legéndy and Salcman 1985, Elias et al. 2007).

FREQUENCY DOMAIN ANALYSIS OF DISCHARGE PATTERN

As mentioned above, the CV ISI might be confusing in cases with periodic bursts. Periodic phenomena are better studied in the frequency domain. The Fourier transform is a mathematical transform that decomposes a continuous signal (a function of time) into its constituent frequencies.[2] The Fourier transform returns a complex-valued (a + jb, j = √–1) function of frequency whose magnitude (absolute value, $\sqrt{(a^2 + b^2)}$, power) represents the power of a sine of that frequency present in the original function and whose argument is the phase offset of that sine. In electrical circuits, power is proportional to the voltage squared, $P = V^2/R$. Power spectrum functions (or densities; PSD) should therefore have units, and the PSD y-axis units are V^2/Hz. We will often normalize our PSDs to a fraction of the total power in the studied domain or by Z normalization of our time function $(Z(t) = (X(t) - mean(X(t))/SD(X(t)))$ to Z^2/Hz.

The results of the Fourier transform in the frequency domain are bounded by the sampling frequency (the highest frequency observed is equal to half of the sampling frequency). On the other hand, the resolution of the power spectrum is set by the size of the analysis window. One should not expect to see frequencies of less than 0.1 Hz if using a one-second window to analyze the data. To expose low-frequency oscillations in spiking activity after 300–6,000 Hz (or similar) band-pass filters, one should convert the signal into one that gives the discharge rate as a function of time. This can be achieved by full-wave rectification (absolute of the band-passed data) and low-pass filtering.

The power spectrum of many natural phenomena, including neural activity, follows the "$1/f^{\alpha}$" rule, often referred to as the "$1/f$" rule for the case of $\alpha = 1$. White noise is the ultimate random noise, with $\alpha = 0$, or $P(f) = $ constant. However, friction and local interactions between the elements of physical objects attenuate high-frequency components, and ideal white noise can exist only in mathematical models. Brownian noise (also known as Brown noise or red noise) is the kind of signal produced by Brownian motion; hence, its alternative name of random walk noise (i.e., from its current point, the particle can move in any random direction; however, there is a memory of one step backward of the location arrived, and therefore random-walk processes, unlike white-noise processes, will not stay near their origins). The term "Brown noise" does not come from the color but from Robert Brown, who documented the motion of particles in water. The term "red noise" comes from the white-noise/white-light analogy; red noise is stronger at lower frequencies (longer wavelengths) at the red end of the visible spectrum. In between the white noise and the red (Brownian) noise is the canonical case of the pink noise, with $\alpha = 1$. In pink noise there is equal energy in all octaves (or any other log bundles; octave is a two-fold increase in frequency) of frequency. In contrast, white noise has equal energy per linear frequency interval. Pink noise falls off at 3 dB (decibels) per octave (10 dB per decade; decade is a tenfold increase in frequency), while red (Brownian) noise falls off at 6 dB per octave (20 dB per decade).

Typical filters used in the physiology lab, such as four-pole Butterworth filters, fall off at 24 dB per octave. Most physical systems, including brain electrical phenomena, generally demonstrate 1/f behavior, with α ranging between 0.1 and 2.5 and more commonly around 1–1.5. However, α may change over the frequency axis, and its estimation might be tricky. The "whitening" procedure—$P(f) = P(f) * f$—named so because it will turn pink into white noise, is sometimes used in physiological studies to enhance the visibility of high-frequency phenomena (e.g., gamma oscillations in LFP and EEG).

PAIRWISE CORRELATIONS ANALYSIS OF SYNCHRONIZATION OF SPONTANEOUS ACTIVITY

Because the brain is a highly connected network, we should study neuronal functional interactions, sometimes called functional (vs. anatomical) connectivity. The cross-correlation function depicts the probability or rate of discharge of a (reference) neuron following a spike of the trigger neurons at time zero (Gerstein and Perkel 1969; Abeles 1982b; Aertsen and Gerstein 1985). Like the autocorrelation function, it will take into account any spike of the reference neuron in the time frame of the cross-correlation function. Unlike the autocorrelation function, the cross-correlation function is not symmetric around time zero, and therefore it should be presented with a time lag before and after the timing of the trigger spikes. Cross-correlation analysis can be applied to spikes, continuous/analog data, and to a mixture of spike and analog data (e.g., spike-triggered average of LFP).

Early studies of cross-correlation functions of simultaneously recorded pairs of neurons were conducted to look for evidence of direct functional connectivity between the studied neurons (Moore et al. 1970; Abeles 1982b). It was assumed that if the trigger neurons provide an excitatory or inhibitory synapse to the reference neuron, we would see an increase or decrease in the discharge rate of the reference (postsynaptic) neurons. The change in the discharge rate of the reference neuron will start a few milliseconds after the discharge of the trigger neuron and will last for the typical duration of excitatory or inhibitory postsynaptic potentials (figure 6.1*A*, *C*). However, the central nervous system is very different from the neuromuscular junction, with a one-to-one relationship between presynaptic and postsynaptic firing. On average, each neuron in the cortex is contacted by ten thousand synapses, and it is naive to expect that one neuron will have an effect on the firing of another that is big enough to detect in the limited time of stable recording of a pair of neurons in a behaving animal. Thus, in real biology we can find two families of cross-correlation functions between the discharge rate of simultaneously recorded pairs of neurons: flat and "common input" functions. The flat cross-correlation functions indicate that the two neurons are independent of each other. A central peak in the cross-correlation functions indicates that the two neurons tend to fire together. This can be the result of a "common input" from a third neuron (figure 6.1*B*, *D*) or, more probably, the result of a scenario in which the two neurons are part of a network with shared connectivity. Sometimes neuron A will fire before neuron B, and sometimes the neurons will fire in reverse order. If the two neurons are part of a network with

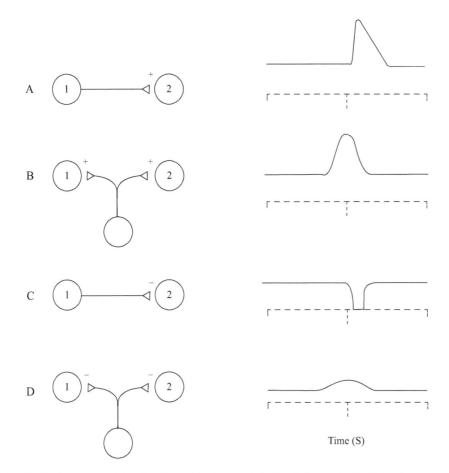

6.1 Idealized cross-correlation functions of basic neural circuits. *Left*, the circuit. *Right*, the cross correlation function between neurons 1 and 2. X-axis time zero indicates the time of a spike in neuron 1. *A*, monosynaptic excitation. *B*, monosynaptic excitatory common input. *C*, monosynaptic inhibition. *D*, monosynaptic inhibitory common input.
Source: Adapted from Moore et al. 1970 and Aertsen and Gerstein 1985.

synchronous oscillations, then periodic peaks will be found in the cross-correlation function (figure 15.3B).

EFFECTIVE CONNECTIVITY IN THE BRAIN AND THE SENSITIVITY OF CROSS-CORRELATION ANALYSIS

Here is my personal story on the heartbreaking gap between the efficacy of brain functional connectivity and the sensitivity of cross-correlation functions. During my time with Moshe Abeles, I learned the art of multiple-electrode recording and of neural network thinking. When I went to Mahlon DeLong's lab, I was looking for a way to use these tools in the basal ganglia. In starting to read the basal ganglia literature, I found that the STN had just changed the color of its suit, from GABA to glutamate.

Traditionally, it was believed that the STN, like all other nuclei of the basal ganglia, use GABA as a neurotransmitter. In line with the inhibitory nature of the STN, a series of electrophysiological studies (Yoshida et al. 1971; Hammond et al. 1978; Larsen and Sutin 1978; Hammond et al. 1983; Rouzaire-Dubois et al. 1983) revealed that STN stimulation hyperpolarized or suppressed the spontaneous firing of neurons in the rodent entopeduncular nucleus (the homologue of the primate GPi).

The story was changed by Steve Kitai, Hitoshe Kita, Yoland Smith, and Andre Parent. Steve Kitai was a man who had no fear. Together with Hitoshi Kita, he wrote a chapter to *The Basal Ganglia II* (Kitai and Kita 1987) with a groundbreaking title: "Anatomy and Physiology of the Subthalamic Nucleus: A Driving Force of the Basal Ganglia." They make it clear that no immunohistochemical study reveals GABAergic markers in the STN and that their in vitro slice recording indicates the excitatory nature of subthalamic efferents (Nakanishi et al. 1987, 1991). Smith and Parent reported that the cell bodies of the STN were closely surrounded by GABA-positive terminals. However, the cell bodies of the subthalamic neurons did not depict GABA immunoreactivity; rather, they displayed an intense immunoreactivity to glutamate (Smith and Parent 1988). These results, verified by other groups (Albin, Aldridge et al. 1989), suggest that STN neurons are the only structure in the main axis of the basal ganglia that utilize the excitatory neurotransmitter glutamate instead of the inhibitory neurotransmitter GABA.

I was young and even more naive at this time than I am now (i.e., no floor effect for naivety). I was hoping to make a change and to resolve the conflict between the biochemistry and intracellular physiology studies and the previous demonstration of a reduction of pallidal discharge rate following subthalamic stimulation. I therefore suggested to Mahlon DeLong that we record the simultaneous activity of neurons in the STN and the globus pallidus and use cross-correlation analysis to reveal the true nature of their monosynaptic interactions. Mahlon was skeptical but, as always, open-minded and allowed me the adventure of two chambers, two electrodes, and cross-correlation analysis (figure 6.2A). After one year of hard work, I had in my pocket a few hundred pairs of subthalamic and pallidal neurons. Annoyingly, all the cross-correlation functions of the spiking activity were flat (figure 6.2B). I also used microstimulation in the STN and recorded the evoked pallidal activity. To add a nail to the coffin of the cross-correlation dream and in line with previous reports, most of the pallidal responses were composed of a decrease in the pallidal discharge rate (figure 6.2C).

With the wisdom of a few more years of researching the basal ganglia and electrophysiology, I understand now that these heroic experiments reveal the minute efficacy of monosynaptic interactions in the brain and the limited sensitivity of cross-correlation methods (Aertsen and Gerstein 1985; Cohen and Kohn 2011). The inhibition of pallidal activity by the STN was probably due to antidromic activation of pallidal axons and lateral inhibition in the pallidum. It provides one more

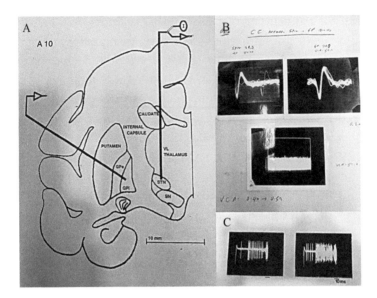

6.2 Early correlation and stimulation studies of the subthalamic nucleus-globus pallidus functional connectivity. *A*, scheme of the experimental setup superimposed on a coronal section at A10 (10 mm anterior to the ear bars) of the monkey brain. The subthalamic nucleus electrode was connected to a recording amplifier or a current stimulator. *B, top,* superimposed waveforms of the spikes of the subthalamic and pallidal neurons; *bottom,* their cross-correlation functions. *C,* effects of subthalamic stimulation at time zero on the spiking activity of the pallidal neuron.
Source: Polaroid pictures from the lab oscilloscope and Digital's PDP11 monitor.

explanation of the clinical effects of deep brain stimulation (DBS) in the STN for patients with advanced Parkinson's disease, but DBS was far in the future in those days. There are many good reasons to doubt negative results of biological experiments. So, at the end of this year, we published them as an abstract of the meeting of the American Society for Neuroscience and put them in the drawer. Revital, my life partner, reminded me of our agreement that in the second year of our Baltimore adventure she would take time for her residency, and I would take care of the kids in the afternoons. Thomas Wichmann joined the lab, and having no more time for long experiments with dual recordings, we switched to "easier and faster" experiments of STN inactivation (Bergman et al. 1990; Wichmann et al. 1994). In science, as in life, luck and good friends are more important than hard work.

CROSS-CORRELATION ANALYSIS OF SPIKE TRAINS
OF NEIGHBORING NEURONS

Cross-correlation functions of two units recorded by the same microelectrodes (figure 5.3) are of special interest because they are probably closer (less than 100 μm apart) and therefore have a higher probability of being affected by common inputs or local circuitry (Smith and Kohn 2008; Rosenbaum et al. 2017). However, as for the autocorrelation function (Bar-Gad et al. 2001a), a problem may arise when recording in an area with high-frequency discharge neurons and a long refractory period (Bar-Gad et al. 2001b). Spike sorting is limited in the detection of simultaneous spikes of two different neurons, so the spikes of one neuron (usually the bigger spikes) may overshadow the spikes of the other neuron. In cases of neurons with Poisson-like activity, this can be corrected, and the independence (or lack of) of the activity can be tested (Bar-Gad et al. 2001b).

COHERENCE: FREQUENCY DOMAIN ANALYSIS OF NEURONAL SYNCHRONIZATION

As for the pattern analysis, we prefer to work in the frequency domain in cases with periodic oscillations. The coherence function depicts the correlation between pairs of neurons as a function of frequency. The coherence function is a complex number $(a + jb, j = \sqrt{-1})$ so that its magnitude-squared value, $\sqrt{(a^2 + b^2)}$, represents the absolute of the correlation coefficient (r, range between zero and one) for a given frequency. The phase difference of the two signals is the argument of the complex number, tan (a/b). We will often use the magnitude-squared coherence function; however, it is affected by common changes in the amplitude of the signal. Phase correlation or the imaginary part of the coherence function (Nolte et al. 2004) might be more sensitive tests of synchronized oscillations in the nervous system.

ANALYSIS OF EVOKED ACTIVITY

Classical electrophysiology has studied the physiological responses to sensory and behavioral events (evoked activity). The seminal work of David Hunter Hubel and Torsten Wiesel on the responses of single neurons in the visual cortex of anesthetized cats, revealing the first steps for information processing of simple visual input, is a good example. Nevertheless, I will only briefly discuss the analysis methods of evoked activity because, first, most of the evoked activity analysis described in this book is first order and straightforward, and second, the analysis of evoked activity can be considered an extension of the previously discussed correlation methods, but with a behavioral event or timed artificial stimulation used as a trigger.

The simplest test is the quantification of the evoked neuronal (spikes or LFP) activity following a sensory cue. Because of the noise and variability of the system, there is a need to average many trials with the same behavioral response. This average, triggered on the sensory cue, is defined as a peristimulus time histogram (PSTH). The PSTH reveals the mean response (latency, polarity, peak amplitude, and duration) of a neuron to behavioral events. In the case of EEG and LFP, we will do a similar calculation of the average analog response to behavioral events.

Pairwise correlation may be affected by the responses of the two elements under study to the same behavioral effect. This can be normalized by the calculation of a shuffle, a shift, or a PSTH predictor that provides us with the average response of the cells to the events. Subtraction of the predictor from the raw cross-correlation function will result in a "pure" functional correlation between neurons (that may differ during the task from the correlation in the intertrial intervals).

We can also compare the similarity of the responses of different neurons to a cue or to a vector of behavioral events. This is called the signal correlation (figure 6.3). Signal correlation can be calculated even between units that are not simultaneously recorded. However, the noise correlation—that is, the shared variability of the responses of two neurons—can be calculated only for simultaneously recorded neurons. The noise correlation is often similar to the spike-to-spike correlation.

The firing pattern and neural synchronization can be modified as a function of time in the trial. We should be careful to discriminate between real changes in pattern and functional connectivity and those that result from the changes in discharge rate. Analytical tools such as the joint peristimulus time histogram (JPSTH) have been suggested and will be used in later chapters of this book.

There is a trade-off between frequency and time in our analysis methods. To obtain an accurate frequency representation of our signal, we need to use a long

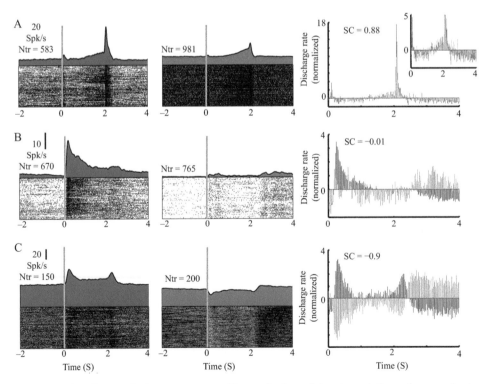

6.3 Examples of raster displays, response profiles, and values of signal correlations of neuronal pairs in the basal ganglia.
Source: Shiran Katabi, unpublished data.

window (to maximize the resolution of our frequency domain analysis). However, in biology we often face nonstationarity and are interested in the evolution with time of the changes in the frequency domain. There are advanced methods to improve the time/frequency trade-off, such as multitapered analysis and wavelets. So far, these methods have not been cost-effective for me. To look at the time evolution of changes in the frequency domain, we will use spectrograms and coherograms, where we calculate the power spectrum or coherence over short windows and display it as a function of time.

ADVANCED ANALYSIS TOPICS

Neuronal activity can be studied at many levels. Here, I am describing more advanced methods that I have not used and that in many cases are beyond my math level.

High order correlations   Figure 6.4 depicts a typical example of four tonically active neurons (TANs, presumably striatal cholinergic interneurons) that were simultaneously recorded in a monkey engaged in an operant conditioning task (Raz et al. 1996). Figure 6.4A shows the stereotypic reduction of discharge rate responses (often bound by excitatory peaks) of the four neurons (Kimura et al. 1984; Graybiel et al. 1994). Figure 6.4B shows the cross-correlation matrix of these neurons. A narrow peak, revealing a tendency for synchronized spikes, is observed for most pairs. The

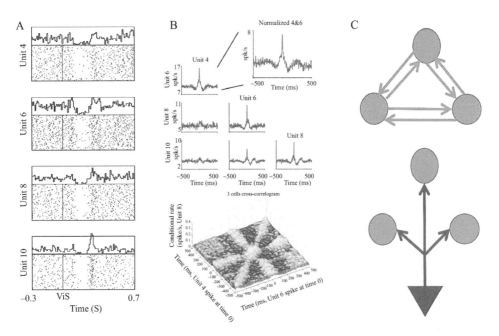

6.4   High-order correlations between striatal tonically active neurons (TANs). *A*, raster display and peristimulus histograms aligned to visual cue presentation at time zero. *B*, pairwise correlation matrix (*top*) and triple correlation (*bottom*) plots. *C*, possible network scenarios.
Source: Copied (with permission) from Raz et al. 1996.

inset is a shuffled cross-correlation function indicating that the observed spike-to-spike correlation is not due to the similar responses. The lower subplots of figure 6.4B show the triple correlation matrix, suggesting that all three neurons are synchronized. However, this triple correlation can reflect two extreme, non-mutually exclusive network scenarios. In the first scenario (figure 6.4C, *upper plot*), there are reciprocal connections between each TAN-TAN pair. In the second scenario, the three neurons receive "common input" from a fourth neuron (or a population of neurons); however, there is no lateral connectivity in the TAN network. Clearly, both scenarios, and any combination of them, could lead to the pairwise and triple correlations shown in figure 6.4B.

The first attempts at network analysis were conducted by George Gerstein and Ad Aertsen, who suggested the gravity analysis of parallel spike trains (Gerstein and Aertsen 1985). Using maximum entropy models, Elad Schneidman and colleagues have shown that weak correlations between pairs of neurons in the retina coexist with the strongly collective behavior of ten or more neurons (Schneidman et al. 2006; Tkačik et al. 2014).

Causality analysis   The careful reader of this book will often find the mantra "correlation does not imply causation." However, as with any logical fallacy, identifying that the reasoning behind an argument is flawed does not necessarily imply that the resulting conclusion is false. Mathematical methods like the Granger causality test and dynamic causal modeling (DCM) are often used in neuroscience and basal ganglia research. These methods often reveal a bidirectional or inconsistent causality relationship. Nevertheless, I would recommend that the reader become familiar with this new field of science.[3]

In summary, advanced analysis and data acquisition tools are not risk-free. A fool with a tool is still a fool. We are all fools when it comes to questions regarding our data and hypotheses. We should make sure to exclude noise from our analysis, to employ multiple comparison correction methods, and to better understand the advantages and limits of our data acquisition and analysis methods (Bennett et al. 2009; Lyon 2017).

CHAPTER SUMMARY

- You know something about the brain and the basal ganglia when you can measure and express it in numbers. We will use temporal and spectral domain data analysis techniques to describe the physiology of the basal ganglia in health and disease.
- Spontaneous neural activity is characterized by its intensity (discharge rate for spiking activity and the power of EEG and LFP), pattern, and synchronization.
- Patterns of neural activity can be characterized in the time domain (e.g., autocorrelation function), TIH of ISIs, CV of ISIs, burst or pauses detection, or in the frequency domain (e.g., power spectrum).

- Pairwise synchronization of neural activity can be measured in the time domain by a cross-correlation function and in the frequency domain by a coherence function. The similarity of the responses of two neuronal elements is assessed by their signal correlation.
- Evoked activity is characterized by a change in intensity (e.g., discharge rate), pattern, and functional connectivity (JPSTH) as a function of time.
- Advanced analysis tools can be (cautiously) used. Of interest are high-level correlation and causality analysis methods.

II

COMPUTATIONAL PHYSIOLOGY OF THE HEALTHY BASAL GANGLIA

7

PHYSIOLOGY OF THE BASAL GANGLIA

Information reaches neurons through the (excitatory and inhibitory) synaptic activation of their dendrites. This synaptic input is integrated or resonated in the cell body to generate action potentials or to decrease the intrinsic discharge of neurons. Therefore, I will focus here on spiking activity and local field potentials (LFP; proxy for synaptic inputs) in the basal ganglia network. Figures 5.2, 7.1, and 11.8 depict typical extracellular recorded spontaneous spiking activity in the basal ganglia network.[1]

SPONTANEOUS DISCHARGE RATE

The spontaneous discharge rate of striatal (projection) medium spiny neurons (MSNs) is very low, around 0.5–2 Hz. This discharge rate is similar to that found in the cortex, and it enables the cortical and striatal MSNs to encode behavioral events by transient elevation of their discharge rate. A second population of striatal neurons, frequently detected in vivo by means of extracellular recording, is the tonically activated neurons (TANs). As their name implies, unlike the MSN's phasic activity the TANs discharge tonically, at 3–7 Hz, even in the absence of attended behavioral stimuli (Kimura et al. 1984, Mizrahi-Kliger et al. 2018; figures 6.4 and 8.4). Several studies indicate the overlap between TANs and striatal cholinergic interneurons. However, the striatal low-threshold spiking interneurons exhibit autonomous firing that in vitro resembles the discharge of cholinergic interneurons (Beatty et al. 2012). We will therefore define the striatal TANs as putative striatal cholinergic interneurons.

The discharge rate of subthalamic neurons is higher than that of the striatal neurons and ranges between 20 and 30 Hz. The picture changes dramatically when we reach the GPe, GPi, and SNr. Here, most of the units discharge almost continuously at a rate of 40–70 Hz (GPi, SNr > GPe). Action potentials have a metabolic cost, so one wonders why biology is ready to pay this price. We will revisit this question, but here

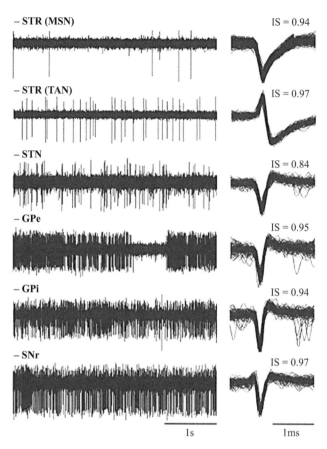

– STR (MSN)   IS = 0.94

– STR (TAN)   IS = 0.97

– STN   IS = 0.84

– GPe   IS = 0.95

– GPi   IS = 0.94

– SNr   IS = 0.97

1s   1ms

7.1 Typical extracellular recorded activity in the basal ganglia network. *Left*, four seconds of activity; *right*, overlapping spike waveforms of the recording on the left. IS, isolation quality; STR (MSN), striatum, medium spiny neurons; STR (TAN), striatum, tonically active neurons; STN, subthalamic nucleus; GPe and GPI, external and internal segments of the globus pallidus; SNr, substantia nigra reticulate. *Source*: Copied (with permission) from Deffains et al. 2016.

we should keep in mind that paradoxical, spontaneous high-frequency discharge can be found elsewhere—for example, the simple spikes of the cerebellum Purkinje cells.

SPONTANEOUS DISCHARGE PATTERN

The discharge pattern of GPe neurons is different from the GPi and SNr. Long (mean duration equals 0.6 s) complete cessation of activity (pauses) frequently (10/minute) interrupts the tonic discharge of GPe neurons (figure 7.2). GPe neurons and pauses are not created equal (DeLong 1971; Elias et al. 2007; figure 7.2*B*). Up to now, we have failed to find a link between the timing of behavioral or external events and GPe pauses. The distribution of the pause follows Poisson (random in time) distribution. In contrast with the lack of precise temporal correlation between GPe pauses and behavioral events, there is a clear and strong relationship between arousal level

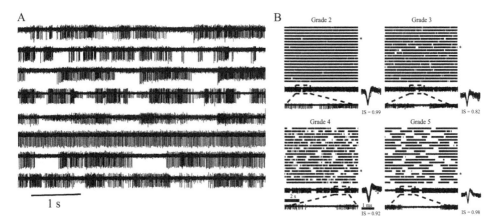

7.2 The GPe pauses. *A,* simultaneous eight-electrode recording of neuronal activity with typical pauses in the external segment of the globus pallidus. Note that at the fourth trace, two different GPe cells are recorded by the same electrode. *B,* diversity of GPe pauses. Activity of four pallidal cells (different cells than shown in (*A*). Tendency to pause is graded from 1 to 5. *Top,* twenty lines of ten second raster display; *bottom,* row data of the line marked with a star. IS, isolation quality.
Sources: A. Unpublished results of Avital Adler and Mati Joshua. B. Copied (with permission) from Elias et al. 2007.

and frequency and duration of GPe pauses. The pauses will be less frequent and shorter when the monkey is engaged in a behavioral task. Their frequency and duration will increase as the monkey switches to a quiet awake state, napping, shallow (N1/2) sleep, and deep (SWS or N3) sleep (figure 12.7). The frequency and duration of GPe pauses will again decrease during the activated state of rapid eye movement (REM) sleep. Why GPe neurons pause is still a mystery.

Another difference between the GPe and the output nuclei of the basal ganglia is the distribution of neuronal type. The morphology and the physiology of GPi and SNr neurons are quite homogenous. In contrast, GPe neurons can be classified as high-frequency (>20 Hz) discharge with pauses (HFD-P; 85 percent of GPe neurons; figures 7.1 and 7.2) and as low-frequency (<20 Hz) discharge with bursts (LFD-B; 15 percent of GPe neurons; figures 5.2, 11.8, and 12.6). A similar dichotomy of GPe neurons has been revealed in early Golgi studies and recently in studies of the rodent GPe discriminating between prototypic and arkypallidal GPe neurons (Mallet et al. 2012; Mastro et al. 2014; Abdi et al. 2015; Hernández et al. 2015). It is not clear yet if the primate LFD-B neurons are the same as the rodent arkypallidal cells, providing feedback projections to the striatum, or if they are GPe interneurons. Finally, in real life not all GPe HFDs pause nor do all GPe LFDs burst.

SYNCHRONIZATION OF SPONTANEOUS DISCHARGE

Thirty years ago, we pioneered multiple-electrode recording of the basal ganglia of behaving monkeys. The first experiments were carried out in the GPe and GPi and led to an absolutely unexpected finding—the cross-correlation histograms of the spiking activity of simultaneously recorded pallidal neurons were completely flat (Nini et al. 1995; Raz et al. 2000; Nevet et al. 2004; figure 7.3). Previous multiple-electrode studies have usually been carried out in the cortex. These studies of the spontaneous activity of pairs of cortical neurons frequently revealed central "common input" peaks in the cross-correlation functions (figure 6.1B and D). The finding of no correlation between the spiking activities of pallidal neurons was therefore very surprising. The basal ganglia networks characterized by a massive reduction in the number of neurons along the cortex-striatum-pallidum axis and significant overlap of the pallidal dendritic disks (table 2.1 and figure 2.5) were expected to display "common input" and correlated activity.

Typically, our electrodes are separated by one-half to a few millimeters. Correlated activity in the cortex decays as a function of distance, and therefore it was critical to show that the absence of correlation of pallidal activity was not due to recording remote cells. Luckily enough, occasionally we can record two different units with the same electrode (figure 5.3). After overcoming the technical issue of shading effect, Izhar Bar-Gad and other students in the lab have shown that even pallidal pairs recorded by the same electrode—namely, at <0.2 mm distance—reveal independent activity (Bar-Gad et al. 2003).

The finding of flat cross-correlation, or in other words the independent activity of neurons in the pallidum and the SNr, has been verified by generations of talented graduate students and postdoctoral fellows (e.g., Asaph Nini, Aeyal Raz, Gali Havetzelt-Heimer, Alon Nevet, Shlomo Elias, Josh Goldberg, Marc Deffains). I believe it should be considered one of the characteristic features of the basal ganglia network. In line with our studies, the group of Atsushi Nambu identified multiple pallidal

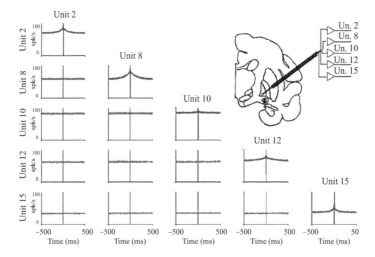

7.3 The correlation matrix of five simultaneously recorded pallidal units in the normal nonhuman primate. Autocorrelation (*blue, outer diagonal*) and cross-correlation (*red*) functions are displayed. *Source*: Copied (with permission) from Raz et al. 2000 and Nini et al. 1995.

neurons receiving inputs from the forelimb regions of the primary motor cortex and supplementary motor area. Most pallidal neurons exhibited task-related firing-rate changes, whereas only a small fraction showed small and short correlated activity during the task performance (Wongmassang et al. 2020). The finding of uncorrelated discharge of pallidal neurons was the main driving force for the model of dimensionality reduction in the main axis of the basal ganglia (chapter 11).

However, not all neuronal pairs in the basal ganglia depict flat cross-correlation functions. Mano Kimura and Ann Graybiel have shown that TANs recorded at different domains of the striatum have a similar increase-decrease-increase discharge rate response (Graybiel et al. 1994). Multiple-electrode recording of TANs activity in our lab (Raz et al. 1996; Raz et al. 2001; Adler et al. 2013a, b) revealed that the spiking activities of most TANs pairs are correlated (figure 6.4B). This was true even when the correlation was calculated only for the intertrial interval epoch or after normalizing the cross-correlation function with a shift predictor. These physiological findings provide further support for the actor/critic model of the basal ganglia and will be discussed further in chapter 10. Even in the main axis (actor) of the basal ganglia, one can find correlated pairs. Spiking activities of neighboring cortical cells are correlated. Avital Adler of our research group has shown that the same is true for MSN pairs and MSN-FSI pairs (figure 5.5). Similar results were reported by Thomas Wichmann, this author, and Mahlon DeLong (1994) for subthalamic pairs. This is in line with the notion of dimensionality reduction (information compression) along the main axis of the basal ganglia and will be further discussed in chapter 11.

SYNCHRONIZATION OF GPe PAUSES AND SPIKING ACTIVITY

The GPe pauses (figure 7.2) are one of the big mysteries of the basal ganglia of nonhuman primates and humans. Our first pairwise correlation studies have not revealed any correlation between the GPe spiking activity and the pauses, as well as between pauses of different cells (figure 7.4).

Nevertheless, Eitan Schechtman and colleagues (2015) performed a statistically higher-powered study by using the pause, rather than the neuron, as the elementary unit of the study. This study revealed a slight decrease of the GPe spiking activity (figure 7.5A) before the pause onset and a slight tendency for pauses of different cells to occur together (figure 7.5B). The reduction of the GPe discharge rate at the onset of the GPe pause reflects a network effect since it is not abolished by removing the effect of other simultaneously occurring pauses (figure 7.5C).

BASAL GANGLIA–EVOKED ACTIVITY: POLARITY AND AMPLITUDE

The neurons of the basal ganglia can be divided into those with a low/medium discharge rate, as in the striatum and the STN, and neurons with a high discharge rate, such as most of the neurons in the GPe, GPi, and SNr. As expected, the low-discharge

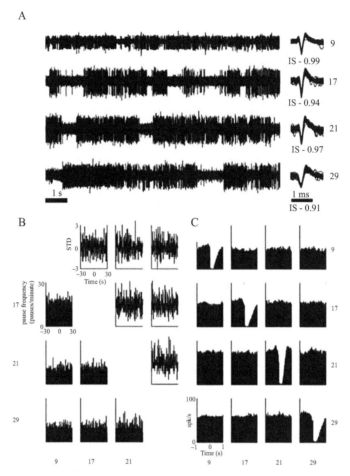

7.4 Neural activity and pauses of neurons in the external segment of the globus pallidus are not synchronized. *A*, simultaneous recording from four electrodes. *B*, pause-to-pause correlation matrix. *C*, pause-to-spike correlation matrix. In the diagonal the correlation between pause and spiking activity of the same unit are shown.
Source: Copied (with permission) from Elias et al. 2007.

neurons encode behavioral events mainly by increases in their discharge rate. The high discharge rate of the GPe, GPi, and SNr neurons and the abundance of GABAergic transmission in the basal ganglia may seem a good start for a neural network that will respond to behavioral events with a decrease of discharge rate (inhibition). Nevertheless, all studies (including ours) of the polarity of the responses of pallidal neurons to behavioral events have revealed that most (60–70 percent) of the changes in discharge rate following a variety of behavioral events (movements, cues, and rewards) are composed of increases in discharge rate (figure 12.5). Moreover, the average amplitude of the increases in discharge rate was greater than the average amplitude of the decreases. Thus, the population average was positive and revealed an increase in discharge rate compared to the spontaneous activity.

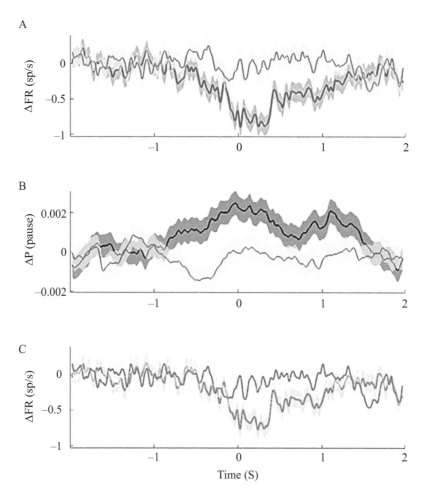

7.5 Pauses of the neurons in the external segment of the globus pallidus are driven by network activity. *A*, mean firing rate across repetitions of start of pause at zero time. *B*, change in pause probability. *C*, change in firing rate after pause removal. Dark lines and shading, the original data; light lines and shading, the shifted data.
Source: Copied (with permission) from Schechtman et al. 2015.

There is one major exception to this trend of more and stronger increases than decreases in the discharge rate of the high-frequency-discharge group of basal ganglia neurons. In 1983 Okihdo Hikosaka and Bob Wurtz published a seminal series of four papers in the *Journal of Neurophysiology* (*JNP*; at that time a series of three back-to-back *JNP* papers was the goal of any hard-core electrophysiologist) describing the response of SNr neurons to saccadic eye movement tasks. Hikosaka and Wurtz (1983 a–d) reported that SNr neurons decrease their spike activity before (memory) saccade onset. Interestingly, unlike the striatal and pallidal responses to upper-limb movements, the SNr saccadic-related activity was present only when the monkey was engaged in saccade tasks; no change in activity was observed when the monkey was making saccades spontaneously—for example, in the dark.

The lack of movement-related responses might be explained by the reward sensitivity of basal ganglia neurons. However, our preliminary studies revealed that many SNr neurons respond to saccades with an increase in the discharge rate. Other groups have reported that the SNr responses are not confined to only a decrease in discharge rate for task-related saccades. Finally, the SNr responses reported by Hikosaka and Wurtz are not as modest as the typical GPi/SNr response. They resemble the GPe pauses and are composed of a decrease in discharge rate to a complete stoppage of firing (i.e., peak amplitude of fifty to eighty spikes/second). Thus, the results of Hikosaka and Wurtz in the SNr differ from our results for the polarity and the amplitude of the response of the basal ganglia high-frequency neurons. There are three possible, not mutually exclusive, solutions to the contradictory results regarding the polarity and the amplitude of the high-frequency-discharge neurons of the basal ganglia.

The first is a bias in the selection of units. The practice of electrophysiologists in the twentieth century was to descend carefully with the electrode until a well-isolated unit was reached and to test the responses of this unit to the behavioral paradigm. Only if the unit responded as expected would the experimenter continue to record and study the unit with many repetitions of the behavioral task. In our recording of the pallidum, we have observed such units with complete suppression of discharge in response to behavioral events (Elias et al. 2007). However, these cases were the exceptions, and the majority of the pallidal neurons recorded by our group responded with small amplitude (five to fifteen spikes/second) modulation of their discharge, more by elevation (figure 12.5).

Second, the SNr might be different than the GPi. Everywhere in this book, we take the approach that "the GPi and SNr are parts of the same structure, accidentally divided by the internal capsule." This idea is driven by the similar anatomy and physiology of GPi and SNr neurons. However, it is not completely true. The main difference between SNr and GPi is their relationship with the SNc dopamine. Dopamine is released not only in the striatum but also in the midbrain by a somatodendritic release (Rice and Patel 2015). This SNc somatodendritic release is probably used for communication between dopamine cells and for regulation of the discharge dopaminergic cell by dopamine autoreceptors. However, it can affect the very close SNr neurons enriched with dopamine receptors.

Finally, as indicated by David Robinson, the oculomotor system is not a scaled-down caricature of the skeletomotor system (six muscles/eye vs. more than six hundred named skeletal muscles in the human body). In the oculomotor system, there are no, or a minimal number of, muscle spindles, probably because the load (eyeball) is fixed, and vision provides the feedback. Unlike the multijoint muscles of the skeletomotor system, the eye muscles can be described as a single joint and have minimal redundancy in their direction of effect. Thus, the need for surround inhibition and stabilization, as John Mink's action-selection model of the basal ganglia may suggest (figure 9.5), is less crucial in the oculomotor system.

BASAL GANGLIA–EVOKED ACTIVITY: TEMPORAL STRUCTURE

Basal ganglia–evoked activity in response to behavioral events can be transient or persistent. Figure 7.6 depicts the average population responses of GPe, putamen TANs, and MSNs to different cues predicting reward, aversive (air puff), and neutral outcomes at the end of the visual cue. Whereas the GPe and the MSN populations reveal a persistent average response, the TANs' response is very short.

SYNCHRONIZATION OF BASAL GANGLIA–EVOKED ACTIVITY

The similarity of the responses of neurons can be evaluated by the signal or response correlation measures (figure 6.3). Figure 7.7 depicts the distribution of the response correlation of pairs of neurons in the different nuclei in the basal ganglia. In chapter 10 we will use this finding of a skewed distribution of correlation values for the TANs and the dopaminergic neurons, versus symmetric distribution for the GPe, GPi, and SNr, to support our claim of division of the basal ganglia to critic and actor subnetworks.

Classical correlation analysis assumes that the synchronization of neural activity is fixed over time. However, the functional connectivity between neurons may be modulated as a function of time. Figure 7.8 depicts the average dynamics of functional connectivity between pairs of neurons in the basal ganglia. While TAN-TAN functional connectivity is fixed over the trial epochs, the functional connectivity (their tendency to fire together) of pairs of dopaminergic neurons and pairs of MSN neurons increased for a short duration after the presentation of a cue predicting future reward. Thus, functional connectivity between pairs of neurons can be dynamically modulated by behavioral events.

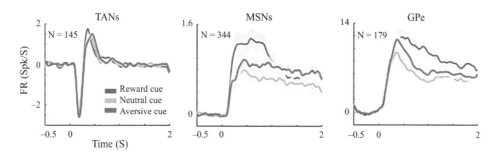

7.6 Transient and persistent responses in the basal ganglia. *Left to right*, average population responses of striatal tonically active neurons (TANs), medium spiny neurons (MSNs), and globus pallidus external segment (GPe) neurons.
Source: Modified (with permission) from Adler et al. 2012.

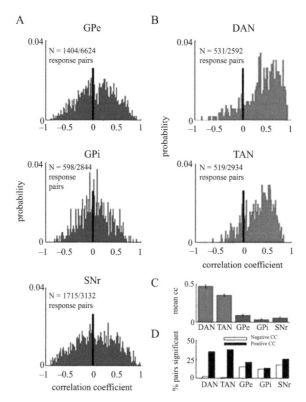

7.7 Distribution of response correlation values in the basal ganglia. *A, blue,* main axis pairs. *B, red,* pairs of critic neurons. *C,* mean of the correlation values of the different populations. *D.* Fraction of neuron pairs with positive and negative significant correlations. GPe, globus pallidus external segment; GPi, globus pallidus internal segment; SNr, substantia nigra reticulate; DAN, dopaminergic neurons (recorded from the substantia nigra compacta); TAN, tonically active neurons.
Source: Copied (with permission) from Joshua, Adler, Prut et al. 2009.

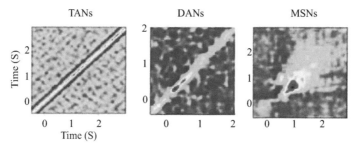

7.8 Dynamic modulation of synchronization in the basal ganglia. Average joint peristimulus histograms (JPSTH) of TAN-TAN pairs (*left*), DAN- (dopaminergic neurons) DAN pairs, and putamen MSN-MSN pairs, showing dynamic modulation of the functional connectivity between the DAN-DAN pairs and the MSN-MSN pairs but not between the TAN-TAN pairs. Cue was given at *T*=0 and outcome at *T*=2 seconds. Note the slightly different timescale for the DAN JPSTH. The color scale (yellow/red indicate stronger connectivity) is adjusted for each structure.
Source: Modified (with permission) from Joshua, Adler, Prut et al. 2009 and Adler, Finkes, et al. 2013.

CHAPTER SUMMARY

- Basal ganglia neurons can be divided into low-frequency and high-frequency discharge neurons. The average discharge rate of striatal medium spiny (projection) neurons is one to two spikes/second, while GPe, GPi, and SNr neurons fire at a rate of fifty to seventy spikes/second. Subthalamic neurons fire at twenty to twenty-five spikes/second. Exceptions exist. For example, striatal interneurons usually fire faster than striatal projection neurons, and GPe low-frequency discharge neurons fire at a rate of fewer than twenty spikes/second.

- The discharge pattern of most basal ganglia neurons is independent in time (like a Poisson process but with a refractory period). The high-frequency discharge GPe neurons exhibit spontaneous pauses (awake state typical duration of 0.6 s and a frequency of ten per minute). The frequency and duration of pauses is inversely correlated with the arousal level.

- The discharge of pairs of neurons in the GPe, GPi, and SNr is uncorrelated. The signal correlation (measuring the similarity of evoked responses of pairs of neurons) is widely distributed around zero for pairs of high-frequency discharge neurons in the GPe, GPi, and SNr.

- The spontaneous spiking and evoked activity of pairs of striatal projection neurons, TAN pairs, and SNc dopaminergic neurons are correlated.

- Most of the responses of basal ganglia neurons to behavioral events consist of an increase in discharge rate.

- Basal ganglia neurons depict transient and persistent responses in response to behavioral events. Striatal TANs and SNc dopaminergic neurons have transient responses, while the populations of neurons in the striatum, GPe, and GPi/SNr have sustained (persistent) responses.

8

BASAL GANGLIA PHYSIOLOGY: THE ROADS LESS TRAVELED

"Two roads diverged in a wood, and I—, I took the one less traveled by, and that has made all the difference." Robert Frost and other poets advocate taking the road less traveled.[1] Real life in the biology/medical science community of the twenty-first century is less romantic. There is a closed loop between grant support and publication in high-impact journals. Both the members of grant committees and the editors of high-impact journals do not encourage scientists today to travel roads too far away from the mainstream.[2] Additionally, some of these less-taken roads are heavily affected by popular writing and "pseudoscience" and are being closed due to taboos and conflicting areas of our society. In this chapter I will discuss some of these less-traveled roads in the basal ganglia world.

THE BASAL GANGLIA AND THE SLEEP/WAKE CYCLE

Spontaneous basal ganglia activity can be the result of two, not mutually exclusive, processes. First, spontaneous activity may be due to intrinsic activity (e.g., pacemaker activity). Second, spontaneous activity might be due to cumulative responses to unknown external and internal events. Studying basal ganglia activity during sleep, when the brain (at least those areas that are beyond the primary sensory cortices) is dissociated from the periphery, might enable us to assess the relative contributions of the intrinsic and driven activities to the generation of spontaneous activity.

Sleep study is not as simple as may be thought. Sleep is not a homogenous state in which we close the windows to the outside. Sleep is an active state, subdivided into different stages (figure 8.1; N1/2, shallow sleep; N3, slow-wave sleep [SWS], deep sleep; and rapid eye movement, paradoxical, dream, REM sleep).[3] Neuromodulators change their activity during sleep, probably also affecting the intrinsic activity of

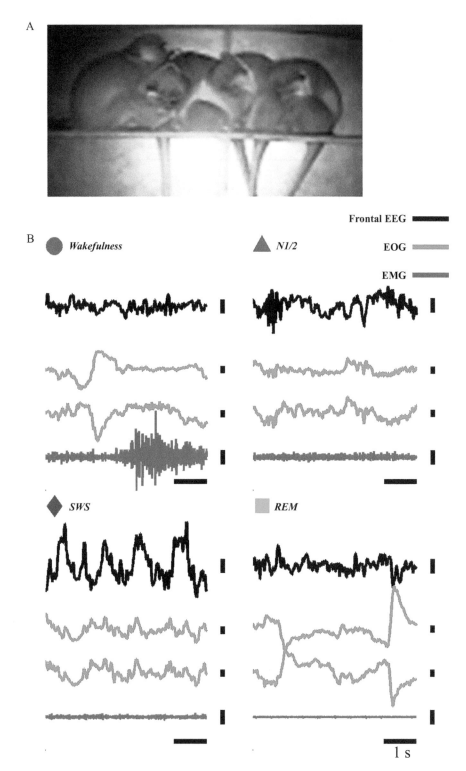

8.1 Monkeys' sleep posture and sleep stages. *A*, group and erect sleeping posture of vervet monkeys in the animal facility of the Hebrew University of Jerusalem. *B*, typical polysomnography traces during sleep stages. *Horizontal bars*, 1 s; *vertical bars*, 50 μV.

Source: Copied (with permission) from Mizrahi-Kliger et al. 2018 (supplementary information).

neurons. We have therefore decided to study the correlates of different sleep stages in the basal ganglia.

Basal ganglia sleep physiology has been neglected over the many years of physiology research of the basal ganglia of nonhuman primates. Mahlon DeLong started the new era of extracellular recording in the basal ganglia of the behaving monkey in Ed Evarts's National Institutes of Health (NIH) lab. Evarts came from the field of sleep physiology (figure 5.1). Therefore, it is not surprising that Mahlon was soon looking at the basal ganglia of sleeping monkeys. Mahlon published his observations as a one-page abstract in the *Physiologist* in 1967 and then, following the example of his mentor, shifted his research to the green field of movement control. The small community of basal ganglia researchers followed DeLong and turned a blind eye to the sleeping basal ganglia. The neglected research on the basal ganglia during sleep is surprising. The arousal effect of amphetamine and similar drugs, blocking the reuptake of dopamine and effectively increasing its amount in the brain, has been known for many years. Most of the brain's dopamine is in the basal ganglia. Therefore, the stimulating effects of amphetamine are probably mediated by the basal ganglia. Sleep disorders are highly common in Parkinson's disease. Thus, it is not clear why sleep has been neglected; maybe basal ganglia researchers have the same biases and prejudices as others and treat sleep as a void state of the brain. In any case, when Aviv Mizrahi-Kliger came to the lab he noticed this gap and together with Alex Kaplan initiated our studies of the basal ganglia during sleep. As expected, the spontaneous discharge rate and pattern of most basal ganglia neurons was significantly affected by the awake/sleep stages (figure 8.2).

Slow-wave sleep (SWS) is characterized by slow (delta, 0.5–2 Hz), synchronized oscillations in the thalamocortical circuits. The discharge of basal ganglia neurons is entrained to these slow oscillations, but even at this deep-sleep stage, in which many other parts of the brain are synchronized, the basal ganglia retain their independent activity, and no correlation is detected between pairs of neurons in the basal ganglia (figure 8.3).

Whether dopamine release is modulated by sleep is still debated. Sleep physiology has traditionally been investigated in cats because of their long sleep duration (more than fifteen hours/day). Classical extracellular recording of feline (cat) midbrain dopaminergic neurons failed to reveal any significant modulation of the discharge spiking activity of these neurons by the transition from wake to sleep and by the sleep stages (Trulson et al. 1981; Miller et al. 1983; Steinfels et al. 1983). Although a few papers have pointed toward a change in the discharge pattern (Dahan et al. 2007), the paradigm has remained that dopamine neurons and therefore brain dopamine do not play a major role in awake/sleep control. This is in contrast to cholinergic neurons in the pons and nucleus basalis of Meynert, noradrenergic neurons of the locus coeruleus, serotoninergic neurons of the dorsal raphe, and histaminergic neurons of the tuberomammillary nucleus that significantly slow down their

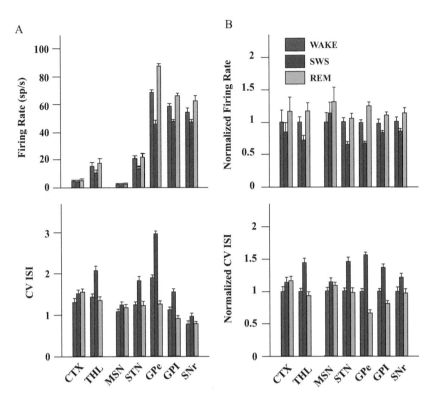

8.2 Discharge rate and pattern of thalamus and cortex as well as of basal ganglia neurons over wake and sleep stages. *A*, average values of discharge rates and coefficient of variation of the interspike intervals (CV ISI). *B*, rate and CV ISI normalized by the wake values. CTX, cortex; THL, thalamus; MSN, medium spiny neurons of the striatum; STN, subthalamic nucleus; GPe, GPi, external and internal segments of the globus pallidus; SNr, substantia nigra reticulata.
Source: Modified (with permission) from Mizrahi-Kliger et al. 2018.

discharge rate during NREM sleep. The transition from NREM to REM sleep does not affect the discharge of the noradrenergic, serotoninergic, and histaminergic neurons, but the cholinergic neurons increase their discharge rate back to awake level. Table 8.1 provides a textbook description of the changes in the activity of cortical neuromodulators over the awake/sleep stages.

The concept that no modulation of dopamine takes place during the sleep stages has been recently challenged (Eban-Rothschild et al. 2016; Fifel et al. 2018). These studies reported that ventral tegmental area (VTA) neurons displayed a low amplitude but significant circadian modulation with increased firing rates during the active phase and that the excitation/inhibition of VTA dopaminergic neurons promotes awake/sleep behavior. The differences between the classical feline physiological studies and the recent optogenetic studies might be caused by species differences and the oversensitivity of optogenetic techniques (Wei et al. 2020). For example, classical extracellular recording of primate midbrain dopaminergic neurons and striatal cholinergic interneurons has provided strong evidence that these neurons are activated not by movement but only by reward-related events (Kimura et al. 1984; Romo and

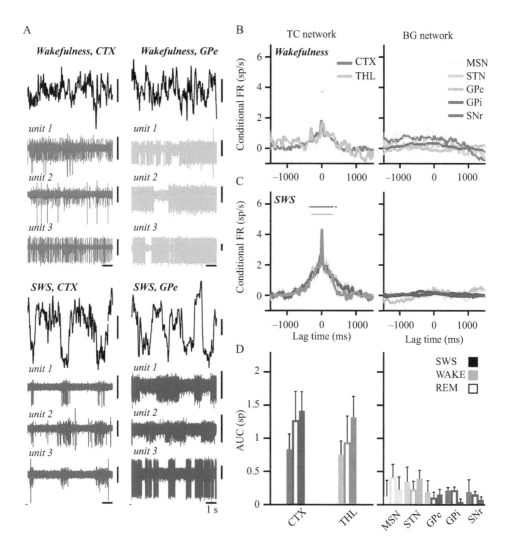

8.3 Synchronized slow oscillations in the thalamus/cortex versus unsynchronized slow oscillations in the basal ganglia during SWS. *A*, an example of LFP and spiking activity recorded from the cortex (*left*) and the GPe (*right*) in the awake (*top row*) and SWS (*bottom row*) states. *B*, average cross-correlation histograms of thalamus/cortex (*left*) and basal ganglia neurons (*right*) in the awake (*top row*) and SWS (*middle row*) states. *Bottom row*, the average area under the curve (AUC) of the cross-correlation histograms. CTX, cortex; THL, thalamus; MSN, medium spiny neurons of the striatum; STN, subthalamic nucleus; GPe, GPi, external and internal segments of the globus pallidus; SNr, substantia nigra reticulata. *Source*: Copied (with permission) from Mizrahi-Kliger et al. 2018.

Schultz 1990; Schultz and Romo 1990). Recent optogenetic studies in mice claim that these neurons are strongly activated by movement (Dodson et al. 2016; da Silva et al. 2018; Howe et al. 2019; Rehani et al. 2019). Thus, it seems that the optogenetic studies in mice detect responses to movement, while those of classical spiking activity do not. Could it be that the optogenetic calcium-sensitive sensors report subthreshold activity? In this case it might be that movement information reaches the midbrain dopaminergic neurons and striatal cholinergic interneurons but does not entrain their

Table 8.1 Sleep textbook description of cortical neuromodulator activity during the awake and sleep stages.

Cortical neuromodulator	Awake	NREM	REM
Acetylcholine	↑	↓	↑
Noradrenaline	↑	↓	↓
Serotonin	↑	↓	↓
Histamine	↑	↓	↓
Dopamine	↑	↑	↑

Notes: Up arrows and down arrows represent increase and decrease discharge rate, respectively. NREM, non–rapid eye movement sleep (N1, N2, and N3); REM, rapid eye movement (paradoxical) sleep.

spiking activity. Accordingly, in Parkinson's disease the MSNs are bombarded by afferent beta oscillations that can be observed in striatal LFP (figure 15.11A); nevertheless, their spiking activity is not entrained to this beta rhythm (figures 15.11B and 15.4A).

Unfortunately, extracellular recording of the spiking activity of dopaminergic neurons is not an easy job, so in our study of the basal ganglia during sleep we did not include SNc and VTA neurons. But we recorded striatal cholinergic neurons—the TANs. Figure 8.4 shows that as reported for feline midbrain dopaminergic neurons, the TAN discharge rate and pattern, as well as the TAN-TAN pairwise correlation, are not modulated by awake, NREM, and REM sleep stages. The dopamine-acetylcholine balance theory and our recording of SNc dopaminergic neurons and striatal TANs (figures 10.4 and 11.4) suggest that SNc dopaminergic neurons and TANs mirror their spiking activity. The jury is still out and waiting for good-quality recording of SNc and VTA dopaminergic neurons in sleeping primates. I predict that they will not be significantly modulated. The changes in the activity of neurons in the main axis of the basal ganglia (figures 8.2 and 8.3) probably reflect the changes in the thalamo-cortical activity, rather than in the activity of the basal ganglia neuromodulators.

RIGHT/LEFT BRAIN AND BASAL GANGLIA ASYMMETRY

In every Jewish wedding ceremony, we break the glass in memory of the destruction of Jerusalem and the temple about two thousand years ago and repeat King David's saying in the Book of Psalms: "If I forget thee, O Jerusalem, let my right hand forget its skill. Let my tongue cleave to the roof of my mouth, if I remember thee not; if I set not Jerusalem above my utmost joy" (137:5–6). These words reflect the old knowledge of association between right hemiparalysis and aphasia.

Modern neurophysiology studies traditionally overlook brain asymmetry. My personal oversight of right/left brain asymmetry started during my first physiological

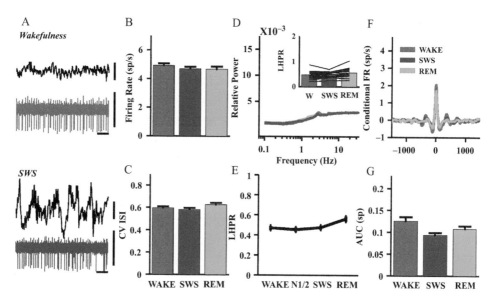

8.4 The discharge rate, pattern, and synchronization of striatal tonically active neurons (TANs) are not affected by awake/sleep stages. A, typical example of striatal TAN activity during wakefulness and SWS, recorded simultaneously with cortical frontal EEG (top). Horizontal bars, 1 s; vertical bars, 100 μV. B, average TAN firing rate. C, average coefficient of variation (CV) of the distribution of interspike intervals (ISIs). D, average relative power spectra. E, low (0.1–2 Hz) to high (9–15 Hz) power ratio (LHPR). F, averaged spike-to-spike correlation histograms. G, AUC for the cross-correlation histograms for TANs during wakefulness, SWS, and REM sleep.
Source: Copied (with permission) from supplementary information of Mizrahi-Kliger et al. 2018.

studies and has continued through most of my academic life. One of the first decisions in monkey physiology experiments is between the right or left location of the recording chamber. It is easy to train a monkey to complete complex arm movements with both hands. I therefore usually repeat the answer I got from my mentors: "Right/left brain asymmetry is pseudoscience and not relevant, so flip a coin for the selection of the hemisphere to record from." It is "pseudoscience" because it leads to concepts such as "males are left brained; females are right brained." It is "not relevant" because "some functions have been displaced to be handled by the right hemisphere only in humans that have developed language in the speech areas of the left cortex. However, the brain of monkeys and all other animals with no language (and note the difference between language and communication) are right/left symmetric."

I was wrong. Right/left brain asymmetry exists very early in the animal kingdom.[4] The huge difference in the size of the left and right habenula of the lamprey (the oldest existing vertebrate) and the zebra fish is a good example. But these are the exceptions. In most vertebrates, including humans and nonhuman primates, the morphology of the right and left hemisphere is very similar. The brain's anatomical right/left symmetry is surprising if we consider the body asymmetry. Although much of humans' external anatomy is bilaterally symmetric, many internal structures develop asymmetrically. Situs solitus is the normal right/left position of the thoracic and abdominal visceral organs.[5] This departure from symmetry in the human body is boosted by a leftward fluid flow (nodal flow) in the embryonic node cavity that is generated by the clockwise rotational movement of a few hundred cilia in the Hensen's node. Thus, 50 percent of people with primary ciliary dyskinesia (also known as immotile cilia syndrome or Siewert-Kartagener syndrome) have situs inversus (a congenital condition in which the major visceral organs are reversed from their normal positions).

In humans, morphological left/right brain asymmetry is minimal compared to the body's left/right asymmetry. The right/left brain asymmetry is at the functional level, with the majority of humans having a right dominant hand (praxis) and most language areas located in the left hemisphere (Gerrits et al. 2020).[6] However, this functional right/left brain asymmetry is probably independent of the molecular cilia machinery leading to situs solitus. In a study of people with primary ciliary dyskinesia, only 15 percent of the individuals with situs inversus and 14 percent of the individuals with situs solitus were left-handed.

The functional asymmetry of the human hemispheres is not limited to language and hand dominance. The left hemisphere is often associated with declarative functions and the right with spatial attention, face recognition, and emotional functions. Alternatively, the valence hypothesis maintains that negative emotions (and avoidance behavior) are served by the right hemisphere, while the positive emotions (and approach behavior), are controlled by the left hemisphere. Probably due to the hostile environment of our evolution, most basic human emotions are negative. For example, four of Paul Ekman's basic emotions (sadness, fear, anger, and disgust) are negative, while only two (happiness and surprise) are positive. Thus, even the valence hypothesis attributes most human emotions to the right hemisphere. Finally, even the roads less traveled may intersect in the wood. Cortical areas fall asleep at different times. The sleep onset process (as judged by increased delta oscillatory activity) follows an anterior–posterior direction, and the left hemisphere falls asleep earlier than the right hemisphere (Casagrande and Bertini 2008a, b).[7]

Human pathology and diseases often open unique windows into understanding the functions of the brain. Although often neglected, the asymmetry of symptom onset and severity in Parkinson's disease is a highly important diagnostic feature and is strongly linked to differential clinical manifestation. Parkinson's disease starts on one side of the body, with weak preference to the dominant side (Yust-Katz et al. 2008; van der Hoorn et al. 2012). Thus, if you see a patient with similar intensity of right/left tremor (and >7 Hz postural/action tremor that is attenuated by a glass of good wine), you should consider the differential diagnosis of essential tremor. Nuclear-imaging methods targeting different dopamine biomarkers reveal asymmetric dopamine loss in the posterior putamen at this early stage and might assure patients of their diagnosis (for neurologists the diagnosis is certain by the clinical symptoms). With disease progression, the motor symptoms (akinesia/bradykinesia, muscle rigidity, and tremor) worsen on both sides; however, the side on which the disease started usually remains more severe until the end stages of the disease, where the ceiling effect equalizes right and left.

Putting aside the onset and severity of Parkinson's motor symptoms, there is no qualitative difference (e.g., more rigidity or more tremor) between patients with more severe right/left striatal dopamine depletion and left/right parkinsonism. However, Parkinson's nonmotor symptoms—that is, emotional, cognitive, and sleep deficits—are affected by the laterality of the dopamine depletion and motor symptoms. As may be expected by the different functions of the left and right cortical hemispheres, patients with left-hemisphere dopamine depletion (and motor symptoms on the right) typically experience problems with language-related tasks and verbal memory, whereas patients with right-hemisphere dopamine depletion (and more severe motor symptoms on the left) more often perform worse on tasks of spatial attention, visuospatial orienting, and memory (Blonder et al. 1989; but see Cubo et al. 2010). In line with the other ideas of division of labor between the right and left hemispheres, other groups have reported that patients with greater dopamine loss in the left striatum showed reduced novelty seeking (or reduced approach behavior), whereas patients with reduced dopamine in the right striatum displayed reduced harm avoidance (Tomer and Aharon-Peretz 2004; Porat et al. 2014).

Renana Eitan, Ruby Shamir, and other members of our team recorded spontaneous spiking activity and responses to vocal nonverbal emotional stimuli in the subthalamic nucleus (STN) in Parkinson's patients during deep brain stimulation (DBS) surgeries (Eitan et al. 2013). Emotive auditory stimulation evoked activity in the ventral nonoscillatory region of the right STN. These responses were not observed in the left ventral STN or in the dorsal regions of either the right or left STN (figure 8.5). Thus, the ventral nonoscillatory regions of the STN are asymmetrically associated with nonmotor functions. In line with emotional function of the right hemisphere, the right STN is associated with emotional processing.

There are still many unknowns in the story of the left and right brains and basal ganglia. I feel that the division of labor between the two brain hemispheres started long ago in the history of biological evolution and, like many other items, arrived at its current peak in primates, especially humans. It seems that the right cortex and subcortical structures such as the basal ganglia are more engaged in negative emotional functions and avoidance behavior, as well as with spatial attention and face recognition. The left hemisphere controls communication (throughout the whole animal kingdom) and language (in humans), as well as positive emotions and approach behavior. Currently, ablation and DBS therapy treat the left and the right basal ganglia equally. Future therapies will hopefully do better.

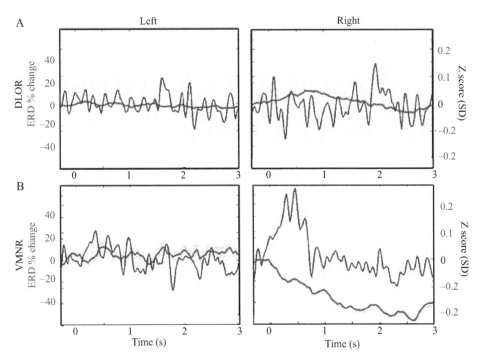

8.5 Subthalamic responses to emotive auditory stimulation are restricted to the right ventral subthalamic nucleus. DLOR, dorsolateral oscillatory (motor) region; ERD, event related de-synchronization; VMNR, ventral-medial nonoscillatory (nonmotor) region.
Source: Copied (with permission) from Eitan et al. 2013.

BIOLOGICAL SEX AND THE BASAL GANGLIA

Like sleep and left/right asymmetry, biological sex issues are overlooked in the typical monkey neurophysiology lab, partly because of fear of pseudoscience ("Men are from Mars, women are from Venus") and partially because of the fear of crossing the borders of politically correct behavior. As for the selection of which hemisphere to record from, the selection of male/female subjects in the monkey lab is quite random. Males are bigger and will perform more trials to get their daily water or caloric needs but are more aggressive. Female monkeys are less aggressive (although our vets are still busy suturing the results of fighting in our female monkey group housing), but they are smaller and their hormonal stages fluctuate.

In most cultures, gender and biological sex used to follow a binary definition. This is clearly not correct anymore, and many individuals exist outside the classical binary groups. Biological sex is defined by the chromosomes (female XX, male XY), reproductive organs (ovaries, testes), gonadal hormones (estrogen, testosterone), and secondary sex characteristics. Gender refers to the cultural differences expected (by society and culture) of men and women according to their biological sex. I insist on "different but equal" and will limit myself to the effects of biological sex on basal ganglia functions.

As for the right and left hemispheres, there is a hot debate regarding morphological differences between male and female brains (Joel and Vikhanski 2019; Rippon 2019). In the basal ganglia world, a recent study reported that males showed significantly larger volumes for the globus pallidus and the putamen, but no differences were found in the male/female volumes of the caudate and the nucleus accumbens (Rijpkema et al. 2012). In many cases, later studies refute initial findings such as males having more volume in the left hemisphere and females having more volume in the right hemisphere. The MRI studies of Daphna Joel and her team led to the "mosaic" hypothesis that suggests high variability in the degree of "maleness" and "femaleness" of different features within a single brain (Joel and Vikhanski 2019).

What about functional differences between the male and the female brain? The chain of the sex hormones (or the hypothalamic-pituitary-gonadal axis) starts at the release of gonadotropin-releasing hormone (GnRH) from the preoptic area of the hypothalamus. The GnRH is carried by the hypophyseal portal bloodstream to the anterior pituitary gland (hypophysis), where it stimulates the synthesis and release of the gonadotropic hormones that include luteinizing hormone and follicle-stimulating hormone. The gonadotropic hormones induce the release of sex hormones, androgens (such as testosterone), and estrogens (such as estradiol), which are steroid hormones (figure 8.6) synthesized primarily in the testes and ovaries, respectively. On average, levels of testosterone in adult males are about seven times as great as in adult females. Estradiol levels in premenopausal women depend on the menstrual cycle and can be as low as for males or peak to ten to forty times the levels in males. After menopause women's estradiol levels are similar to those in men.

The effects of sex hormones are not limited to the reproductive and secondary sex organs. Estrogen receptors are abundant in the prefrontal cortex, dorsal striatum, nucleus accumbens, and hippocampus (Del Río et al. 2018). Similarly, androgen receptors are present at high concentrations in the hypothalamus, amygdala, nucleus accumbens, bed nucleus of the stria terminalis, and septal nuclei (Celec et al. 2015). The relationships among changes in the levels of female sex hormones, mood and cognition during the menstrual cycle, over the life span, or due to hormonal therapy (e.g., contraceptive drugs) attract the attention of the public (Westly 2012). It seems there is solid evidence that estrogen hormones affect both dopamine and serotonin transmission (Jacobs and D'Esposito 2011). Here, I can only second the NIH call to consider, collect, characterize, and communicate sex-based data.[8]

AGE EFFECTS IN THE BASAL GANGLIA

Every textbook of neurology and movement disorders will maintain that Parkinson's akinesia/bradykinesia and rigidity are due to the degeneration of dopamine neurons

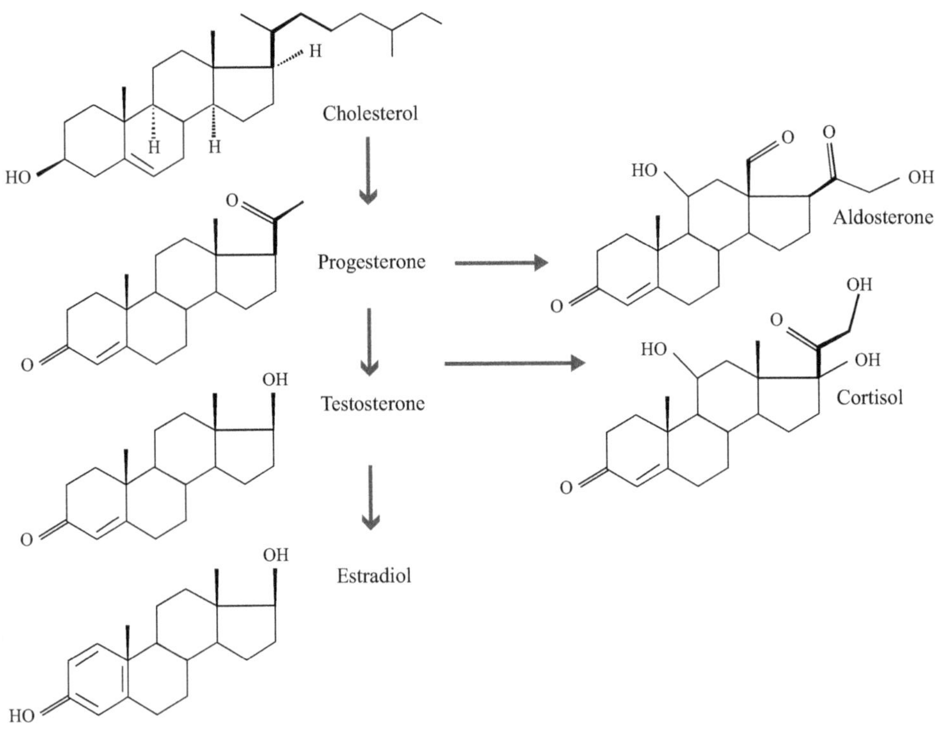

8.6   The biosynthesis of the sex hormones. Related steroid hormones are also shown.

in the substantia nigra pars compacta (SNc) and the resulting loss of dopamine in the dorsal striatum (the dopamine loss starts and is more severe in the putamen than in the caudate). The loss of dopamine neurons is gradual and slowly progresses over the years.

The first method to answer the question of age and related decrease in the number of SNc dopaminergic neurons was to use stereological estimates of autopsy data (Pakkenberg et al. 1991; Ma et al. 1999; Cabello et al. 2002; Rudow et al. 2008; Giguere et al. 2018). Considerable interstudy and interindividual variations exist. Nevertheless, these studies reveal a linear, rather than exponential, decrease in the number of dopaminergic neurons starting at early adulthood and steadily continuing with age (figure 8.7*A*). Data are less solid regarding changes during childhood and adolescence. Most probably, dopamine levels reach their life peak during adolescence, enabling maximal neural plasticity and learning during this critical period (Larsen and Luna 2018). Aging effects on dopaminergic neurons are gradual (Branch et al. 2014). Positron emission tomography (PET) and single-photon emission-computed tomography (SPECT) can be used to assess the dopamine system in bigger populations. As for the number of midbrain dopamine neurons, most of the striatal dopamine markers (figures 8.7*B–D*) show a close to linear decline over the years in healthy human controls (Karrer et al. 2017).

The reasons for the degeneration of dopaminergic neurons are still debated. Aging-related factors such as general brain atrophy and gonadal hormones may play a role.

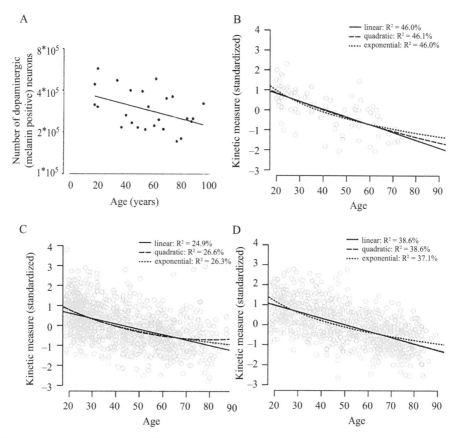

8.7 A reduction in the number of dopaminergic neurons and biomarkers as a function of age. *A*, number of midbrain dopaminergic neurons. *B–D*, meta-analysis of dopamine markers reveals decline in D1, D2 dopamine receptors and transporters with age. A linear mode fit to the data has higher correlation than quadratic and exponential fits.
Sources: A. Adapted from Cabello et al. 2002. B–D. Adapted from Karrer et al. 2017.

The gradual decline of dopamine transmission with age is probably one of the main factors leading to the typical changes in behavior with regard to younger men and women's tendency toward risk-seeking versus their parents' harm-avoidance policy. The effects of ageing and dopamine decline on the computational physiology of neurons in the basal ganglia are still unknown. Are the discharge rate, pattern, and synchronization of the spiking activity in the basal ganglia affected by normal ageing? Like the other subjects discussed in the chapter, this is a pathway not yet taken.

EVOLUTION AND COMPARATIVE BIOLOGY OF THE BASAL GANGLIA

After exploring the riskier basal ganglia pathways, I am back to (relative) safety, and below I will discuss the evolution and the comparative biology (anatomy and physiology) of the basal ganglia. Although basal ganglia homologues are reported in invertebrates (Strausfeld and Hirth 2013), most researchers agree that the cell-rich basal ganglia came in with the appearance of the vertebrates about six hundred million years ago (mya; Reiner et al. 1998; Grillner and Robertson 2016; Murray et al. 2017; Cisek 2019). The gross anatomy of the basal ganglia of many nonmammalian animals does not follow the schemes given in figure 1.1. Nevertheless, Sten Grillner and colleagues established a detailed similarity of input from the cortex/pallium and thalamus of intrinsic organization and projections of output nuclei in the basal ganglia of lamprey, the phylogenetically oldest group of vertebrates, and primate basal ganglia. Similar homology with the primate basal ganglia have been reported for birds (Medina et al. 1999; Doupe et al. 2004; Gale and Perkel 2010; Fee and Goldberg 2011).

The transition of the dorsal pallium to the neocortex, the development of the corticospinal tract, and the corticothalamic reciprocal innervation probably appeared only with mammals about 250 mya (Cisek 2019). The major projections of the basal ganglia were directed to the tectum (today, the superior colliculus) and the tectospinal and the reticulospinal pathway for the first half of their existence. The cerebellum is about the same evolutionary age, or slightly younger than the basal ganglia. Unlike the basal ganglia that affect the contralateral body side, the cerebellum receives information and projects it to the ipsilateral body side. I hope that one day we will have a coherent understanding of the basal ganglia's and cerebellum's different up and down connectivity. Francis Crick (1988) was probably right in saying, "Biologists must constantly keep in mind that what they see was not designed, but rather evolved."

The most common animal species used today for basal ganglia research are rodents and primates. The cellular biochemistry and morphology of the rodent and the primate basal ganglia are very comparable. Nevertheless, there are some critical differences between the gross anatomy, clinical effects of dopamine depletion, and discharge physiology between rodents and primates. In primates the dorsal pallidum is divided into external and internal segments (GPe and GPi). Rodents have a clear homologue of the GPe—the globus pallidus (GP)—however, the entopeduncular nucleus (EP), which is often suggested as the rodent homologue of the GPi, is a different story. Unlike the GPi, which is a closed nucleus with clear boundaries, the EP cells are scattered within the internal capsule. The substantia nigra pars reticulata (SNr) alone, rather than the GPi/SNr, is the main output of the rodent basal ganglia. Rodents are axial animals that usually turn toward their target with their body axis. It could be that the GPi is engaged with the control of nonaxial movements, while the SNr is engaged more with axial (and gaze in the primate) orientation.

The importance of the dopamine system is greater in the primate than in the rodent. Severe depletion of dopamine in the dorsal striatum will lead to severe akinesia/bradykinesia and muscle rigidity in the contralateral body side of all primate species tested and, additionally, to low-frequency tremor in some monkey species (e.g., in vervet monkeys). Severe dopamine depletion of the right or left dorsal striatum will cause minimal deficits (e.g., asymmetric walk) in the spontaneous activity of rodents; however, rigidity and tremor are not self-evident. To expose the dopamine asymmetry, or its pharmacological rescue, researchers test the rotational behavior induced by dopamine agonists or reuptake inhibitors given to the unilateral dopamine-depleted rodent.

Physiologically, one of the characteristics of the primate pallidum is the high (50–80 Hz) discharge rate of most pallidal neurons and the spontaneous pauses of many GPe cells. A recent study by Dana Cohen, Izhar Bar-Gad, and their colleagues revealed that the firing rate is significantly lower in rats than in primates (Benhamou et al. 2012). Interestingly, a very high discharge rate has been reported in the avian homologue of the globus pallidus (Goldberg et al. 2010). Pauses of firing were observed when the birds sang alone, in a less stereotypical manner, and it was suggested they enable the generation of variability in singing behavior (Woolley et al. 2014).

CHAPTER SUMMARY

• Sleep stage affects the activity of neurons in most structures (with the exception of striatal cholinergic interneurons and striatal projection neurons) of the main axis of the basal ganglia. Discharge rate is reduced and the tendency of neurons to discharge in bursts is increased during NREM sleep. Basal ganglia neurons remain uncorrelated even during deep SWS (N3).
• Right/left asymmetry of brain hemispheres is observed in many biological systems and is reflected in basal ganglia physiology and pathophysiology.
• Gonadal hormones, especially estrogen, affect basal ganglia physiology.
• Dopamine transmission gradually declines with aging.

9

BOX-AND-ARROW RATE MODELS OF THE BASAL GANGLIA

Albin, Young, and Penney (1989) were the first to offer a two-arm box-and-arrow model of the basal ganglia. They were followed by Bergman, Wichmann, and DeLong (1990) and by Gerfen et al. (1990; figure 9.1). The three versions use box-and-arrow diagrams to shed light on the mechanism of hypokinetic and hyperkinetic movement disorders. Because their general frames are similar, they will be lumped together as the D1/D2 direct/indirect pathways box-and-arrow model of the basal ganglia.

D1/D2 DIRECT/INDIRECT PATHWAYS BOX-AND-ARROW MODEL OF THE BASAL GANGLIA

The main assumptions of the model are as follows:

1. All neurons in one structure (e.g., cortex, striatum) have one function and can be lumped into one box.
2. All the projection neurons of a structure use a single transmitter that can excite/inhibit or increase/decrease the discharge rate of their target structure.

   2a. Glutamate is an excitatory transmitter, increasing the discharge rate of its postsynaptic neurons.

   2b. Gamma-aminobutyric acid (GABA) is an inhibitory transmitter, reducing the discharge rate of their target structure.

   2c. Dopamine is different and can be excitatory or inhibitory depending on the nature of its postsynaptic receptors. Dopamine is excitatory to neurons with D1 dopamine receptors and inhibitory to neurons with D2 dopamine receptors.

Figure 9.2 depicts the details of the D1/D2 direct/indirect pathways box-and-arrow model. Green arrows are for glutamatergic/excitatory connections. Red arrows

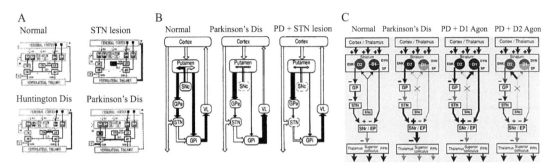

9.1 The original box-and-arrow models of the basal ganglia. PD: Parkinson's disease; Agon: agonists. *Sources:* A. Adapted from Albin, Young, and Penney 1989. B. Adapted from Bergman et al. 1990. C. Adapted from Gerfen et al. 1990.

are for GABA/inhibitory connections. Midbrain dopamine innervation of the striatum is depicted in black arrows, along with its excitatory/inhibitory effects on D1/D2 (green/red) medium spiny neurons (MSNs), respectively. The model states that

- All cortical areas send glutamatergic projections to the input stage of the basal ganglia—the striatum.
- Dopamine excites D1-direct and inhibits D2-indirect striatal MSNs.
  - □ Excitation of D1 MSNs leads to inhibition of the GPi/SNr (direct GABAergic pathway).
  - □ Inhibition of D2 MSNs leads to excitation of the GPe, which inhibits the STN. The inhibition of the STN leads to reduced excitation (i.e., inhibition) of the GPi/SNr (indirect net excitatory pathway).
  - □ Thus, an increase in dopamine in the striatum leads (through both the direct and the indirect pathways) to GPi/SNr inhibition.
- Inhibition of the GPi/SNr leads to disinhibition of thalamocortical motor areas, and the increased activity of the motor cortices leads to movements.

Thus, the model shows that

- An excess of dopamine (as in levodopa-induced dyskinesia) leads to hyperkinetic movement symptoms.
- A lack of dopamine (as in Parkinson's disease) leads to hypokinetic symptoms (akinesia and bradykinesia). To see this, reverse the arguments above, starting with the saying that lack of dopamine inhibits D1-direct and excites D2-indirect MSNs.

The model got its huge popularity first because it explains the hypo/hyper clinical symptoms associated with a lack/excess of dopamine and because it correctly predicts (personal disclosure: because it was in line with the experimental results) the increased discharge rate in the STN in the MPTP primate model of Parkinson's disease. The model's behavioral predictions were further confirmed by optogenetic and

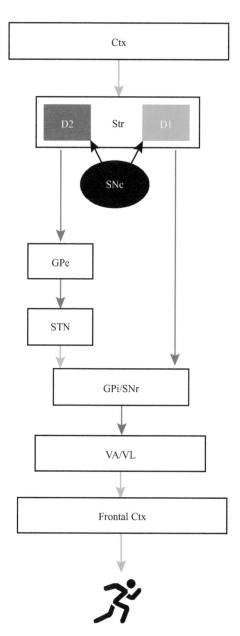

9.2 The D1/D2 direct/indirect pathways box-and-arrow model of the basal ganglia. D1 and D2 are for striatal (Str) medium spiny neurons with D1 and D2 dopamine receptors. Ctx, cortex; GPe, GPi, globus pallidus external and internal segments; SNc, SNr, substantia nigra pars compacta and reticulata; STN, subthalamic nucleus; VA/VL, ventral anterior/lateral thalamic nuclei. *Red/green fonts and arrows*, inhibitory and excitatory effects, respectively.

chemogenetic manipulation of the activity of D1 and D2 MSNs (Shen et al. 2008; Kravitz et al. 2010; Alcacer et al. 2017). Notably, the model predicts the amelioration of parkinsonian (akinetic) symptoms following transient (by direct injections of muscimol, $GABA_A$ agonist; Wichmann et al. 1994) and permanent (ablation, by direct injection of the axon-sparing toxin ibotenic acid) STN inactivation (Bergman et al. 1990) and by radio-frequency lesion (Aziz et al. 1991).

COMPUTATIONAL PHYSIOLOGY OF THE D1/D2 DIRECT/INDIRECT PATHWAYS BOX-AND-ARROW MODEL OF THE BASAL GANGLIA

From the point of view of the basic questions of a computational model (chapter 4), the D1/D2 direct/indirect box-and-arrow model of the basal ganglia suggests that

- The computational goal (*What does the system do?*) of the basal ganglia is to modulate the excitability (vigor) of the motor apparatus.
- The computational algorithm (*How does the system do what it does?*) of the basal ganglia network is that of collaboration or competition between D1/direct and D2/indirect pathways.
- And the computational implementation (*How is the system physically realized?*) in the basal ganglia is realized through D1/D2 modulation of striatal excitability and discharge rate, leading through the direct/indirect pathways to modulation of the excitability and discharge rate of the basal ganglia output nuclei and the motor cortical fields.

THE D1/D2 SAGA

The D1/D2 direct/indirect models (figures 9.1 and 9.2) assume the population of striatal MSNs is divided into two classes, MSNs with D1 dopamine receptors that project to the GPi/SNr and a second, non overlapping, population of MSNs with D2 dopamine receptors that project to the GPe. Due to the opposite dopamine effects, the activity of the two populations of MSN neurons is expected to mirror each other.

Jim Surmeier and colleagues (1993) challenged the dogma and asked in a 1993 *Trends in Neuroscience* (*TINS*) paper: Are neostriatal dopamine receptors colocalized? Surmeier's original data were based on RNA studies, but later he changed his mind and turned into a strong supporter of D1/D2 segregation. RNA and protein levels might be different, and the proteins (receptors) are the more important parameter. However, other groups (Nicola et al. 2000; Kupchik et al. 2015) working mainly in the ventral striatum continue to claim that a significant fraction of MSNs coexpress D1 and D2 markers and that dopamine D1-D2 receptor heteromers exist in striatal MSNs (Perreault et al. 2011).[1]

The first decade of the twenty-first century witnessed the exponential growth of optogenetic and chemogenetic methods. The early studies using D1- or D2-labeled

MSNs usually confirmed the basic assumptions of the classical D1/D2 direct/indirect pathways models. Optogenetic activation of D1 or D2 MSNs provided the causality link between the activity of the two populations of striatal projection neurons and the motor symptoms of Parkinson's disease (Kravitz et al. 2010, 2012; Lee et al. 2016).

However, recent studies have revealed anatomical and functional connectivity between D1 and D2 MSNs. D1 MSNs were found to project not only to the GPi/SNr but also to the GPe (Wu et al. 2000; Gittis and Kreitzer 2012). Additionally, D1 and D2 MSNs often interact (Tepper et al. 2004; Planert et al. 2010, Burke et al. 2017; Hjorth et al. 2020). Recording of D1 and D2 MSN activity in behaving animals by the groups of Rui Costa and Mark Schnitzer (Cui et al. 2013; Tecuapetla et al. 2016; Parker et al. 2018) has revealed coactivation and a complex coordinated activity pattern of D1 and D2 MSNs. Thus, D1 and D2 MSNs do not mirror each other in behaving animals.

THE TWO PALLIDAL SEGMENTS: FRIENDS OR FOES?

In the D1/D2 direct/indirect model of the basal ganglia, the GPe to GPi connectivity has a net inhibitory action (GPe inhibits the STN that excites the GPi). If GPe activity decreases (as during dopamine depletion), GPi activity is expected to increase and vice versa. This opposite relationship between the GPe and GPi discharge rate is augmented by the effects of dopamine on the striatum. Thus, the GPe and GPi should behave as foes, with opposite trends of changes in their discharge rates.

Two of our recent studies have not confirmed this GPe-GPi "foe" prediction. Marc Deffains recorded the activity of GPe and GPi neurons during a classical-conditioning task (Deffains et al. 2016). Neurons in both pallidal segments responded by both increasing and decreasing their discharge rate during the cue and the delay epoch. The neurons of both structures tended to increase their discharge rate during the cue and the delay. One might argue that our results do not rule out the possibility that there are GPe-GPi pairs with mirror activity. However, then we would expect opposite (or equal) fractions of increase/decrease responses in the two structures.

In the second experiment, Aviv Mizrahi-Kliger and Alex Kaplan recorded basal ganglia neurons over the sleep/wake cycle. Figure 8.2 depicts the average discharge rate and pattern of the neurons in the different basal ganglia structures during the different awake/sleep stages. During SWS, both pallidal segments reduced their discharge rate by approximately one-third and increased their CV ISI by approximately one-third (i.e., became more bursting) compared to the awake state. Similar trends were observed by Avital Adler when she compared GPe and GPi activity during a task and during "napping," when the monkey stopped performing and closed his eyes in the middle of the task (Adler et al. 2010). Thus, changes in the discharge rate of pallidal neurons over wake/sleep states were in the same direction.

THALAMO/CORTICAL VERSUS DOPAMINE DRIVE OF THE STRIATUM AND THE BASAL GANGLIA NETWORK

Of course, this is not the end of the D1/D2 direct/indirect model (although it might be the beginning of the end, as Winston Churchill said after the victory in El Alamein during World War II). The similar trends in discharge rate of the two pallidal segments are probably due to similar activation of the D1 and D2 MSNs by their thalamic and/or cortical afferents. The conflicting results presented above can reflect the duality of the striatal inputs (figure 9.2). On one hand, both populations of D1 and D2 MSNs are activated by their thalamo and cortical glutamatergic input. On the other hand, D1 and D2 MSNs might be oppositely affected by drastic modulation of the striatal dopamine tone. Coactivation of D1 and D2 MSNs might be due to increased cortico-thalamic inputs with no modulation of the dopaminergic tone. The similar changes in the discharge rate of the two pallidal segments may reflect the discharge rate of their striatal afferents. However, the cumulative physiological data call for updating the basic assumptions of the D1/D2 direct/indirect model of the basal ganglia.

ANATOMICAL UPDATES OF THE D1/D2 DIRECT/INDIRECT PATHWAYS BOX-AND-ARROW MODEL OF THE BASAL GANGLIA

A successful scientific model needs:

1. To explain previous knowledge in a simple way.
2. To predict the results of future experiments.
3. To have good public relationships.

The D1/D2 direct/indirect box-and-arrow model has achieved at least two of the above requirements. The physiological questions raised in the previous paragraphs could be neglected. However, from the beginning, and even more with the passing years, it has been evident that the D1/D2 direct/indirect model does not describe the reality of the anatomical connectivity of the basal ganglia network.

Even at the time of the model's birth, it was well known that the STN heavily projects back to the GPe and not just to the GPi/SNr. Similarly, the cortico-STN hyperdirect pathway, as named by Atoshi Nambu, was neglected (even if was marked on the plotting board by Bergman, Wichmann and DeLong; figure 9.1*B*). About half of the excitatory inputs to the striatum are coming from the thalamus. These thalamic projections from the intralaminar thalamic nuclei—the centromedian and parafascicular (CM-PF) nuclei—were also neglected. Finally, the output nuclei of the basal ganglia—the GPi and the SNr—do not only project to the ventral lateral and ventral anterior thalamic nuclei, respectively. In the thalamus, pallidal projections are also directed toward the intralaminar nuclei, as well as to the dorsomedial nucleus of the thalamus. Thus, the basal ganglia output also affects the arousal level through the modulation of the intralaminar thalamic nuclei, as well as the associative areas of the prefrontal cortex through the modulation of the dorsomedial thalamic nucleus. The GPi and SNr project also to brain stem motor nuclei, such as the pedunculopontine nucleus (PPN) and the superior colliculus (SC), respectively.

Progress in anatomical techniques has revealed new connections in the basal ganglia. First, a direct path from the GPe to GPi was described in the rodent by Hitoshi Kita (Kita and Kitai 1994; Kita 2001) and found to be a major one also in the non-human primate (Hazrati et al. 1990; Smith et al. 1994; Shink et al. 1996; Sato et al. 2000). Second and most surprisingly, a subgroup of pallidal neurons, the arkytypical neurons, were found by Pete Magill and colleagues to project back to the striatum. Finally, it was found that D1 MSNs also project to the GPe, and lateral interactions exist between D1 and D2 MSNs.

Students of basal ganglia have tried to incorporate most of this data into more and more complex models of the basal ganglia. Figure 9.3 depicts one such comprehensive box-and-arrow model of basal ganglia structures and functional connectivity (again green/red arrows symbolize glutamate/GABA connectivity. Midbrain dopamine projections to the striatum are omitted because of graphic constraints).

THE THREE-LAYER MODEL OF THE BASAL GANGLIA

Neurology and neurosurgery residents, as well as neuroscience undergraduate and graduate students, have to memorize the classical D1/D2 direct/indirect box-and-arrow model (figure 9.2) for their board or end-of-semester exams. They honestly describe this as a nightmare. The updated detailed box-and-arrow model of the basal ganglia (figure 9.3) is a nightmare also for dedicated researchers of the basal ganglia. This is not only because of the overload of boxes and arrows. It is impossible for most of us to see the forest through the many trees of this model. Neither can we do simple predictions regarding the effects of a lack or excess of dopamine on the activity in the frontal cortex and the like.

A more simplified three-layer box-and-arrow model is emerging (Kita 2007; Deffains et al. 2016; figure 9.4, *again green/red for excitatory/inhibitory connectivity*). The model is built of all the boxes and arrows of the detailed model (figure 9.3) but in different positions. The major change is the removal of the STN from its position as a relay nucleus in the indirect pathway. Rather, the STN is considered as a second input structure in the basal ganglia that, like the striatum, receives (hyper) direct innervation from the cortex and the intralaminar thalamic nuclei—the CM-PF. The second major change is the positioning of the GPe as the central nucleus of the basal ganglia. The basal ganglia input nuclei project to the GPe, which in turn project back to the STN and the striatum (Sato et al. 2000). Finally, the STN, the striatum, and the GPe project to the output structures of the basal ganglia—the GPi and the SNr. These

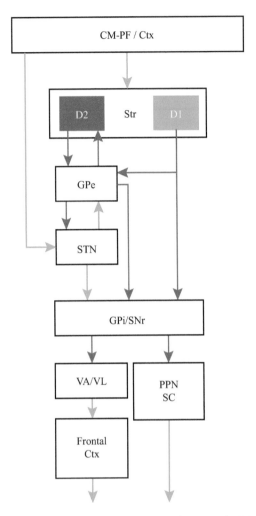

9.3 Comprehensive box-and-arrow model of the basal ganglia network. CM-PF, centromedian para-fascicular thalamic nuclei; Ctx, cortex; D1 and D2, striatal (Str) medium spiny neurons with D1 and D2 dopamine receptors; GPe, GPi, globus pallidus external and internal segments; PPN, pedunculopontine nucleus; SC, superior colliculus; SNc, SNr, substantia nigra pars compacta and reticulata; STN, subthalamic nucleus; VA/VL, ventroanterior/lateral thalamic nuclei. *Red/green fonts and arrows*, inhibitory and excitatory effects, respectively.

basal ganglia output structures project to the cortex (through the thalamic motor nuclei) and brain stem (PPN and SC) motor structures.

The main advantage of the three-layer box-and-arrow model of the basal ganglia is its simplicity, on one hand, while accounting for most of the major connections in the basal ganglia, on the other hand. The model is not complete. The corelease of transmitters by a single neuron (like Substance P and enkephalin by D1 and D2 GABAergic MSNs, GABA and glutamate by midbrain dopaminergic neurons) and the somatodentritic release, volume transmission, and extrasynaptic effects of neuromodulators (Rice and Patel 2015) are neglected. Nevertheless, the model focuses

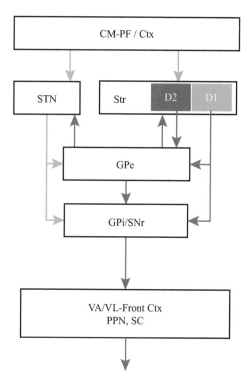

9.4   The three-layer model of the basal ganglia. CM-PF, centromedian parafascicular thalamic nuclei; Ctx, cortex; D1 and D2, striatal (Str) medium spiny neurons with D1 and D2 dopamine receptors; GPe, GPi, globus pallidus external and internal segments; PPN, pedunculopontine nucleus; SC, superior colliculus; SNc, SNr, substantia nigra pars compacta and reticulata; STN, subthalamic nucleus; VA/VL, ventroanterior/ lateral thalamic nuclei. *Red/green fonts and arrows*, inhibitory and excitatory effects, respectively.

our attention on the GPe as a central nucleus of the basal ganglia. Finally, as any good model does, it raises new questions, such as which structure is playing a more important role in shaping the activity of downstream basal ganglia structures—the big striatum or the small but excitatory STN? We will return to these questions in chapter 18.

THE PITFALLS OF BOX-AND-ARROW MODELS OF THE BASAL GANGLIA

Box-and-arrow models of the basal ganglia have played a critical role in understanding basal ganglia physiology and pathophysiology, as well as in establishing new surgical therapies (ablation and DBS) for common basal ganglia disorders. However, we should recognize the major pitfalls of these models.

- First, if the basal ganglia were so simple as to follow "box-and-arrow" operational rules, we would probably be so simple that we could not.[2] More seriously, we have to remember that box-and-arrow models are oversimplified. For example, in the one box of human striatum, there are 10^8 medium spiny projection neurons (table 2.1).

Each one of these striatal neurons receives 10^4 synapses from the cortex, thalamus, SNc dopaminergic neurons, and other striatal cells. In turn, each striatal neuron emits one hundred to one thousand collateral synapses and projects to downstream basal ganglia nuclei. Additionally, there are $5 * 10^6$ interneurons in the striatum, leading to a complex striatal microcircuitry.

- Second, does dopamine only modulate striatal and motor cortices' excitability and tonic discharge rate? What about dopamine modulation of the corticostriatal synaptic plasticity? Do dopamine and the basal ganglia play a role in learning?

- Finally, what about the dynamics of basal ganglia activity and behavior? Box-and-arrow models are static by nature and do not relate to oscillations and tremor.

FOCUSING VERSUS SCALING: ACTION SELECTION IN THE BASAL GANGLIA

The box-and-arrow models of the basal ganglia described above were given from the point of view of scaling of motor vigor (e.g., movement speed) as the main goal of basal ganglia processing. An alternative model, probably already proposed by Denny-Brown (1962), states that the basal ganglia act as a "clearing house," selecting the most appropriate action for a given situation. The model suggests that the basal ganglia select which movements should be carried out and suppress unwanted movements in response to competing sensory stimuli. This model was further developed by John Mink (figure 9.5A). Mink suggested that the selection of specific movements requires activation of the direct pathway, while the inhibition of competing movements requires activation of the indirect pathway (Mink 1996). The indirect pathway supplies a blanket inhibition in the GPi, out of which the direct pathway carves the intended movement. The dense lateral inhibitory network of the striatum (figure 2.2A, B) and the resulting winner-take-all dynamics may help in ensuring that a single action is selected.

To overcome the slow conduction along the indirect pathway (small-diameter striatal axons, multisynaptic), the "hyperdirect" cortex to the STN pathway (Nambu et al. 2002) can be recruited to supply a fast route to the GPi/SNr (figure 9.5B). Parkinson's akinesia is therefore attributed not to underscaling and suppression of the motor vigor. Rather, Parkinson's akinesia is the result of an overactive inhibitory blanket that does not allow the selection of even one possible action. Similarly, hyperkinetic disorders (e.g., the involuntary tics of Tourette's patients) are due to aberrant action foci escaping the inhibitory blanket.

COMPUTATIONAL PHYSIOLOGY AND PITFALLS OF ACTION-SELECTION MODELS OF THE BASAL GANGLIA

The focusing/action-selection model takes one step beyond the box-and-arrow scaling model of the basal ganglia. The scaling model of the basal ganglia depicts the different layers of the basal ganglia network (striatum, GPi/SNr, thalamus/cortex)

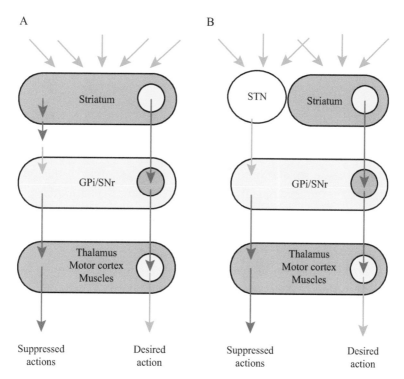

9.5 Action-selection models of the basal ganglia. *Red and green arrows*, inhibitory and excitatory connections, respectively. *Yellow and blue areas*, areas with increased and decreased discharge rates, respectively. Action selection is mediated by competition between (*A*) indirect and direct pathways and (*B*) hyperdirect plus indirect and direct pathways.

with millions of neurons as one box. The action-selection model divides these boxes into two or many foci. In considering the basic questions of a computational model (chapter 4), the action-selection model of the basal ganglia suggests that

- The computational goal (*What does the system do?*) of the basal ganglia is to select which movements should be carried out in response to competing sensory stimuli and to suppress unwanted movements.
- The computational algorithm (*How does the system do what it does?*) of the basal ganglia network is a spatial competition between the D1/direct and the hyperdirect plus D2/indirect pathways in a Mexican-hat operation pattern with strong "center" excitation (or disinhibition) flanked by broad "surround" inhibition.
- The computational implementation (*How is the system physically realized?*) in the basal ganglia is a narrow focus for the direct pathway and broad distribution for the indirect/hyperdirect pathways. The dopamine modulation of striatal lateral connectivities leads to widening or narrowing of the basal ganglia direct pathway focus or focuses.

Still, the action-selection model can't explain the consistent findings of late timing of pallidal activity in relation to the initiation of movements (Mink and Thach 1991). The action-selection model provides an intuitive framework for the binary selection of one of two possible choices. However, our actions are based on the activation and suppression of so many muscles that a model of two alternative choices seems too simplistic.

CHAPTER SUMMARY

- The D1/D2 direct/indirect box-and-arrow model of the basal ganglia assumes that dopamine excites D1 MSNs and inhibits D2 MSNs. The D1 MSNs provide direct inhibition, whereas the D2 MSNs provide indirect excitation of GPi/SNr, the output nuclei of the basal ganglia that inhibit the thalamic-cortical networks.
- The computational goal of the basal ganglia network according to the D1/D2 direct/indirect pathways model of the basal ganglia is to scale the activity in the motor cortex and the motor vigor of the subject.
- The D1/D2 direct/indirect model predicts that too much dopamine will lead to a reduction in the GPi/SNr discharge rate, disinhibition of the motor cortex, and hyperkinetic disorders. Striatal dopamine depletion (as in Parkinson's disease) would lead to an increase in the discharge rate of the GPi/SNr, inhibition of the motor cortices, and hypokinetic (akinesia) motor disorders.
- The complex connectivity of the basal ganglia suggests a three-layer model. In this model the striatum and the STN are the input nuclei of the basal ganglia, receiving inputs from all cortical areas and from the intralaminar thalamic nuclei. The striatum and the STN project to the central structure of the basal ganglia—the GPe—and to the basal ganglia output nuclei—the GPi/SNr. The GPe project back and forward to basal ganglia input and output structures.
- The action-selection model of the basal ganglia proclaims that the direct pathway of the basal ganglia provides a focus of activation of the selected action. All other actions are inhibited by the blanket of the indirect or the hyperdirect pathways.
- According to the action-selection model of the basal ganglia, too much dopamine will lead to many wide-open activation foci in the basal ganglia and hyperkinetic (e.g., tics) disorders. Following striatal dopamine depletion, no focus is opened, and no action can be selected.

10

REINFORCEMENT-LEARNING ACTOR/ CRITIC MODELS OF THE BASAL GANGLIA

Box-and-arrow models of the basal ganglia dominate textbooks of neurology and neurosurgery. In textbooks of neuroscience, however, one can find another family of models of the basal ganglia—reinforcement-learning models.

PRINCIPLES OF REINFORCEMENT LEARNING

Reinforcement learning (RL) is a machine-learning algorithm that makes possible learning the optimal behavioral policy (maximal cumulative rewards over time) from experience (and not from a know-all teacher) in a stochastic and changing environment. The environment provides the behaving agent with rewards and punishments—positive and negative rewards, respectively.[1] Reinforcement learning is a very successful branch of applied computer science, and its technological use (from computer games to autonomous driving) is fast expanding. It was inspired by psychological learning theory (e.g., Rescorla-Wagner model of conditioning), and in return it gave neuroscience one of the best frameworks for understanding the neural mechanism of learning and behavior.

Figure 10.1 depicts the main parts and functions of reinforcement learning. The machine-learning agent (*gray rectangle*) interacts with the environment (world) and adjusts its behavior according to negative and positive received rewards. The agent acts on the world, and as a result it may get a reward (negative or positive) and face a new state. The agent optimizes its behavior by means of a division of labor between its two parts: the actor and the critic (teacher). The actor connects (probabilistically) the two vectors of possible states and possible actions according to the current behavioral policy. The critic uses the acquired rewards and the expected value of the previous state to calculate the prediction error (PE; the mismatch between the predicted and actual value of the previous state) to adjust the critic stored value of

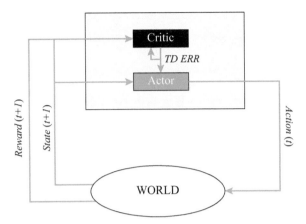

10.1 The main components of reinforcement learning. The agent (*gray rectangle*) is composed of two parts: critic and actor. The agent calculates the prediction temporal-difference error (TD ERR) and modifies the behavioral policy of the actor and its state value estimation.

that state and to change the actor's behavioral policy. The PE is also called the temporal difference error (figure 10.1, *TD ERR*) since it represents the difference between the two temporal values of the state $(PE = V(t) - V(t-1))$. If the PE is positive (reality better than prediction), then the previous state-to-action association is increased (reinforced); if the PE is negative (reality worse than prediction), the previous state-to-action association is decreased. The behavioral policy would not be changed in cases in which the reality does not differ from the predictions. Thus, reinforcement learning ensures the optimal adjustment of behavioral policy with no prior knowledge (in that case, we will start with equal probability for all possible state-to-action pairs) of the world. It can also start with prior knowledge and adjust it to the changing environment.

There are several critical factors in reinforcement learning, including:

- The learning gain (α and β for state value and policy, respectively). What is the optimal gain of learning from PEs? If the gain is high, we will learn faster; however, we will then be prone to distraction from every small glitch and noise. Learning gain is therefore usually kept at low values. Some algorithms aiming at both fast and stable learning start with high learning gain and decrease it with time.
- The temporal (future) discount rate (γ). The value of a state is the agent's current estimate of the cumulative reward she will get in the future with the assumption of no change in behavioral policy. For most of us, publication within a month from now in a journal with an impact factor of ten is worth more than publication twenty years from now in the most prestigious journal with an impact factor of twenty-five (for the lucky ones who do not have to fight with professional editors and hostile referees, think about $10 now versus $1,000 in twenty years). This is the temporal discount rate, which is dependent on many factors, including personality

and maybe even disease states. An agent with close to zero temporal discounting (γ) is very shortsighted, and any future reward does not count. An agent with $\gamma = 1$ treats the very remote future as the next moment. Most adult human and good reinforcement-learning agents prefer intermediate discount rates ($0 < \gamma < 1$).

- The eligibility (past, backward updating) trace (λ). The eligibility trace sets how far we look back to the past when we update the state values. Thus, one-step temporal-difference (TD) methods that update just one step to the past have $\lambda = 0$. On the other hand, Monte Carlo methods use $\lambda = 1$; namely, they have infinite memory and update all previous states. In between are intermediate eligibility trace values ($0 < \lambda < 1$) that are often better than either extreme method.

- Policy versus value iterations. Naturally, policy and state values affect each other. Thus, we can optimize our value estimation for a specific policy; however, the values of the same states would be changed for a different policy. In the same vein, the optimization of the policy is dependent on the state values. The real goal of the reinforcement-learning agent is to find the optimal policy with accurate estimation of state values. However, given the condition of the changing environment, this might not be achieved in real conditions. A possible solution is to alternate between policy and value optimization iterations—that is, fixing the values while optimizing policy and then fixing policy while optimizing state value estimation. Another approach is critic/actor architecture, in which the agent is simultaneously modifying both values and policy according to the PE.

A critical issue in reinforcement learning is the trade-off between exploration and exploitation. The simplest action selection rule is to select the action (or one of the actions) with the highest estimated action value. This greedy method always exploits current knowledge to maximize immediate reward. However, the greedy method doesn't make possible the sampling of unknown or apparently inferior actions. In a dynamic environment, unexplored options or previously low-grade options might be found to be better than the current highest estimated action. A simple alternative to the greedy method is the ε-greedy method. The agent behaves greedily most of the time, but from time to time (with a small probability ε, epsilon) selects an action at random. This method enables exploration of the inferior or never-visited options and therefore ensures the finding of the optimal solution. The method drawback is that when it explores, it chooses equally among all inferior actions. This means that it is as likely to choose the worst-appearing action as it is to choose the next-to-best action. The soft-max policy varies the action probabilities as a graded function of their estimated value. The greedy action is still given the highest selection probability, and all the others are ranked, weighted, and selected according to their value estimates. The soft-max behavioral policy uses θ as the "temperature" controlling the risk behavior of the agent. With a high temperature value, the probability of picking all actions is the same and does not depend on their estimated value (gambling, exploring behavior). On the other hand, low temperature values ensure that only the high-value actions would be selected (greedy behavior). Medium value of the soft-max temperature will lead to a "probability matching"-like behavior, in which the probability of choosing an action is proportional to its estimated value.

REINFORCEMENT LEARNING IN EQUATIONS

For those of us who prefer to see equations, we can express the essence of reinforcement learning with the following:

The state value function: $V(s_t) = r_{t+1} + \gamma r_{t+2} + \gamma^2 r_{t+3} \cdots \gamma^n r_{t+n+1} = r_{t+1} + \gamma V(s_{t+1})$.   10.1

The temporal prediction error: $\delta_t = r_{t+1} + \gamma V(s_{t+1}) - V(s_t)$.   10.2

The state value updating rule: $V(s_t) \leftarrow V(s_t) + \alpha\,\delta_t$.   10.3

The policy updating rule: $p(s_n, a_n) \leftarrow p(s_n, a_n) + \beta\,\delta_t$.   10.4

The soft-max policy: $p(s_n, a_n) = e^{\wedge}(Q(a_n)/\theta)/\Sigma\ e^{\wedge}(Q(a)/\theta)$,   10.5

where V is the value of a state, γ is the future discounting factor, δ is the prediction error, r is the reward, α is the learning rate of the value of a state, and β is the learning rate of the behavioral policy. Note that we are using a TD($\lambda=0$) algorithm and update (left-pointing arrow) only the value of the last state. $p(s_n, a_n)$ is the probability of choosing action a_n being in a state s_n. Q is the value of performing an action. θ is the soft-max temperature.

I have provided only a brief and simplistic introduction to TD(0) reinforcement learning. There are many alternatives to the scheme given here, such as Q and SARSA learning, off and on learning policies, and more. The interested reader should consult some of the more professional literature (e.g., Sutton and Barto 2018).

REINFORCEMENT LEARNING IN THE BASAL GANGLIA NETWORKS

In chapter 9 we discussed the box-and-arrow rate models of the basal ganglia. We concluded that the computational goal of the basal ganglia according to these models is (static) regulation of the motor vigor. The goal is achieved by the effect of dopamine modulation on the excitability of striatal D1 and D2 MSNs and competition between the direct and indirect pathways on the resulting excitability and discharge rate of GPi/SNr and motor cortex neurons. The static nature of the box-and-arrow model does not allow fast dynamic behavioral processes, such as plasticity and oscillations at the neuronal level or learning and tremor at the behavioral level. The reinforcement-learning actor/critic models of the basal ganglia underscore the role of dopamine in modulation of the efficacy of the corticostriatal synapse and behavioral motor learning. Therefore, unlike the box-and-arrow models they emphasize the role of the basal ganglia in phenomena with fast dynamics.

Figure 10.2 depicts the basal ganglia as a critic/actor reinforcement-learning network. The basal ganglia are built of an actor-main axis that connects between cortical areas, encoding the current state of the agent and the brain motor centers. The midbrain (SNc) dopaminergic neurons constitute the critic (the teacher). Most basal ganglia reinforcement-learning models and literature are focused on the dopamine/striatum interactions and neglect the physical details of the basal ganglia actor. Here, we will use our knowledge of the functional connectivity of the basal ganglia and cortical networks to close the loop with the brain motor centers and the environment (world; figure 10.2). The main axis of the basal ganglia is shown for both the D1/D2 direct/indirect pathways and the three-layer models of the basal ganglia (figure 10.2*A*, *B*). In both cases the main axis of the basal ganglia connects all cortical areas, encoding the current state of the biological agent and the cortical (Ctx) and brain stem (BS) motor centers.

The reinforcement-learning models of the basal ganglia are our first step toward state-to-action loops models of the basal ganglia. A hidden assumption of the box-and-arrow models is their vertical (figures 9.2–9.5), hierarchical, top-down structure. These box-and-arrow models follow the classical view that actions are initiated by perception, or even by our internal free-will agent. The state-to-action loops models second John Dewey, who said, "What we have is a circuit, not an arc or broken segment of a circle. This circuit is more truly termed organic than reflex, because the motor response determines the stimulus, just as truly as sensory stimulus determines movement. Indeed, the movement is only for the sake of determining the stimulus, of fixing what kind of a stimulus it is, of interpreting it" (Dewey 1896). State-to-action loops models of the basal ganglia and the brain do not have start and end stations; rather, behavior is continuously looping between sensing, perception, state of the behaving agent, and actions. One aim of the actions is to provide a better estimate of the current state (active sensing; Anderson 2011; Ahissar and Arieli 2012; Cisek 2019).

To be able to play the role of the basal ganglia critic, the midbrain dopaminergic neurons should

- Homogeneously encode the prediction error (PE)
- Widely broadcast this PE to the striatum
- Modulate the state-to-action probability matrix of the main axis of the basal ganglia

Below I will show that these features have been found in the basal ganglia network.

ENCODING OF REWARD PREDICTION ERROR BY DOPAMINERGIC NEURONS

Wolfram Schultz has pioneered in the extracellular recording of the activity of dopaminergic neurons in the behaving monkey. Driven probably by the consensus that basal ganglia are part of the motor system, he and Ranulfo Romo were looking for

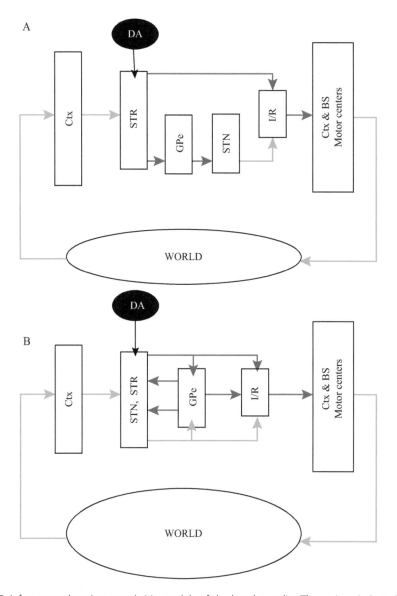

10.2 Reinforcement-learning actor/critic models of the basal ganglia. The main axis (actor) of the basal ganglia is depicted according to the D1/D2 direct/indirect (*top*) and the three-layer (*bottom*) models of the basal ganglia. *Red arrow*, excitatory connections; *green arrow*, inhibitory connections. Ctx, cortex; DA, dopamine neurons; GPe, external segment of the globus pallidus; I/R, internal segment of the globus pallidus and substantia nigra reticulata; STN, subthalamic nucleus; STR, striatum.

a correlation between the spiking activity of dopaminergic neurons and arm movements (Romo and Schultz 1990; Schultz and Romo 1990). Careful reading of their 1990 seminal series of papers reveals their frustration. They failed to find consistent correlation with arm movements and mainly found activation of the dopaminergic neurons around the time of reward and, according to the consensus, attributed it to the facial movement of licking. But like the Jewish king Shaul (1 Sam. 9), sometimes you go to look for donkeys and find a kingdom. Careful analysis of this data, supported with the insight of reinforcement learning, led Schultz, Dayan, and Montague in 1997 to hypothesize that the dopaminergic neurons do not encode reward per se but rather encode the reward PE.[2] Figure 10.3 is their "best typical example," which has been shown at almost any presentation of the biology of reinforcement learning that I have attended in the last twenty years. These raster displays (dots represent spikes, lines represent trials; data are aligned to time of the relevant behavioral event) and their peristimulus histograms (top black bars showing the average activity per time bin) reveal that dopaminergic neurons display

- 4–7 Hz spontaneous activity (see the one second before the behavioral events) that probably encodes no mismatch between prediction and reality.
- Short (approximately fifty to one hundred milliseconds) bursts of activity following unpredicted reward (figure 10.3A).
- This burst of dopamine spike does not encode the reward per se since after the monkey was trained to associate an auditory click with a later delivery of reward, the dopamine burst of activity happened following the auditory click (conditioning stimulus; CS), and no change in activity is noticed one second later when the predicted reward has been given (figure 10.3B). The auditory click is given following a random intertrial interval and therefore cannot be predicted. Thus, a burst of dopamine spikes encodes states in which reality is better than predicted.
- Finally, if the auditory click is played but the predicted reward is omitted, the dopaminergic neurons reduce their discharge at the expected time of the predicted reward (figure 10.3C), encoding a reality worse than predicted.

In an amazing set of experiments, Schultz and his colleagues, as well as other research groups, showed that dopamine neurons fulfill the requirements of a critic in reinforcement-learning theory. For example, phenomena such as temporal discounting, blocking procedure, and probability coding (figure 10.4A, B) were correlated with the spiking activity of dopaminergic neurons. However, the encoding of negative (worse than predicted) events by dopaminergic neurons is still debated. Thus, in the context of a probability operant conditioning task, Genela Morris (Morris et al. 2004) reported a monotonic increase in the response of dopaminergic neurons to cues predicting future reward with a higher probability (figure 10.4A, B, left), an opposite trend for the actual reward delivery (figure 10.4A, B, middle column), but no robust encoding of the reward omissions (figure 10.4A, B, right).

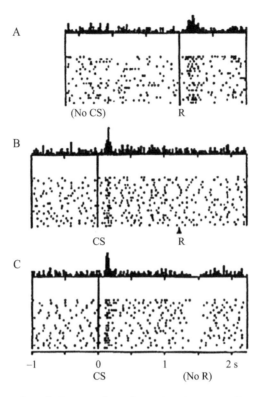

10.3 Encoding of reward prediction error by a dopaminergic neuron. Raster display of spikes of a dopaminergic cell aligned to the reward (R), conditioning stimulus (CS), and reward omission (no R). *A*, unpredicted reward. *B*, predicted reward. *C*, omission of predicted reward.
Source: Modified from Schultz et al. 1997.

We got similar results in a second set of experiments, carried out by Mati Joshua and Avital Adler, using cues that encode the probability of both future reward and an (aversive) air puff (figure 11.4; Joshua et al. 2008; Joshua, Adler, and Bergman 2009).

There could be several reasons for the discrepancy regarding the encoding of a negative PE. First, it may be due to the different behavioral paradigm. Schultz's monkeys were overtrained with positive reward cues, and the omission of a reward was not done routinely and was therefore highly unexpected. On the other hand, our monkeys were trained with a probabilistic task, and therefore the omission of a reward might be better expected. The floor effect of a zero discharge rate and the low (4–7 Hz) spontaneous discharge rate of dopaminergic neurons limit their ability to encode negative events versus encoding positive (better than expected) events in which "the sky is the limit."[3] Paul Glimcher and colleagues (Bayer et al. 2007) therefore suggested that the amplitude of the negative PE is encoded by the duration of the reduction of discharge rate. Mati Joshua has failed to find this phenomenon in our recording (Joshua, Adler, and Bergman 2009). Okihide Hikosaka suggested that different populations of dopaminergic neurons located in the ventral tegmental area encode aversive events by reduction of their discharge rate (Matsumoto and

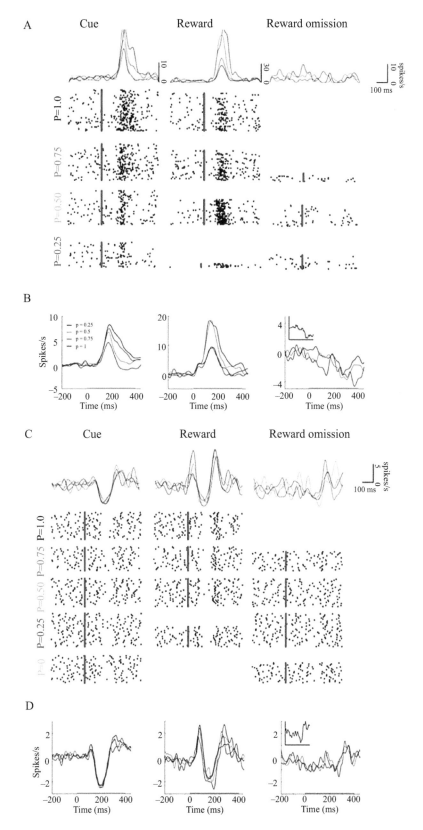

10.4 Encoding of reward probability by dopaminergic neurons. *A*, peristimulus histogram (PSTH) and raster display of the response of a single dopaminergic neuron to cues predicating future reward at different probabilities (*left*), reward outcome (*middle*), and reward omission (*right*). *B*, population average of dopaminergic neuron PSTHs. *C*, *D*, as *A*, *B*, but for striatal TANs.
Source: Modified (with permission) from Morris et al. 2004.

Hikosaka 2009). Finally, it could be that dopaminergic neurons are not alone, and other basal ganglia neuromodulators (e.g., the striatal cholinergic interneurons) take part of the critic load and encode negative PE events (chapter 11).

Reinforcement-learning basal ganglia students generally agree that tonic (background) dopamine activity encodes no significant difference between prediction and reality. However, the mechanisms of this tonic activity, as well as the meaning of slow modulation of its level, are hotly debated. A parsimonious mechanism for the tonic activity of the dopaminergic neuron is that it is created by the summation of many small events in which prediction is slightly better than expectations. Any modulation of the tonic level of dopamine activity reflects the frequency and amplitude of these miniature events. On the other hand, there is a school of basal ganglia students who support the idea of different mechanisms for tonic and phasic activity of dopamine neurons. This school will point toward different timescales and ramping anticipatory activity of dopaminergic neurons (Schultz 1998; Fiorillo et al. 2003; Bromberg-Martin et al. 2010a, b; but see Niv et al. 2005), dissociable regulation of tonic and phasic dopamine activity (Floresco et al. 2003), and the different role of tonic and phasic dopamine activity in the pathophysiology of dopamine-related disorders (Grace 1991, 2016).

HOMOGENEOUS DOPAMINE INNERVATION OF THE STRIATUM

A simple condition that schoolteachers and parents often require is to be consistent. Situations in which one teacher encourages playing in the schoolyard while another outlaws it should be avoided. Thus, simply thinking, if the dopamine neurons act as a critic of the basal ganglia network, one would expect them to have correlated activity, wide axonal arborization, and a large volume of effect.

The spontaneous spiking activity of dopaminergic neurons is weakly correlated. However, the similarity of their responses is outstanding. This can be evaluated by the measure of response or signal correlation (figure 6.3). Figure 7.7 depicts the distribution of the response and signal correlation of pairs of dopaminergic neurons (DANs) and pairs of TANs (critics) versus GPe-GPe, GPi-GPi, and SNr-SNr pairs. While the distribution is wide in both populations, it is centered on zero for the pairs in the main axis and strongly skewed to the right (positive values of correlations) for pairs of the basal ganglia critics. Scientists are of divided mind regarding the spectrum between lumpers and splitters. Thus, the homogeneity versus diversity of dopamine neurons is hotly debated. Modern science methods are better equipped for splitting than lumping (the fact that we failed to find a significant difference between two populations of neurons does not imply that they are equal). Thus, many recent studies of rodent dopaminergic neurons emphasize the importance of different groups and clusters of these neurons (Engelhard et al. 2019). However, see Nao Uchida for the view that rodent dopaminergic neurons are a homogeneous group (Gershman and Uchida 2019).

The dopamine critic message should be distributed to all components of the actor, or at least to all neurons in one of the layers of the chain connecting the state to action. Indeed, the axonal arborization of a single dopaminergic neuron in the striatum is huge (figure 10.5). The total length of the axonal arborization of a single human dopaminergic neuron is estimated to exceed four meters and to give rise to 0.5–1 million synapses (Pissadaki and Bolam 2013; three orders of magnitudes more than the number of synapses emitted by a single neuron in the main axis of the basal ganglia; table 3.1).

The homogeneity of the dopamine message to the striatum is augmented by the structure and function of dopaminergic synapses in the striatum. Anatomically, the dopaminergic boutons are not always opposed by a postsynaptic structure (Moss and Bolam 2008). More importantly, the release of dopamine in the striatum is organized not as a classical synaptic transmission mode but rather as a "volume transmission"

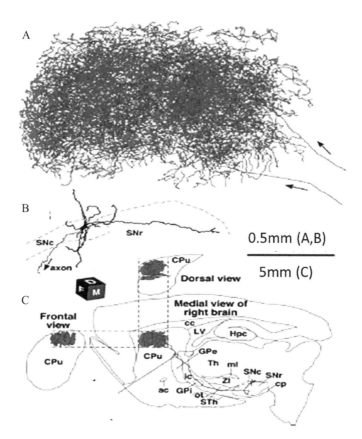

10.5 The broad distribution of axonal arborization of a single dopaminergic neuron. Camera lucida reconstruction of a dopaminergic neuron from the substantia nigra compacta. The axon fibers in the caudate-putamen (CPu) (A) and local dendrites (B) are projected onto a parasagittal plane. In (C) the axial and coronal views of the axonal arborization are shown in line with the parasagittal view. Red and blue lines in the striatum indicate the axon fibers located in the striosome and matrix compartments, respectively.
Source: Adapted from Matsuda et al. 2009.

mode (Rice and Cragg 2008). Thus, dopamine released by one neuron diffuses and mixes with dopamine released from other neurons.

Early electron-microscope studies have revealed a unique microcircuit structure in the striatum. The synapses of the glutamatergic axons are located at the head of the spine of the striatal MSN, whereas the synapses of the dopaminergic neurons are located at the neck of the spine (Freund et al. 1984). This suggests that dopamine can regulate corticostriatal transmission at the resolution of a single spine. However, a recent quantitative study (Moss and Bolam 2008) has revealed that the probability of a dopaminergic release site to touch a spine neck is not larger than the probability of apposing any other subcellular element. Dopaminergic synapses are randomly and very densely distributed in the striatum in accordance with volume transmission of striatal dopamine.

DOPAMINE MODULATES THE EFFICACY OF THE CORTICOSTRIATAL SYNAPSES

The third prerequisite of the critic/actor reinforcement model of the basal ganglia is that dopamine should modulate the state-to-action probability matrix of the main axis of the basal ganglia. Jeff Wickens and colleagues (Reynolds et al. 2001) provided proof of this principle by showing that dopamine modulates the efficacy of the corticostriatal synapse when coupled with postsynaptic depolarization (figure 10.6).

Needless to say, much more must be learned about the dopamine-modulated corticostriatal synaptic physiology. We do not expect the efficacy of all connections between the cortex and the striatum to be augmented by dopamine release. This will not lead to a meaningful learning of a behavioral policy. One possibility is a triple Hebbian learning rule. A change in the efficacy of a corticostriatal synapse would be conditioned on a presynaptic spike, followed by a postsynaptic MSN spike and dopamine transient release (Bar-Gad et al. 2003). The biological eligibility trace should be explored since the efferent spikes would precede the reward delivery. Finally, a mechanism must exist for a decrease in the efficacy of the corticostriatal synapses in cases where not all three Hebbian elements are activated in the proper timing. This may be achieved by a decrease in the synaptic efficacy if one or two, but not three, of the conditions are met. Importantly, such learning rules would sustain an overall homeostasis of the corticostriatal synaptic machinery.

WHO IS THE TEACHER OF THE TEACHER?

Classical machine-learning actor/critic models assume that the actor is informed about the current state and that the information about the current state and reward is transmitted to the critic (figure 10.1). The reinforcement-learning actor/critic models of the basal ganglia depicted in figure 10.2 intentionally neglect the inputs to

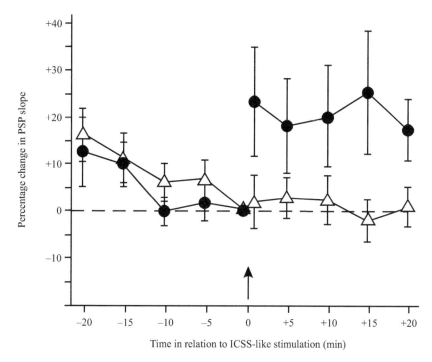

10.6 Dopamine modulates the efficacy of the corticostriatal synapse. The slope of the corticostriatal postsynaptic potential (PSP) following postsynaptic stimulation and intracranial self-stimulation (ICSS) at time zero is potentiated (*black circle*). The synaptic potentiation is suppressed by the D1 dopamine antagonist (*empty triangles*).
Source: Adapted from Reynolds et al. 2001.

the dopaminergic neurons. However, the dopaminergic neurons should be informed about the state and reward to calculate the temporal prediction error.

The synaptic afferents of midbrain dopaminergic neurons are highly diverse (Watabe-Uchida et al. 2012), and there have been many attempts to find the identity of the main drivers of these neurons. Several options have been suggested. Ann Graybiel has done a superb job of extending the neurochemical definition of the striosomes to the identification of their critic properties (Friedman et al. 2015). In a very nice series of papers, Okihide Hikosaka has demonstrated the role of the lateral habenula in providing information about negative events to midbrain dopaminergic neurons (Matsumoto and Hikosaka 2007). However, no head-to-head study has examined the relative contribution of the inputs to midbrain dopaminergic neurons, and no consensus exists regarding the identity of the main drivers of the dopaminergic neurons.

The basal ganglia are usually divided into three domains: associative/cognitive, motor, and limbic/emotional. At least at the level of the input structure—the striatum and the STN—these domains can be identified topographically (e.g., caudate, putamen, and ventral striatum for the cognitive, motor, and limbic domains, respectively), although with fuzzy and intermixed boundaries. Here, I would like to

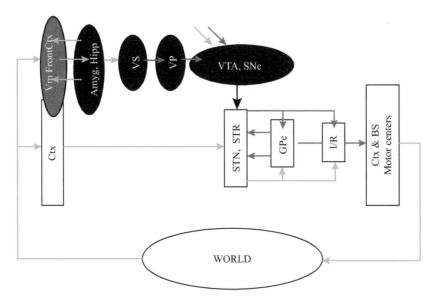

10.7 Actor/critic model of the basal ganglia depicting the major drivers of midbrain dopaminergic neurons. *Red arrows*, excitatory connections; *green arrows*, inhibitory connections. Amyg, amygdala; BS, brain stem; Ctx, cortex; GPe, external segment of the globus pallidus; Hipp, hippocampus; I/R, internal segment of the globus pallidus and substantia nigra reticulata; SNc, substantia nigra compacta; STN, subthalamic nucleus; STR, striatum; VmFrontCtx, ventromedial frontal cortex; VP, ventral pallidum; VS, ventral striatum; VTA, ventral tegmental area.

note the possible overlap between limbic/mood and reward information. Imaging studies by John O'Doherty and colleagues (2004) have suggested the ventral striatum is part of the critic network of the basal ganglia. The ventral pallidum (VP) is located below the GPe, receives significant innervation from the ventral striatum, and in turn projects to the ventral tegmental area. VP neurons can be classified into critic-like and actor-like assemblies (Kaplan et al. 2020; Ottenheimer et al. 2020). I therefore suggest (figure 10.7) that limbic/reward state information is streamed through the orbitomedial frontal cortex (VmFrontCtx) to the amygdala and the hippocampus and then to the ventral striatum, the VP, and the midbrain dopaminergic neurons. But this is only one of many possible scenarios, as indicated by the additional excitatory and inhibitory connections depicted at the top of figure 10.7.

COMPUTATIONAL PHYSIOLOGY OF THE REINFORCEMENT-LEARNING ACTOR/CRITIC MODEL OF THE BASAL GANGLIA

From the viewpoint of determining basic questions about a computational model (chapter 4), the reinforcement-learning actor/critic model of the basal ganglia suggests that

- The computational goal (*What does the system do?*) of the basal ganglia is the learning of an optimal behavioral policy that would maximize the cumulative rewards of an agent interacting with a noisy and nonstationary world.

- The computational algorithm (*How does the system do what it does?*) of the basal ganglia network is a reinforcement-learning algorithm.
- The computational implementation (*How is the system physically realized?*) in the basal ganglia is by actor/critic, main axis/dopamine neurons, architecture, and triple Hebbian modulation of the efficacy of the corticostriatal synapse.

Learning is probably the most important role of the brain, and neural correlations of learning can be found in almost any neural structure studied. I suggest that corticocortical networks are concerned with declarative, rule-based learning (e.g., the name of the guest that just entered your office and the MATLAB routine you can use to analyze your data). The cerebellum is part of the extrapyramidal motor system that makes possible implicit (nonconscious) learning. Cerebellar implicit learning is more strongly supervised by the error functions (my reaching movement missed the target by 10 cm). The basal ganglia make possible weakly supervised learning driven by negative and positive rewards. The basal ganglia only "know" if we have received a reward for our previous actions. Brain networks are highly interconnected. Therefore, we should not be surprised at finding correlates of all learning domains in all brain structures. Moreover, in cases of disease, brain networks will compensate for malfunctions of the network. Still, the major goal of the basal ganglia is the implicit, reward-driven reinforcement learning of optimal motor behavioral policy.

PITFALLS OF THE REINFORCEMENT-LEARNING ACTOR/CRITIC MODEL OF THE BASAL GANGLIA

The reinforcement-learning actor/critic models of the basal ganglia have played a critical role in the development of computational thinking regarding the basal ganglia and reward physiology. However, we should recognize the major pitfalls of these models.

- The limits (floor effect of zero discharge rate) and less robust encoding of negative prediction errors by dopamine neurons have an adverse impact on the ability of the basal ganglia to optimize behavioral policy in a world with both positive rewards and punishments (negative rewards).
- The model assumes that dopamine is only affecting synaptic plasticity and learning. The learning rate should be slow; otherwise, learning might be erratic. This is not in line with the ultrafast physiological (figure 10.8*A*), behavioral (figure 10.8*B*), and clinical effects of fast-acting postsynaptic dopamine agonists and antagonists.
- The goal of reinforcement learning actor/critic is maximization of cumulative pleasure. This is good for an ultracapitalist bank owner. However, most of us would like to believe that we are driven by more than the single goal of maximizing cumulative pleasure.

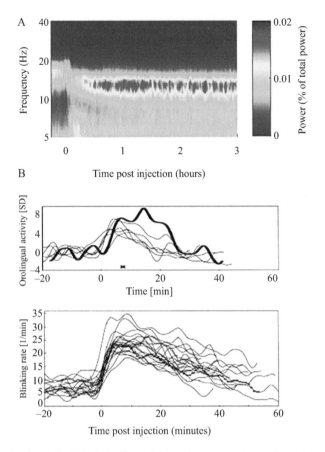

10.8 Fast behavioral and physiological effects of dopamine antagonists and agonists. *A*, average spectrogram of local field potentials recorded at the external segment of the globus pallidus following IM injection of 1 mg/kg of haloperidol at time zero. *B*, orofacial (lingual and blinks) induced by IM injection of 0.1 mg/kg apomorphine at time zero.
Sources: A. Unpublished results of Lily Iskhakova and Pnina Rappel. B. Copied (with permission) from Nevet et al. 2004.

CHAPTER SUMMARY

- Reinforcement learning makes possible the optimization of behavioral policy by interactions with the environment and without a "know-it-all" teacher.
- Reinforcement learning can be achieved by a division of labor between actor and critic. The actor implements the behavioral (state-to-action) policy, and the critic adjusts it according to the prediction error.
- The basal ganglia can be viewed as a reinforcement-learning network. The main axis of the basal ganglia is the actor, and midbrain dopaminergic neurons are the critic.

11

MULTIPLE-CRITICS, MULTIOBJECTIVE OPTIMIZATION MODELS OF THE BASAL GANGLIA

Table 11.1 summarizes the outline and the open questions concerning the models of the basal ganglia discussed so far.

There is an old Jewish joke about two neighbors who were fighting over a financial dispute and took their case to the local rabbi. The rabbi heard the first litigant's case, nodded his head, and said "You're right." The second litigant then stated his case. The rabbi heard him out, nodded again, and said "You're also right." The rabbi's wife was justifiably confused. "But, Rebbe," she asked, "how can they both be right?" The rabbi thought about this for a moment before responding, "You're right too!"

I thought for a moment about the statements and problems of the two models of the basal ganglia and decided to adopt the same approach—"Yes, both are right."

MULTIOBJECTIVE OPTIMIZATION

Reinforcement learning is an online algorithm aiming at the optimization of one parameter—the cumulative signed amount of positive and negative rewards. Despite Richard Bellman's curse of dimensionality and given a stationary world, a reinforcement-learning agent will achieve an optimal behavioral policy (namely, no other behavioral policy would do better in maximizing the cumulative rewards). However, life is more complex, and biological agents have more than one objective.

One way of overcoming the multiobjective problem is to reduce it to one objective problem, by either merging or scaling all parameters to the single axis of positive to negative rewards. Thus, both punishment (physical or mental pain) and metabolic cost will be lumped together (with appropriate scaling, if available) to be considered a negative reward. However, there are some theoretical answers (more in the economic thinking) to the issue of multiobjective optimization (Emmerich and Deutz 2018; Gunantara 2018). Vilfredo Pareto has suggested that a manifold (frontier) of

Table 11.1 Outline and open questions of the two families of basal ganglia computational models. The competitive D1/D2 direct/indirect pathways and the action-selection models are considered different models of the same generation (1.0 and 1.1).

Version	Model	Dopamine effect	Open questions
1.0	D1/D2 direct/indirect pathways box-and-arrow model	Excitability of striatal projection neurons	1. One box $= 10^8$ neurons? 2. Dynamics (learning, tremor)?
1.1	Action-selection box-and-arrow model	Widening of the focus of action selected	1. Late timing of pallidal activity 2. Dynamics (learning, tremor)?
2.0	Actor/critic reinforcement-learning model	Efficacy of the corticostriatal synapse	1. Floor effects of zero discharge rate limit the encoding of negative events by the dopaminergic teacher 2. Fast effects of dopamine agonists and antagonists? 3. Is maximization of cumulative pleasure the only goal of life?

optimal solutions can be achieved by multiobjective optimization. Pareto's optimality is a situation that cannot be modified to make any one objective better off without making at least one other objective worse off. The Pareto frontier divides the possible space of parameters into two domains, one that can be achieved and another that cannot. Multiobjective optimization is achieved when the agent is located on the Pareto frontier. However, many optimal solutions sit on Pareto's frontier, and from the mathematical/economical point of view, they are equally optimal, and the decision of which to choose is subjective or in accord with other constraints of the environment.

For example, your factory is able to manufacture two kinds of goods: butter or guns; however, you have limits set by the physical properties of the machinery, number of workers, and so on. If you manufacture more guns, you make less butter, and vice versa. The production-possibility trade-off reveals that the optimal solution (the Pareto frontier) is the purple curve in figure 11.1. Point x is impossible. If your factory is located at point A, you are not optimal, and you should aim at pushing your workers and machines to the blue curve (points B, C, or D). However, when you are on the Pareto frontier, moving from B to D will give you more butter but fewer guns, and all points on the Pareto frontier are equally optimal. My two cents: make butter, not guns, and move to point C, and even lower, on the frontier.

A single critic—for example, the basal ganglia dopaminergic critic—encoding pleasure prediction error can optimize only one parameter. Below we suggest that multiobjective optimization of behavioral policy may be achieved by multiple critics in the basal ganglia network.

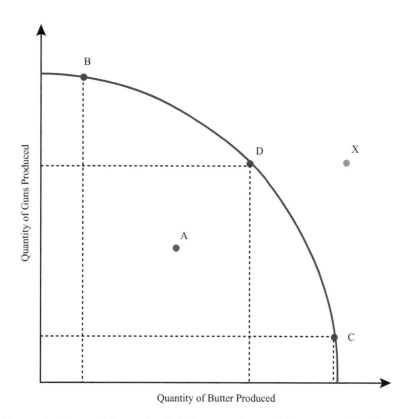

11.1 The production possibility trade-off for butter and guns. A is not optimal, and x cannot be achieved. B, C, and D are on the Pareto frontier and optimal.

THE CHOLINERGIC CRITIC

"Two are better than one." If the floor effect of zero discharge rate limits the ability of midbrain dopaminergic neurons to encode events that are worse than predicted, the basal ganglia can use an additional teacher to supply the missing information. The immediate suspects would be the cholinergic interneurons of the striatum.

First, for one hundred years, from 1870 until 1970 (or from Charcot to Cotzias), anticholinergic treatment was the main pharmacological weapon in the battle against Parkinson's disease. It was not a great therapy, but it was better than nothing. With the discovery of dopamine replacement therapy, the "dopaminergic-cholinergic balance" hypothesis was in the mainstream of neurological thinking. The hypothesis stated that in the normal/healthy brain there is a balance between (striatal) dopamine and acetylcholine (figure 11.2, *left*). However, in Parkinson's disease, due to the degeneration of dopaminergic neurons, the balance is shifted toward the acetylcholine side (figure 11.2, *right*). One can restore the balance and cure parkinsonian symptoms either by anticholinergic treatment or by dopamine replacement therapy.

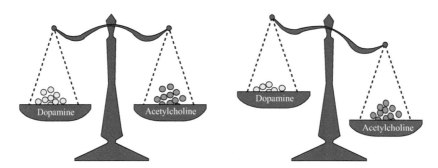

11.2 The cholinergic-dopaminergic balance hypothesis. In the normal (*left*) state, the two neuro-modulators are balanced. In Parkinson's disease (*right*), dopamine neurons degenerate, and the acetylcholine system is overweighted.

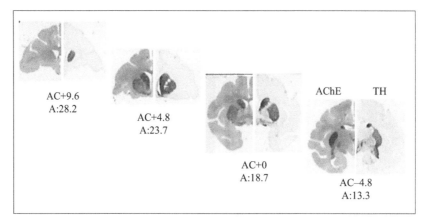

11.3 Cholinergic and dopaminergic markers in the nonhuman primate brain. Rostro-to-caudal coronal sections with immunohistochemistry of acetylcholine esterase (*left*, AChE, cholinergic marker) and tyrosine hydroxylase (*right*, TH, dopaminergic marker) in the brain of a macaque monkey. AC, anterior commissure; A, stereotaxic anterior-posterior coordinate.
Source: Courtesy of Suzanne Haber.

Second, immunohistochemical staining studies have consistently reported a dense staining of the striatum with cholinergic markers (figure 11.3). Apparently, most of these markers are created by the cholinergic interneurons of the striatum and their extensive dendritic and axonal arborization (figure 2.2*D*). However, the pedunculopontine nucleus (PPN) also sends cholinergic innervation to the striatum.

The first description of phasic versus tonically active neurons (TANs) in the striatum, encoding movements versus reward, respectively, was by Minoru (Mano) Kimura, working with Ed Evarts (Kimura et al. 1984). Charlie Wilson, working with Steve Kitai, conducted the leading experiments revealing the subcellular processes of these neurons, as well as the identity of TANs as the striatal cholinergic interneurons (Wilson et al. 1990). Kimura has continued to contribute to the studies of TANs in behaving primates, with his research team in Japan and also with Ann Graybiel, who shifted from being one of the best basal ganglia anatomists to one of the best physiologists (Graybiel et al. 1994). Another dedicated student of TAN physiology in behaving monkeys is Paul Apicella (Ravel et al. 2001). Finally, our group has also recorded the spiking activity of TANs before and after treatment with the MPTP neurotoxin and the development of Parkinson's symptoms (Raz et al. 1996; Morris et al. 2004; Joshua et al. 2008; Deffains et al. 2016). These physiological studies have revealed that striatal TANs modulate their discharge rate in response to reward-related events. The most dominant feature of the TAN response is the depression of their activity, in contrast to the elevation of discharge by the SNc dopaminergic neurons (figure 10.4). The TANs' responses are synchronized with the dopaminergic burst (Morris et al. 2004) and can complement the information encoded by the SNc dopaminergic neurons (figure 11.4).

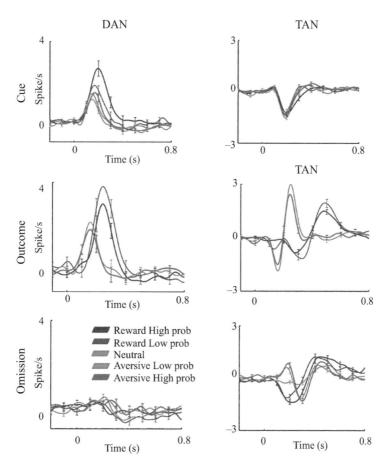

11.4 Population response of the critics of the basal ganglia to probabilistic rewarding and aversive events. Midbrain dopaminergic neurons (DAN, left) and striatal tonically active neurons (TANs, right) to rewarding (blue), aversive (red), and neutral (green) probabilistic cue, outcome and omissions. *Source:* Adapted from Joshua et-al 2008.

BASAL GANGLIA CRITICS MODULATE BOTH STRIATAL EXCITABILITY AND AFFERENT SYNAPTIC EFFICACY

If two are better than one, four are probably better than two. Together with dopamine and cholinergic markers, one can find in the striatum an exceptional level of serotoninergic (5-HT; Parent et al. 2011) and histaminergic (Bolam and Ellender 2016) markers (figure 11.5). Unlike the cortex, noradrenaline is not a major neuromodulator of the basal ganglia.

The four neuromodulators of the basal ganglia enable us to adopt the rebbe's "both are right" solution. I would like to suggest that basal ganglia critics modulate both the excitability of the striatum and the efficacy of the corticostriatal connections. The changes in the excitability are fast (seconds to minutes), whereas the changes in the efficacy of the corticostriatal synapses are slower (days to weeks). As a result of these modulations, we will get immediate changes in the excitability (vigor) of the motor apparatus (Niv et al. 2007; Rigoli et al. 2016; Shadmehr and Alaa 2020), as well as slow changes in the behavioral policy (state-to-action matrix). I further suggest that all four possibilities of the modulation of excitability and plasticity are encoded by the elevation of discharge rate of the different basal ganglia critics (table 11.2).

Table 11.2's left column depicts the domains affected by the basal ganglia neuromodulators: the behavioral and the neuronal effects. In table 11.2A the up arrows and down arrows depict the effect on behavior (motor vigor and state-to-action behavioral policy). The multiple critics, multiobjective model of the basal ganglia is not affected by the detailed model of the main axis of the basal ganglia. The suggested effects on the excitability of MSNs shown in table 11.2B are in line with the D1/D2 direct/indirect rate (scaling) model of the main axis of the basal ganglia. Dopamine for example, will lead to an increase/decrease in excitability and discharge rate of D1 MSNs and D2 MSNs (left and right arrows), respectively. This will lead to a reduced GPi/SNr discharge rate that will disinhibit (increase excitability and discharge rate) the motor cortices and increase motor vigor. Likewise, the changes in the efficacy of the corticostriatal synaptic transmission would be different for D1 and D2 MSNs. The suggested effects will also work for the action-selection (focusing) model in which the surround inhibition is provided by the indirect pathway (figure 9.5A). We will have to modify the suggested effects for the action-selection model in which the surround inhibition is provided by the cortex-STN hyperdirect pathway (figure 9.5B) and the three-layer model of the basal ganglia (figure 9.4).

From the behavioral point of view, an increase in dopamine (e.g., when a cue indicates a future positive reward) would increase both the motor vigor (fast approach to the reward) and the state-to-action probability (reinforcing behavioral policy that led to better than expected situation). If a worse than expected situation happens (you enter the classroom and find the most terrible teacher at the university), it is time for your basal ganglia serotonin to rise. It will induce the same change in the motor vigor as dopamine (so you can escape from the class). But the change in the

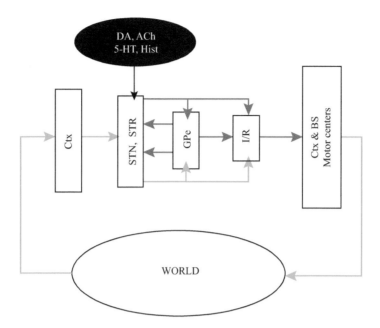

11.5 The multiple critics of the basal ganglia. The critics are DA, dopamine; ACh, acetylcholine; 5-HT, serotonin; and Hist, histamine neurons. The structures of the main axis are BS, brain stem; Ctx, cortex; GPe, external segment of the globus pallidus; I/R, internal segment of the globus pallidus and substantia nigra reticulata; STN, subthalamic nucleus; STR, striatum. *Red and green arrows*, excitatory and inhibitory connections, respectively.

Table 11.2 Dual actions of the critics of the basal ganglia. *A*, suggested changes in the behavioral policy. *B*, suggested changes in neuronal excitability and plasticity according to the D1/D2 direct/indirect model of the main axis of the basal ganglia. DA-dopamine, ACh-acetylcholine, 5-HT-serotonin.

A. Behavior effects	DA	ACh	5-HT	Histamine
Motor vigor (temperature)	↑	↓	↑	↓
State-to-action association	↑	↓	↓	↑
B. Neuronal effects	**DA**	**ACh**	**5-HT**	**Histamine**
Striatal D1/D2 MSN excitability	↑/↓	↓/↑	↑/↓	↓/↑
GPi-SNr excitability	↓	↑	↓	↑
Motor cortex-brain stem centers excitability	↑	↓	↑	↓
Cortex-to-striatal D1/D2 MSN synaptic efficacy	↑/↓	↓/↑	↓/↑	↑/↓
Cortex-to-motor centers multisynaptic efficacy	↑	↓	↓	↑

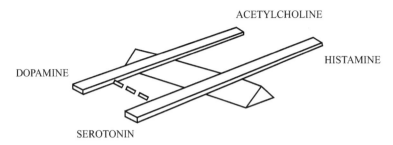

11.6 Early notion of multiple critics of the basal ganglia.
Source: Modified from Barbeau 1962.

plasticity would be different (you will change your behavioral policy to avoid repeating the same mistake again and having to take this class with the awful lecturer). As may be expected from the acetylcholine-dopamine balance hypothesis and the responses of the TANs to behavioral events (figures 10.4 and 11.4), we hypothesize that an increase in acetylcholine will have an effect opposite that of dopamine. Histamine closes the battery and is therefore believed to have an effect opposite that of serotonin. To my surprise, I found that I am stepping in the pathway of a giant, and the complex interactions among the four critics of the basal ganglia were already suggested sixty years ago (figure 11.6).

The multiple-critics, multiobjective model of the basal ganglia can be augmented by the soft-max behavioral policy and its temperature parameter (equation 10.5). In the reinforcement-learning model, dopamine and temperature levels are associated with explorative versus exploitative behavioral policy. However, previously we left aside the question of the control of the temperature. In the multiobjective models of the basal ganglia, the temperature of the soft-max equation can be equated with the excitability of the striatal neurons (table 11.2). The temperature of the soft-max behavioral policy enables the immediate modulation of behavior by changing the location on the multidimensional Pareto frontier. For simplicity let's consider the case of a two-dimensional multiobjective optimization agent that is aiming at maximization of its cumulative reward and the predictability of the future (figure 11.7). Let's further assume that our agent is already located on the Pareto frontier. If the soft-max temperature is low, the agent will shift toward point B, where by slow and careful decisions it will minimize future unpredictability (exploitation), a low-gain, low-risk policy. If the temperature is high, the agent will be a fast-acting gambler and will shift its behavior toward point A, high-risk (low predictability) and high-gain behavior. With more than two basal ganglia critics and objectives, the Pareto frontier is a high-dimensional manifold that enables complex and highly effective behavior.

The suggestion that different basal ganglia critics affect the same actor makes the requirement for coordination of these different neuronal groups critical. The coordination of response timing between TANs and SNc dopaminergic neurons (figures 10.4 and 11.4) is striking. Both dopaminergic and serotoninergic drugs have beneficial

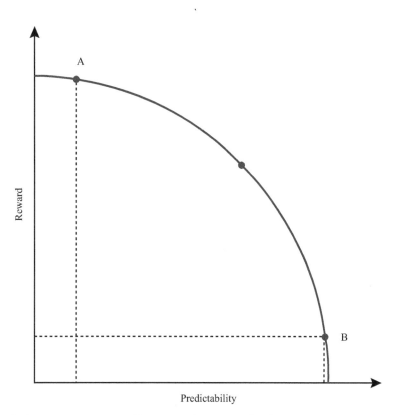

11.7   Multiobjective optimization of future cumulative reward and predictability. *A, B,* high- and low-temperature optimal behavioral policies, respectively.
Source: Modified from Parush et al. 2011.

effects on mental disorders. Many ways exist for coordination of the critics, either by direct synaptic interactions (Straub et al. 2014) or even by modulation of synaptic release (Threlfell et al. 2012; Brimblecombe et al. 2018; Zhang et al. 2018).

The biological effects of the neuromodulators discussed above are restricted to the basal ganglia and should not be confused with the general behavioral effects of drugs affecting these neuromodulators. Cholinergic, serotoninergic, and histaminergic receptors are abundant in so many areas of the nervous system. Therefore, the general effects of their agonists or antagonists might reflect their action outside of the basal ganglia. Finally, table 11.2 should be worked out by future domain-specific experiments.

FREESTANDING (OUTSIDE THE STRIATUM) NEUROMODULATION IN THE BASAL GANGLIA

Up to now, we have discussed neuromodulation and plastic changes in neural activity in the striatum and assumed a hard-wired connectivity in the other parts of the basal ganglia network. This is a good first-order approximation; however, there are many exceptions.

Mahlon DeLong has identified an additional class of pallidal units in the borders of the two pallidal segments (figure 11.8*D*). These border neurons exhibited a slower (on average thirty spike/second) and very regular pattern of discharge. In DeLong's first publication, the discharge of the border cells was equated with the regular discharge of units recorded from the substantia innominata of Reichert or the ventral pallidum (figure 11.8*E*). Later (Mitchell et al. 1987; Richardson and DeLong 1990), DeLong suggested that the border cells are aberrant cholinergic cells of the basal forebrain. Although in the initial report DeLong mentions a stronger tendency of border neurons to respond to rewarding events, more recent publications by DeLong, as well as by the Bordeaux group (Bezard et al. 2001), reported that border neurons are activated by movement.

In the rodent (but probably less in the primate; Eid and Parent 2015), many of the pallidal neurons are cholinergic. These pallidal cholinergic neurons are intermingled with the pallidal GABAergic neurons. They often corelease acetylcholine and GABA and may provide a direct path from the GPe to the cortex (Saunders et al. 2015; Guo et al. 2016). Direct projections from the GPe to the cortex are not in line with the description of the main axis of the basal ganglia as a feedforward network (figures 9.2, 9.4, and 9.5). I will therefore keep the working hypothesis that the cholinergic pallidal neurons are part of the basal forebrain cholinergic and cortex neuromodulation system. Nevertheless, it is tempting to speculate that the pallidal cholinergic innervation is part of the extrastriatal neuromodulation of basal ganglia activity.

The cholinergic teachers of the brain have been extensively studied by Lennart Heimer (Heimer et al. 2008) and many others (e.g., Mesulam et al. 1992; Stephenson et al. 2017). The cholinergic modulation of the brain is achieved by small nuclei in the brain stem and the basal forebrain. Most important are the cholinergic neurons of the nucleus basalis of Meynert that innervate the cortex, the septal nuclei and diagonal band of Broca that innervates the hippocampus, and the PPN that innervates the intralaminar nuclei of the thalamus. Recently, Juan Mena-Segovia and colleagues have shown that the PPN (and the nearby lateral-dorsal tegmental nucleus) provides significant cholinergic input to the striatum (Dautan et al. 2020). Recent studies by Martin Parent revealed that both pallidal segments receive significant and diffuse PPN cholinergic innervation (Eid and Parent 2016).

The dopaminergic innervation of extrastriatal structures such as the GPi and the STN is well established (Eid and Parent 2016). The direct physiological effects of these neuromodulators on pallidal and STN activity have been verified in in vitro and in vivo studies (Galvan et al. 2014). Nevertheless, there is probably at least one order of magnitude difference in the density of dopaminergic, cholinergic, and other neuromodulator biomarkers in the striatum versus extrastriatal basal ganglia structures. Therefore, I will stay with the first-order assumptions that basal ganglia neuromodulation and plasticity take place in the striatum and wait for future studies to teach us the computational physiology of extrastriatal neuromodulation.

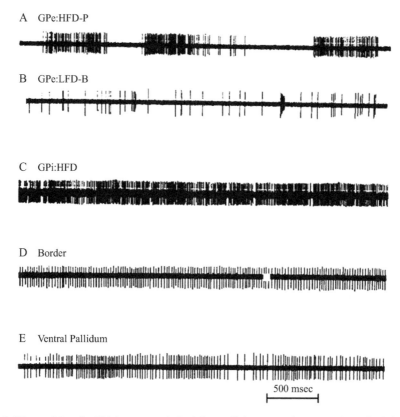

11.8 Spiking activity of pallidal neurons. *A, B,* globus pallidus external segment (GPe) high-frequency discharge with pause (*A,* HFD-P) and low-frequency discharge with bursts (*B,* LFD-B). *C,* globus pallidus internal segment (GPi). *D,* pallidal border. *E,* ventral pallidum, originally denoted as a substantia innominate (SI) neuron.
Source: Modified from DeLong 1971.

COMPUTATIONAL PHYSIOLOGY OF THE MULTIPLE-CRITICS, MULTIOBJECTIVE OPTIMIZATION MODEL OF THE BASAL GANGLIA

Focusing on the basic questions of a computational model (chapter 4), the multiple-critics, multiobjective optimization model of the basal ganglia suggests that

- The computational goal (*What does the system do?*) of the basal ganglia is both the modulation of the motor vigor and the learning of an optimal behavioral policy with the multiple objectives of an agent interacting with a noisy and nonstationary world.
- The computational algorithm (*How does the system do what it does?*) of the basal ganglia network is a multiobjective optimization algorithm (multidimensional soft max).
- The computational implementation (*How is the system physically realized?*) in the basal ganglia is achieved by a division of labor between an actor (main axis) and four critics (dopamine, cholinergic, serotonergic, and histaminergic neurons), affecting the striatal excitability and the efficacy of the corticostriatal synapse.

PITFALLS OF THE MULTIPLE-CRITICS, MULTIOBJECTIVE OPTIMIZATION MODEL OF THE BASAL GANGLIA

The multiple-critics, multiobjective optimization model of the basal ganglia

- Is too complicated and has many free parameters.[1] A lot of future experimental work is needed to improve the model.
- Neglects other parts of the brain that might play significant roles in the behavioral loop of state to action.

CHAPTER SUMMARY

- The basal ganglia have four neuromodulators (critics): dopamine, acetylcholine, serotonin, and histamine.
- The basal ganglia critics affect both the excitability of the medium spiny neurons and the efficacy of the corticostriatal synapses.
- The changes in the motor excitability of the striatum lead to fast modulation of motor vigor.
- The changes in the efficacy of the corticostriatal synapses enable slow learning of an optimal multiobjective behavioral paradigm.
- The four critics of the basal ganglia are coordinated.

12

THE BASAL GANGLIA AS ONE OF MANY STATE-TO-ACTION LOOPS

The basal ganglia represent only a small fraction of the whole brain. Having some understanding of the computational physiology of the basal ganglia, I will take the first step toward a holistic understanding of the computational physiology of the brain. Proceeding from the broad definitions of state and action, we can safely conclude that all brain machinery is devoted to the state-to-action loop. Figure 12.1 depicts several aspects of state and action and also points toward two additional neural networks beyond the basal ganglia that connect state to action. The cortico-cortical (Ctx-Ctx) network starts with the sensory organs that project to the primary sensory cortices that in turn, through the rich and staged cortical network, project to the prefrontal cortex, to the motor cortices, and through the corticospinal (pyramidal) tract to the spinal cord and muscles. The cerebellum receives inputs from the spinal cord and cortex through the pontine nucleus and the mossy fibers and projects back through the vestibulospinal, rubrospinal pathways, as well as to the motor cortical areas through the Vim thalamic nucleus.[1]

Many other networks affect the state-to-action loop—for example, the hypothalamic-pituitary-adrenal (HPA) axis and the autonomic (sympathetic and parasympathetic) system. I have decided to focus on the three networks of the cortex, basal ganglia, and cerebellum because I believe that our central nervous system is motor centric.[2] Evolution's first command is "Thou shalt forward your genes." This is achieved and optimized by motor actions (get food, water, shelter, and sex and run away from danger). Cognition, emotion, and probably even language have evolved to optimize our motor behavior and policy.

BRAIN NETWORKS AND KAHNEMAN'S SYSTEMS 1 AND 2

In his seminal 2011 book *Thinking Fast and Slow*, Daniel Kahneman offers the idea that two different brain systems close the state-to-action loop:

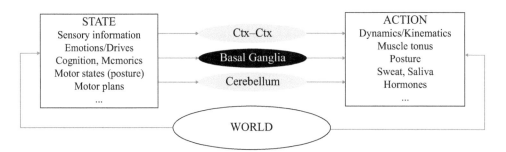

12.1 State-to-action brain networks. Three out of the many brain networks connecting state and action are shown. Ctx, cortex.

- System 1—the frequent, automatic, fast, emotional, stereotypical, procedural, model-free, subconscious (experiencing self) network
- System 2—the infrequent, effortful, slow, logical, calculating, declarative, model-based, conscious (remembering self) network.[3]

I would like to suggest that the basal ganglia and the cerebellum (extrapyramidal) networks constitute Kahneman's system 1. The basal ganglia's implicit learning is driven by reward prediction errors, while cerebellum learning is driven by explicit spatial error. The corticocortical (pyramidal) network constitutes Kahneman's system 2. This explains why Parkinson's patients with dopamine-depleted basal ganglia suffer from a lack and slowness of automatic movements (akinesia, bradykinesia). Auditory or visual cuing helps the patients since the control is shifted to the corticocortical goal-directed brain network. Table 12.1 summarizes our suggestion for anatomical naming of Kahneman's functional systems.

Thus, I suggest that the computational goal of the basal ganglia is to control and optimize our automatic and subconscious actions. Most of our actions are automatic and subconscious. This puts (in my perhaps biased view) the basal ganglia at the top of the brain hierarchy. There are other suggestions (e.g., Williams et al. 2019) for the anatomical correlates of Kahneman's system 1 and 2. While respecting their right to differ, I claim that their approach is perception, cortex, and EEG centric. The other extreme view is basal ganglia centric. Peter Redgrave and colleagues (including me—my apologies) suggested that goal-directed (system 2) behavior is controlled by the rostromedial (caudate) striatum, and habitual (system 1) behavior is mediated by the posterolateral (putamen) striatum (Redgrave et al. 2010). Here I suggest Maimonides's golden mean and to divide the labor of system 1 and 2 between the basal ganglia and the cortex. The two systems are not completely separate (Collins and Cockburn 2020). There are interconnections between the cortical and the basal ganglia networks, and both networks serve a diversity of functions. But, if we take a step back and get a bird's-eye view of the entire brain, we may locate the basal ganglia and the cortex as the anatomical correlates of the

Table 12.1 Kahneman's systems and brain networks.

Kahneman's system	Properties	Brain network
1—"experiencing self" network	Frequent, automatic, fast, emotional, stereotypical, procedural, model-free, subconscious	Basal ganglia and the cerebellum (extrapyramidal) networks
2—"remembering self" network	Infrequent, effortful, slow, logical, calculating, declarative, model-based, conscious	Corticocortical (pyramidal) network

edges of the continuous spectrum between systems 1 and 2 serving the state-to-action loop of behavior.

BOTTLENECK DIMENSIONALITY REDUCTION IN THE BASAL GANGLIA

The space of possible states (what we see, hear, remember, plan, and so on) is much larger than the space of possible actions (limited by the number of joints and muscles). The structure and the physiology of the basal ganglia provide some hints for a dimensionality reduction, or compression of information along the main axis of the basal ganglia.

Many years ago, while studying the subthalamic nucleus (STN) at Johns Hopkins medical school, I witnessed heated debate regarding parallel segregated pathways versus funneling or integration in the basal ganglia. My mentors Garry Alexander, Mahlon DeLong, and Peter Strick, probably influenced by their physiological findings on the somatotopic organization of movement-related neurons in the basal ganglia, argued that the basal ganglia are structured in the form of (five) parallel segregated loops (Alexander et al. 1986, figure 12.2). On the other side stood the French battalion, led by the late Gerald Percheron. They were probably more influenced by the anatomical findings of a huge reduction in the number of neurons along the cortex-striatum-pallidum axis (table 2.1). The large diameter of the pallidal dendritic disk (figures 2.4 and 2.5) suggests that pallidal neurons integrate the information emitted by remote cortical and striatal territories. Gerald Percheron and Michel Filion (from Quebec, so it is still a French-American "war") wrote a short letter to the *Trends in Neuroscience* (*TINS*) journal (then the most respected review journal in neuroscience) titled "Parallel Processing: Up to a Point." This is a good example showing that science would better advance by open debate rather than by "politically correct" avoidance of debate (Percheron and Filion 1991).

Starting my research group in Jerusalem, I decided to enter the segregated/funneling basal ganglia pathways "battlefield" with a new weapon—multiple-electrode recording of the simultaneous spiking activity of several pallidal neurons. If the basal ganglia have the form of segregated pathways, then most pairs of recorded neurons

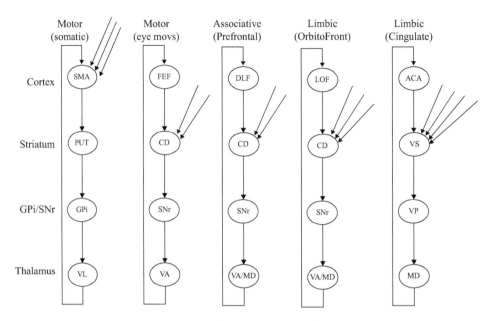

12.2 The parallel segregated pathway model of the basal ganglia circuitry. ACA, anterior cingulate area; CD, caudate; DLF, dorsolateral prefrontal cortex; FEF, frontal eye fields; GPi, internal segment of globus pallidus; LOF, lateral orbitofrontal cortex; MD, medial dorsal thalamic nucleus; PUT, putamen; SMA, supplementary motor area; SNr, substantia nigra pars reticulata; VA, ventral anterior thalamic nucleus; VL, ventral lateral thalamic nucleus; VP, ventral pallidum; VS, ventral striatum. Downward-pointing arrows to striatal structures indicate inputs from other cortical areas.
Source: Adapted from Alexander et al. 1986.

would be part of different functional pathways (especially because even the five pathways of Alexander, DeLong, and Strick should be further subdivided—e.g., the motor pathway into face and upper- and lower-limb pathways) and would be uncorrelated. On the other hand, funneling and integration should lead to many "common inputs" to the recorded neurons, and therefore the neuronal activity of most pallidal pairs would be correlated (figure 12.3, *left*). The correlation studies revealed an unusual black-and-white answer (unusual because most biology results are in the gray domain). Almost all pallidal pairs revealed uncorrelated, independent activity (figure 7.3). The few exceptions of correlated activity of pairs of pallidal neurons were found to be the result of recording problems and artifacts.

I prefer French wine over American hamburgers, and I did believe more in the anatomical description of numbers of neurons and their dendritic field than in the physiological (subjective) findings of somatotopic organization in the pallidum. But hard-core data are more important than personal preferences. I was invited to write a *TINS* review of our multiple electrode results, highlighting the shift from uncorrelated to synchronous oscillations in the pallidum of MPTP-treated monkeys (Bergman et al. 1998; figure 15.3), and I was frustrated by the contrast between my intuitions and findings. When I went back to Israel to start my basal ganglia research group, I was

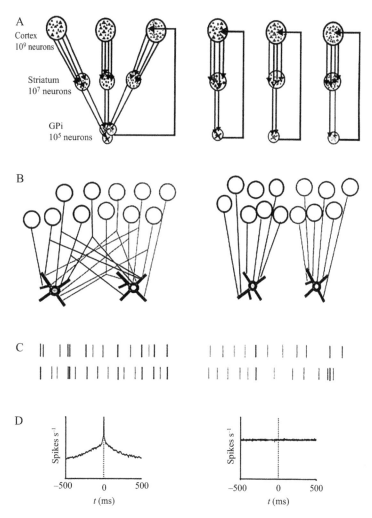

12.3 Physiological predictions of the funneling/integration (*left*) and parallel segregated pathways (*right*) models of the basal ganglia circuitry. *A, B,* gross and detailed anatomical schemes. *C, D,* predicted raster displays and cross-correlation functions of two simultaneously recorded pallidal units. *Source:* Copied (with permission) from Bergman et al. 1998.

lucky to be part of the founding of the Hebrew University Interdisciplinary Center for Neural Computation (ICNC). I did not like the neglect of the basal ganglia practiced there, but I was delighted with the fresh and motivated minds of the students. We attended (and still do) the annual retreats of the ICNC (now called the Edmond and Lily Safra Center for Brain Research; ELSC) in Kibbutz Ein-Gedi, near the Dead Sea. On the way back from one of those meetings, I was sitting in a bus seat near a young ICNC student, Izhar Bar-Gad, and shared with him my frustration at the contradiction between my belief in funneling/integration along the main axis of the basal ganglia and our findings of the noncorrelated activity of pallidal neurons. Izhar looked at me and asked if I had heard about the dimensionality reduction (e.g., principal component analysis; PCA) done by neural networks. I had never heard of it before, and I am still thankful to Izhar for mentioning this amazing conjecture and for further development of this idea (Bar-Gad et al. 2003). Dimensionality reduction is data compression as needed when going from a layer with many neurons to a layer with a small number of neurons. However, to preserve as much of the information of the input layer as possible, we need to get rid of the redundancy (correlation) in the input layer. Thus, dimensionality-reduction neural networks actively decorrelate their discharge to maximize the information capacity of layers with a small number of neurons. The small number of neurons in the GPe and the GPi/SNr layers of the main axis of the basal ganglia can maximize their information capacity by keeping their activity independent. Independent activity means close to zero pairwise correlation. Strong negative correlations, like strong positive correlations, are situations in which the activity of the neuronal pairs is dependent, and the activity of one neuron can be predicted by the activity of its peer.

Dimensionality reduction comes with a price tag. Depending on the amount of redundancy in the input layer and the intensity of the dimensionality reduction, we may lose information following the processes of dimensionality reduction. In any case we will lose redundancy. It might be that the flexibility of the dimensionality reduction in the basal ganglia is the solution for the lost redundancy. If we lose our grandmother cell concept due to the loss of some neurons in the main axis of the basal ganglia, we will be able to recreate it with dynamic reorganization of the network.

Classical dimensionality reduction is tailored to minimize the reconstruction error of the input layer. However, the neural networks of the basal ganglia do not compress the state information in order to reconstruct it. Rather, the basal ganglia extract the most important state features for the next action. Thus, if you see a friend, you usually do not care about the color of your friend's T-shirt or hair, and this information is not needed for your next action (e.g., discussing your recent great experiments) and should not be saved by the basal ganglia dimensionality-reduction networks. This kind of dimensionality reduction, aiming at maximizing the information about different aspects other than the input layer, is called bottleneck dimensionality reduction.[4] The bottleneck dimensionality reduction should be flexible. If it is the first time that your friend has changed his or her hair color to shining blue, then probably your best next action is to tell your friend how much you like the new hairstyle.

How is decorrelation achieved in the main axis of the basal ganglia? The functional connectivity and activity along the main axis of the basal ganglia is probably controlled by Hebbian rules (Bar-Gad et al. 2003). We suggest that the rich lateral inhibitory connectivity in the striatum, as well as in other structures of the main axis of the basal ganglia, is used for active decorrelation processes. Lateral inhibition networks support winner-take-all network modes. The simplest example is of two neurons connected by reciprocal inhibitory connections. The neuron that has won the competition strongly inhibits the other neuron. This inhibited other neuron provides less inhibition to the winner, thus establishing a stable state of one winner. A network of many inhibitory neurons with all-to-all connectivity will be stabilized with one winner (like in the King of the Hill children's game). Tomoki Fukai has nicely shown that depending on the parameters of the connectivity the inhibitory network can exist over a broad spectrum from "winner takes all" to "winner shares all" (Fukai and Tanaka 1997). The high discharge rate of pallidal and SNr neurons suggests that they are closer to the "winner-shares-all" mode.

In summary, pairwise correlations of close and remote neurons in the pallidum are flat. This independent pallidal activity, the numerical reduction in the number of neurons from the striatum to the pallidum and the unique dendritic morphology of pallidal neurons suggesting sparse sampling of broad striatal territories, is in line with the idea of dimensionality reduction in the main axis of the basal ganglia. The jury is still out regarding the nature of this dimensionality reduction (PCA-like, bottleneck, and so on) and the mechanism (lateral inhibition?). The jury is still out even for parallel processing versus funneling/integration in the basal ganglia network (Lee et al. 2020), but I vote for bottleneck dimensionality reduction.

LINEAR AND NONLINEAR NEURONAL INPUT-OUTPUT RELATIONSHIPS IN THE BASAL GANGLIA

In 1948, Alan Hodgkin (of the 1952 Hodgkin-Huxley model of the action potential) set apart two classes of neurons: class I excitability neurons, which increase their discharge rate linearly from zero to thirty to fifty spikes/second in response to increasing net input current (figure 12.4A), and class II excitability neurons, which act in a nonlinear fashion (Hodgkin 1948). Class II neurons do not fire until a certain current threshold is crossed, and then they hop to a high (greater than one hundred spikes/second) discharge rate. After crossing the threshold, the discharge rate of class II neurons increases linearly with the input current; however, the slope of the change in the discharge rate as a function of the input current is modest, and the current frequency (I-f, where I is the input current and f is the discharge frequency) curve may look like a step function (figure 12.4B).

Classical I-f curves of neurons have commonly been achieved through current injections to the soma. However, most of the input to neurons is on the dendrites, and the shape of the I-f curve is a function of the location of the current injection, the general morphology of the dendritic field, and if/which other elements of the dendritic field are activated. Nevertheless, it seems that striatal MSNs resemble class I excitability (very low background discharge rate and maximal rates of around forty spikes/second). The GPe neurons with tonic discharge at fifty to seventy spikes/second and occasional pauses with an abrupt change in the high-frequency discharge to zero (figures 7.2, 7.4, and 12.7) resemble the class II excitability neurons (figure 12.4B, C).

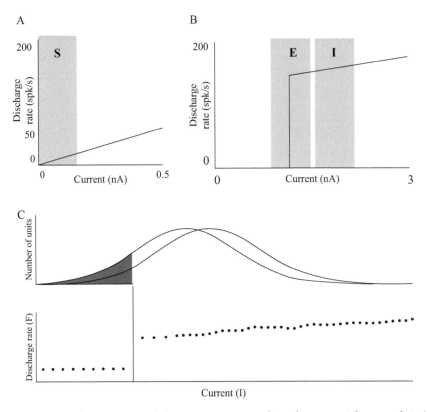

12.4 Two classes of neuronal excitability. Current intensity-firing frequency (I-f) curve of Hodgkin type I (A) and type II (B) excitability neurons. Note the different x-scale. S, E, and I denote the theoretical location of striatal, GPe, and GPi neurons. C, effect of changes in the population discharge rate of GPe class II neurons on the probability to pause.
Source: Modified (with permission) from Goldberg and Bergman 2011 and Schechtman et al. 2015.

Theoretical models have shown that dimensionality reduction can be enhanced in neural networks with a mixture of linear and nonlinear layers. Could this be the reason why the basal ganglia network is composed of more than one layer? What is the input-output relationship of the STN neurons? Do basal ganglia class I and II neurons behave as integrators and resonators, respectively?[5] Can basal ganglia neurons change their dynamic behavior, as do thalamic neurons?[6] The quest for the answers to these questions is left to the next generations of basal ganglia students.

COMPUTATIONAL PHYSIOLOGY OF THE BASAL GANGLIA: THE CURRENT STATUS

In chapter 4 we posited the major questions about the basal ganglia (What are their computational goal and algorithm? How are they physically implemented?). Table 12.2 summarizes the answers to these questions given by the different computational models presented above.

Table 12.2 Computational goals, algorithms, and implantation of the basal ganglia models.

Model version	1.0	1.1	2.0	3.0	4.0
Model name	D1/D2 direct/ indirect box-and-arrow model	Action-selection box-and-arrow model	Actor/critic reinforcement-learning model	Multiple-critics, multiobjective optimization model	Kahneman's system 1 of brain's state-to-action loops
Computational goal (*What do the basal ganglia do?*)	Modulation of the motor vigor	Action selection and suppression of competitive actions	Optimization of behavioral policy	Modulation of the motor vigor and behavioral policy	Modulation of the vigor and policy of automatic movements
Computational algorithm (*How do the basal ganglia do what they do?*)	Converged competition of D1/direct and D2/indirect pathways on excitability of GPi/SNr neurons	Mexican hat spatial profiles of narrow direct vs. broad hyperdirect and/or indirect pathways	Critic/actor network; dopamine encodes pleasure prediction error	Multiple-critics/actor networks; critics encode prediction error for valance and motor vigor	Division of labor between cortical and basal ganglia networks
Computational implementation (*How is the basal ganglia system physically realized?*)	D1/D2 modulation of striatal excitability. Similar divergence of direct/indirect pathways.	Dopamine modulation of the diameter of striatal focus. Narrow/broad divergence of direct/indirect, hyperdirect pathways	Triple Hebbian (dopamine + pre/ post spikes) modulation of corticostriatal synaptic efficacy	Triple Hebbian (critics + pre/ post spikes) modulation of corticostriatal synaptic efficacy	Multiple neural networks connecting brain areas encoding states and actions

COMPUTATIONAL PHYSIOLOGY OF THE BASAL GANGLIA: SECOND-ORDER QUESTIONS

Having provided our current answers to the big computational questions of the basal ganglia, we can now turn to the additional second-order questions:

- Why does the main axis of the basal ganglia use GABA (inhibitory transmitter) as a carrier of information?
- What are the advantages of the high-frequency discharge of neurons in many basal ganglia structures?
- What are the physiological/computational roles of GPe pauses?

I admit that these are teleological questions, and most philosophers of science will claim that they are not pertinent. Evolution may take complex paths and no one claims that all final results of the evolutionary processes are optimal. Nevertheless, the search for an answer might provide us with insight into the physiology of the basal ganglia.

A simple answer would be that the use of an inhibitory transmitter as a carrier of information makes it possible to encode relevant information by means of a decrease in the discharge rate of high-frequency discharge neurons. The GPe pauses are responses whose trigger we still have not identified. However, most of the responses to behavioral events in the central and output layers of the basal ganglia (GPe, GPi, and SNr) are in the form of increases in discharge rate (figure 12.5). Below, we will present some other, perhaps premature, reflections on these questions.

Why do the basal ganglia use GABA as a carrier of information?   A possible explanation for the use of GABA as a carrier of information is the advantage of inhibitory versus excitatory closed-loop neuronal networks. Any positive feedback loop, such as two excitatory neurons reciprocally connected, and a network of excitatory neurons with lateral connectivity (like the cortex) is prone to exponential growth of its activity and epilepsy. A network of inhibitory connections has a built-in safety valve of the floor of zero discharge rate. Take, for example, a network that uses disinhibition as part of its closed-loop structure: A excites B; B inhibits C, which inhibits D. Finally, D is exciting A (as an analog to the cortex, which excites the striatum. The

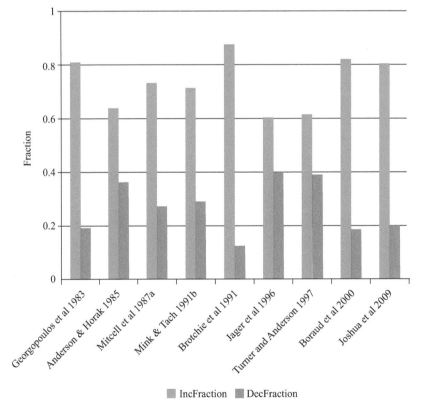

12.5 Meta-analysis of nine studies (*n* = 1,779 neurons) reporting fraction of neurons with increases (Inc) and decreases (Dec) of discharge rate in response to behavioral events.
Source: The relative fraction has been estimated from the figures of Mitchell et al. 1987, Jaeger et al. 1995, and Joshua, Adler, Rosin, et al. 2009. Based on Goldberg and Bergman 2011.

striatum inhibits the GPi that inhibits the thalamus. The thalamus excites the cortex). At some point the increase in discharge rate of A and B would lead to such a strong inhibition of C that the discharge rate of C will be reduced to zero. From that point on, further increases in the discharge rate of A and B would not lead to stronger excitatory feedback to A and to epilepsy.

Lateral inhibition is probably a critical process for active decorrelation in the basal ganglia. In the cortex, the excitatory pyramidal neurons use inhibitory interneurons for lateral inhibition. However, such lateral inhibition entails disynaptic processes, and it is unavoidably slower and less stable than the lateral inhibition in the basal ganglia provided by the lateral axons of the GABAergic projection neurons.

A release of inhibition may lead to faster changes in the discharge rate than an increase in excitation. The "left-foot braking" method is a good example. Think about waiting in your car for the traffic light to turn green so you can speed to the lab to conduct the most critical experiment of your life. You have two options: Option 1, the conventional one, is to push the brake pedal with your right foot and then to shift your right foot to the gas pedal. In option 2, you, like a "professional race car driver," will keep your right foot on the gas pedal but prevent the car from moving by simultaneously pressing the brake pedal with your left foot. When the light turns green, you will release the brake, and since the engine is already at full speed, you will accelerate faster than the conventional driver (please stay safe and leave the left-foot maneuver to trained and professional drivers). It is probably the same story for two neurons that are connected by excitatory or inhibitory connections. The release of inhibition will lead to faster depolarization of the postsynaptic neuron in comparison with direct excitation.

Finally, the basal ganglia project back to the frontal cortex mainly through the ventral anterior and ventral lateral thalamic nuclei. The basal ganglia are unique because they provide the only pathway that inhibits (using GABA) their thalamic targets. All the other thalamic afferents use glutamate (an excitatory transmitter). If the basal ganglia serve as Kahneman's default, automatic, fast, and most frequently used system 1, the corticocortical network, the goal-directed, deliberate system 2 is called into action when system 1 fails. Maybe the major role of the basal ganglia inhibitory innervation of the thalamus is to inhibit the corticocortical network. This inhibition allows the automatic state-to-action behavior (carried through the brain stem motor centers) to dominate our behavior. Interestingly, our thinking pendulum is shifting back to the old notion of the basal ganglia major output directing the brain stem, spinal cord, and motor apparatus (figure 13.1). It is in line with the late responses of basal ganglia neurons relative to movement onset (Mink and Thach 1991). This new/old theory can explain the efficacy of thalamic inactivation (e.g., ablation) therapy to ameliorate Parkinson's tremor and rigidity and the lack of a thalamic sweet spot for akinesia. As we will discuss in chapter 19, one can distinguish between akinesia, the core negative symptoms of Parkinson's disease, and the positive compensatory symptoms, like tremor and rigidity. Akinesia is probably mediated from the basal ganglia to the spinal motor apparatus by the brain stem motor nuclei, while tremor and rigidity are driven by the cerebellar and long-loop reflexes mechanism that goes through the thalamocortical network. I beg forgiveness for swinging the basal ganglia pendulum again and suggest that the basal ganglia output to the thalamus and frontal cortex is a corollary of the output to the brain stem motor nuclei and that its main functional role is to inhibit the corticocortical networks.

In summary, there are very good reasons (safety, speed, function) for the use of GABA as a carrier of information. Of course, we will never know if this is the reason for this practice in the basal ganglia system or whether it is just a glitch of evolution. But, considering all the above-mentioned advantages of GABA as a carrier of information, I wonder why the cortex uses glutamate? Maybe the cortex is a new player in the game of brain evolution and therefore has to offer a "high-gain, high-risk" method, in comparison with the more conservative methods of the older players, like the basal ganglia and the cerebellum.

What are the advantages of a high-frequency discharge rate? There is a metabolic cost (in terms of glucose, oxygen, and ATP used) for the high tonic discharge rate of pallidal neurons (Buzsáki et al. 2007; Hasenstaub et al. 2010; Moujahid et al. 2011; Yi and Grill 2019). Naturally, we are looking for the return of this cost, usually in terms of neuronal information. However, the discussion below should be taken with a bit of caution. First, as mentioned there is the understanding of the convoluted pathway of evolution. Second, the return, and at what level (single neuron, the whole basal ganglia, the organism) it should be calculated, is a complicated function. It also depends on the assumed neural code—for example, rate coding and integrate-and-fire neurons versus temporal coding and resonant neurons (Levakova 2017).

Using GABA as a carrier for information over more than one synapse (disinhibition) sets the need for a tonic discharge rate. Admittedly, the average discharge rate of pallidal neurons is higher than their average response amplitude. However, there is a broad distribution of the discharge rate of pallidal neurons, starting probably as low as twenty spikes/seconds (figure 12.6). Could it be that the distribution of the pallidal discharge rate shifted to higher values to enable even the slow runners of the pallidal group to encode behavioral events by decreasing the discharge rate without reaching the hard floor of a zero discharge rate? If each pallidal neuron counts, as suggested by the ideas of dimensionality reduction and information integration in the pallidum, then the metabolic cost of one pallidal neuron firing at 100 Hz is equal to the metabolic cost of one hundred striatal neurons with one spike/second discharge rate. Maybe the high discharge rate is used to achieve pallidal active decorrelation and to maintain the uniqueness and importance of each neuron.

What are the physiological/computational roles of GPe pauses? Disclosure: for many years the pauses of the GPe (figures 5.2, 7.1, 7.2, 7.4, 11.8, and 12.7) have been at the

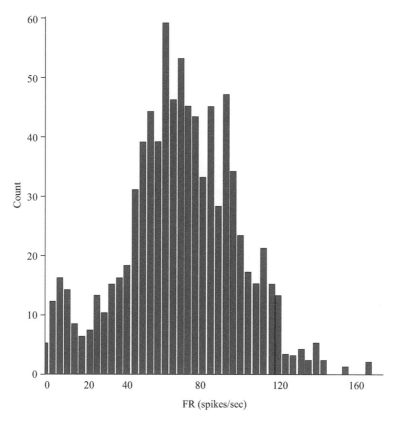

12.6 Firing-rate histogram of GPe cells. Data from 5 monkeys, 852 units, bin width = 4 spikes/second. Two classes of GPe cells can be observed, units with low (<20 spikes/s) discharge rate and units with high (>20 spikes/second) discharge rate. FR, firing rate.
Source: Unpublished results of Inna Vainer Pinkas.

top of my list of natural wonders of the brain.[7] The question at the beginning of this section is one of the first I pose to any student of the basal ganglia. There are several possible answers. The first, hinted by Hikosaka's saccade-triggered pauses of activity in the SNr (Hikosaka and Wurtz 1983a–d), is that the GPe pauses are triggered by events we have not yet identified. However, most of the responses of pallidal neurons to behavioral events are increases in discharge rate (figure 12.5), and even the responses with a decrease in discharge rate are of modest (five to twenty spikes/ second) amplitude. Second, pauses are much more common in the GPe than in the GPi and the SNr (DeLong 1971; Elias et al. 2007). Finally, the frequency and the duration of the pauses are modulated by arousal level, with a reduction in the probability of pausing in a state of high arousal—probably the state with more hidden triggers. Thus, although I accept the possibility that triggers of the GPe pauses might be found in the future, I tend to believe they do not exist.

The other possibility, suggested by Alison Doupe, Mati Joshua, and colleagues (Woolley et al. 2014), is that the pallidal spontaneous pauses are a source of variability ("noise generators"). The pallidal cells might use them to "listen" to their

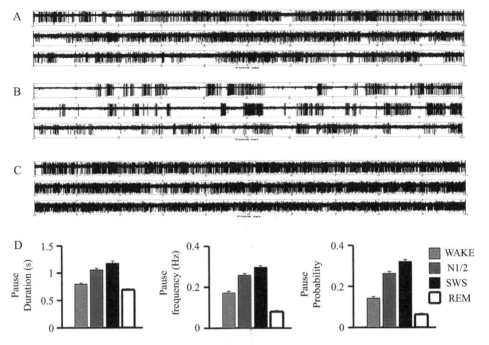

12.7 Pausing activity in the external segment of the globus pallidus (GPe) over the sleep/wake cycle. *A*, example of ten-second simultaneous recording of three GPe neurons in the awake state. *B*, same as (*A*) but for slow-wave sleep (SWS) state. *C*, same as (*A*) but for rapid-eye movement (REM) sleep. D. Population (*n* = 321 neurons) analysis of GPe pause properties.
Sources A–C. Unpublished results of Aviv Mizrahi-Kliger and Alex Kaplan. D. Modified (with permission) from Mizrahi-Kliger et al. 2018.

afferents without the interference of dendritic backpropagating spikes and to enable explorative behavior. Exploratory behavior is more appropriate during quiet awake states. On the other hand, if you are in the middle of a "fight-or-flight" scene (either in awake reality or during your dreams, as in REM sleep; figure 12.7), it is better switch to exploitation behavioral mode.

With these speculations we end part II, in which we introduced the main concepts of the computational physiology of the basal ganglia. Table 12.2 can be used to remind the reader of the main take-home concepts of this part. In part III, we will use this background to turn our attention to the computational physiology of basal ganglia–related disorders and their therapies.

CHAPTER SUMMARY

• The basal ganglia network is one of many brain networks connecting state and action. The basal ganglia network is Kahneman's default, fast, and automatic system 1; the cortex is system 2.

- Anatomical studies support the two extreme views of the main role of the basal ganglia: parallel segregated pathways versus funneling and integration of information.
- Information processing along the main axis of the basal ganglia can be characterized as motor-centered (bottleneck) dimensionality reduction (information integration). The main axis preserves maximal information about the features of the current state that is relevant for future actions.
- Lateral inhibition along the main axis of the basal ganglia provides a mechanism for active decorrelation.
- The possible reasons for the basal ganglia to use an inhibitory transmitter (GABA) as a carrier of information between the different basal ganglia nuclei are 1) as a safety valve against an avalanche of positive feedback loops, 2) as an effective method for fast active decorrelation, 3) for fast action, and 4) for inhibition of the corticocortical system 2 network.

III

COMPUTATIONAL PHYSIOLOGY OF BASAL GANGLIA DISORDERS AND THEIR THERAPY

13

PARKINSON'S DISEASE AND OTHER BASAL GANGLIA–RELATED HUMAN DISORDERS

The previous part was devoted to the basic and theoretical science of the computational physiology of the healthy basal ganglia. This next part is shaped according to my desire for translational science and for making a difference in the life of patients. I cannot think of a better description of what this concerns than that provided by Lev Zasetsky (1920–1993), who suffered a severe brain injury in World War II. His case is described by Aleksandr Luria in his book *The Man with a Shattered World: The History of a Brain Wound*, first published in1972.

In the introduction to the book, Lev Zasetsky wrote:

Perhaps someone with expert knowledge of the human brain will understand my illness, discover what a brain injury does to a man's mind, memory and body, appreciate my effort, and help me avoid some of the problems I have in life. I know there is a good deal of talk now about the cosmos and outer space, and that our earth is just a minute particle of this infinite universe. But, actually, people rarely think about this; the most they can imagine are flights to the nearest planets revolving around the sun. As for the flight of a bullet, or a shell or bomb fragment, that rips open a man's skull, splitting and burning the tissues of his brain, crippling his memory, sight, hearing, awareness—these days people don't find anything extraordinary in that. But, if it's not extraordinary, why am I ill? Why doesn't my memory function, my sight return? Why does my head continuously ache and buss? It is depressing, having to start all over and make sense of a world you've lost because of injury and illness, to get these bits and pieces to add up to a coherent whole.

May we learn from Lev Zasetsky first the ugliness of war, and then that so many people outside are waiting for us—the students of neuroscience, neurology, and psychiatry, gifted with intelligence and social security—to devote ourselves to advancing translational medical science.

THE PYRAMIDAL VERSUS THE EXTRAPYRAMIDAL SYSTEMS

The basal ganglia might be small and less popular than other brain structures such as the cortex, cerebellum, and hippocampus, but nevertheless, the basal ganglia account for a significant share of human suffering due to neurological and psychiatric disorders. In the past the basal ganglia were classified as part of the motor system, composing, together with the cerebellum, the extrapyramidal motor system. The extrapyramidal system lies parallel to the pyramidal system connecting the upper motor neurons of the motor cortices and the lower (alpha) motor neurons in the spinal cord (figure 13.1). The diseases associated with the basal ganglia and the cerebellum are classified as "extrapyramidal movement disorders." These disorders are characterized by abnormal movements, changes of muscle tonus, postural

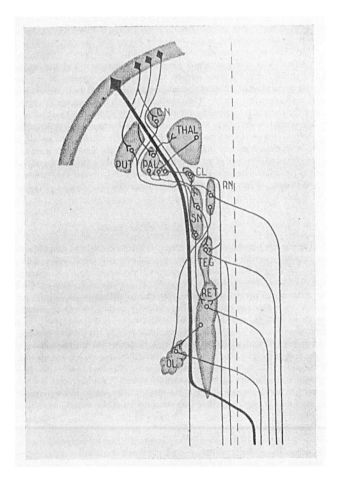

13.1 Classical description of parallel down-going pyramidal and extrapyramidal systems. *Black*, the pyramidal tract; *gray*, the basal ganglia and the extrapyramidal tracts. Put, putamen; Cn, caudate; Pal, pallidum; Thal, thalamus; SN, substantia nigra; RN, red nucleus; Teg, tegmentum; Ret, reticular formation; OL, olive.
Source: Copied from Denny-Brown 1946.

abnormalities, and tremor. In a sharp contrast, lesions of the pyramidal system, as happened in cerebral palsy and following a middle cerebral artery stroke, are associated with the paralysis of distal body parts (limbs, digits, and speech).

Classically, the pyramidal system starts from the upper-motor (layer 5 pyramidal) cells of the motor cortices, travels through the lateral corticospinal tract, decussates at the brain stem pyramids, and continues to innervate the alpha-motor neurons of the spinal cord and to create phasic contractions of muscles. The extrapyramidal (or medial) tracts (i.e., the tectospinal, rubrospinal, vestibulospinal, and reticulospinal tracts) transmit the cerebellum and basal ganglia commands to the interneurons of the spinal cord and affect the body posture and muscle tonus.[1] The pyramidal system probably represents the efferent pathway of the corticocortical system 2, and the extrapyramidal tracts are the efferent pathway of Kahneman's default system 1. In line with the extrapyramidal system as the default experiencing system, the number of fibers in the extrapyramidal tracts is much higher than in the corticospinal tracts (in the human, $20 * 10^6$ vs. $1 * 10^6$, respectively).

Basal ganglia–related movement disorders are often classified as hypokinetic and hyperkinetic movement disorders.[2] Parkinson's disease, the most common basal ganglia and movement disorder, is the prototypical example of a hypokinetic disorder. Parkinson's disease patients are bradykinetic or akinetic—that is, their spontaneous, automatic movements (like emotional facial expression, frequency of blinks, and accompanying upper-arm swing movements during walking) are slow (bradykinesia) or absent (akinesia).[3] The hyperkinetic disorders include the chorea (dance-like, involuntary movements) of Huntington's disease, the flailing limb movements of hemiballismus due to subthalamic infarct, and the sustained or repetitive muscle contractions of dystonia. Human diseases are often not in agreement with the classical teachings of textbooks of medicine and neurology. Thus, hyperkinetic phenomena such as dystonia (in many cases leg dystonia is the initial motor presentation of Parkinson's disease), muscle rigidity (increased tonus), and tremor are integral parts of Parkinson's disease.[4]

Last, but not least, brain disorders are still classified as neurological or psychiatric disorders. This old classification is based on finding, or lack of finding, structural changes associated with the disorders. Thus, diseases such as Parkinson's disease (with a histological finding of Lewy bodies, loss of melanin-containing pigmented midbrain dopaminergic neurons, and depletion of dopamine in the striatum) are classified as neurological symptoms. On the other hand, disorders for which pathological and imaging techniques have failed to find a constituent abnormality in the brain are referred to as psychiatric disorders. The public attitude and care of these two populations of brain disorders follow this borderline, with much less attention to and support for psychiatric patients. I hope that all readers of this book will join the battle against this biased and unfair classification of brain disorders. First, it is very clear that "absence of evidence is not evidence of absence." The failure to find consistent brain pathology associated with psychiatric disorders only reflects our premature state in finding better, more precise, higher-resolution tools. Indeed, few of today's neurological disorders (e.g., dystonia) were classified as psychiatric disorders until recently. Second, "pure" neurological and even movement disorders such as Parkinson's disease are known today to include cognitive and emotional psychiatric disorders as an integral part of their clinical spectrum.

Below, I will briefly summarize the current knowledge regarding the basal ganglia–related disorders. Because of my background, I will mainly discuss Parkinson's physiology here. Please note that medical knowledge is rapidly changing, and the information is given for educational purposes only.

PARKINSON'S DISEASE

The symptoms (tremor, akinesia) of Parkinson's disease are mentioned in ancient history. However, it was James Parkinson, who in his 1817 "Essay on the Shaking Palsy" (based on six patients, two of whom he had just observed in the street) suggested that akinesia (palsy), tremor (shaking), posture and gait deficits, and even autonomic functions such as drooling are all of one entity.[5] Parkinson, an enlightened person and surgeon (not a physician), was limited by the medical practice and knowledge of his time. Thus, he did not mention muscle rigidity, probably because physicians and surgeons of the early nineteenth century did not perform routine physical examination of their patients. Neither did he offer a treatment for the patients with the "shaking palsy" beyond the medical tools of his time, such as bloodletting from the neck vessels.[6]

After Alzheimer's disease, Parkinson's disease is the second most common neurodegenerative disease and affects 0.3 percent (three out of one thousand) of the total population, with a current world prevalence of five to ten million patients (GBD 2016 Neurology Collaborators 2019). Parkinson's disease is usually a disease of the elderly. The mean age of first diagnosis is sixty years, and the frequency increases with aging. One percent of humans older than sixty suffer from Parkinson's disease, and the frequency probably increases to 2 percent and 4 percent for the population older than seventy and eighty years. Parkinson's disease is slightly more common in males than females and in Caucasians than in black people. A small fraction (approximately 5 percent) of all Parkinson's patients is due to strong (monogenetic) genetic inheritance. These patients usually develop Parkinson's disease earlier than the major population of Parkinson's disease and may display a slightly different spectrum of symptoms. During the twentieth century, most Parkinson's patients were considered sporadic or idiopathic (the nice medical term for a disease we do not understand the reasons for). Today, it is well known that idiopathic Parkinson's disease is associated with a network of mutations; some of them (e.g., LRRK2 and GBA in Jewish Ashkenazi people) are more common and have stronger effects than many other genes.

Parkinson's disease is commonly and easily diagnosed according to four major cardinal motor symptoms:[7] Akinesia/bradykinesia (lack and slowness of automatic movements), rigidity (muscle stiffness), low-frequency (4–7 Hz) tremor at rest, and postural and gait deficits. These symptoms follow the degeneration of a significant fraction of SNc dopaminergic neurons and depletion of dopamine in the striatum (starting at the posterior putamen). The worsening of the symptoms (with the exception of tremor) is correlated with the progression of the degeneration of dopamine neurons and depletion of the dopaminergic innervation of the striatum (initially spreading anteriorly in the putamen and then medially to the caudate).

Why do midbrain dopaminergic neurons die? This is the one-million-dollar question that if answered should lead to true therapy, or at least a slowdown in the progression of the disease. The jury is still out. Previous hypotheses include the toxin-like effects of dopamine and its metabolites (suggesting that, evolutionarily, unicellular organisms use neurochemicals to repel their predators, and these same toxic neurochemicals are used in the multicellular organism for cell-to-cell communication); oxidative stress due to mitochondrial (complex 1) malfunction or iron accumulation; and even the metabolic overload of the five hundred thousand boutons (synaptic connections) a single dopaminergic cell has to "feed." Probably, many roads lead to Rome, and the different mechanisms suggested above are not mutually exclusive.

Recently, most of the attention has been directed to alpha-synuclein. The major role of this protein is still debated, but it is very ubiquitous in the central nervous system and is the main protein found in Lewy bodies. Lewy bodies are intracellular aggregations easily recognized by light microscopy. The pathological diagnosis of Parkinson's disease is conditioned on the presence of Lewy bodies, and their amount and distribution correlate with the progression of Parkinson's disease. It is therefore widely accepted today that Parkinson's disease is due to misfolding of alpha-synuclein that can spread from one neuron to another like Creutzfeldt–Jakob's prions (Max 2007). The very popular Braak's hypothesis (Braak et al. 2003; Del Tredici and Braak 2020) suggests that Parkinson's pathology starts in the gut (or even in the olfactory system) and ascends to the dorsal nucleus of the vagus nerve (in the pons). The midbrain and the dopaminergic neurons are affected only in the third and fourth stages of the disease. Afterward, the Lewy body pathology progresses to the cortical mantle. Exceptions exist (e.g., Horsager et al. 2020), but Braak's hypothesis provides a new framework that states that Parkinson's is neither a pure motor disorder nor a pure dopaminergic disorder. The early pathology at the level of the nose, gut, and pons accounts for the common description of hyposmia or anosmia (partial or complete loss of the sense of smell), constipation, and sleep deficits (especially REM sleep behavior disorders, when the patient starts to enact his/her dreams) in the prodromal stage of Parkinson's disease (i.e., before the development of the motor symptoms).[8] Braak's hypothesis is also in line with the finding that the vast majority of Parkinson's patients develop dementia within twenty years from diagnosis.

OTHER BASAL GANGLIA–RELATED DISORDERS

The list of basal ganglia–related disorders is really a big one and includes typical "movement disorders" but also classical psychiatric disorders.

Parkinson-plus and parkinsonism syndromes   In the "movement disorders" theater, we can find the Parkinson-plus syndromes, including multiple-system atrophy, progressive supranuclear palsy, and corticobasal degeneration (MSA, PSP, and CBD, respectively), all of which include damage to midbrain dopaminergic neurons, as well as to other brain centers. Thus, they all show the symptoms of Parkinson's disease in addition to other symptoms. Most of these Parkinson-plus syndromes run a faster course than Parkinson's disease, and patients usually develop dementia within a few years of the initial diagnosis. Interestingly, MSA, as well as Lewy body dementia (a form of Parkinson's disease in which the dementia precedes the motor symptoms) are related to alpha-synuclein mutations.

Parkinson-plus syndromes may be considered as one example of parkinsonism. Parkinsonism is any condition that causes a combination of the movement abnormalities seen in Parkinson's disease. There are many causes of parkinsonism. Among them are drug-induced symptoms, most commonly from the typical neuroleptic (D2 antagonist) treatment of schizophrenia and other mental disorders and environmental toxins like exposure to pesticides and herbicides (more common in those living out of the city), and to manganese (mine workers). Parkinsonism can be observed following infectious, vascular, or traumatic damage to the brain and the basal ganglia. Finally, the incidence of parkinsonism increases with age. The amount of dopaminergic neurons and the dopamine innervation of the striatum in the human brain decline with age in each of us (figure 8.7). Some of us have a lower number of dopaminergic neurons to start with, accelerated degeneration of dopaminergic neurons (e.g., due to genetics, chronic exposure to toxins, and so on) or have been affected by a single or a few events (e.g., head trauma) that have led to a more severe reduction in brain dopamine.

Encephalitis lethargica and postencephalitic parkinsonism   This book was written during the time of the 2020–21 COVID-19 (coronavirus) pandemic. The cumulative reports of a spectrum of brain disorders associated with COVID-19 infection (Paterson et al. 2020) may remind the reader of the one-hundred-year-old story of the H1N1 influenza pandemic probably leading to encephalitis lethargica (an infectious brain disease with excessive sleeplike states, distinct from tsetse fly-transmitted sleeping sickness). Encephalitis lethargica is also called von Economo's disease, after Constantin Freiherr von Economo, who first described the symptoms, pathology, and histology of the disease in 1918.[9]

The epidemic of encephalitis lethargica spread around the world between 1915 and 1926 (parallel to World War I). It is estimated that nearly five million people were affected (about the global number of Parkinson's patients worldwide today), a third of whom died in the acute stages. Many of those who survived the acute phase developed the chronic syndrome of post-encephalitic lethargica, or postencephalitic parkinsonism, a Parkinson-plus condition.[10] Molecular biology techniques were unavailable in these early days. Today, most researchers attribute the encephalitic lethargica to the H1N1 influenza infection. However, this is still debated. For example, the H1N1 epidemic was between 1918 and 1920, while encephalitis lethargica continued until 1927, when it disappeared as abruptly and mysteriously as it first appeared.

The encephalitis lethargica pandemic changed our world. Like COVID-19 it teaches us that "minor" viral disease might be severely complicated by the body's autoimmune response. Immune activation in the olfactory system might eventually lead to the misfolding of alphasynuclein and the sequential development of Parkinson's disease. Viral infection might also interfere with alpha-synuclein clearance. Neurotropic viruses can obstruct protein clearance to maintain optimal viral protein levels, rendering infected host cells unable to counterbalance alpha-synuclein accumulation (Lema Tomé et al. 2013; Cohen et al. 2020). It points toward the close relationships between protein misfolding pathology, parkinsonism and sleep disorders (Max 2008).

There is speculation that Adolf Hitler may have had encephalitis lethargica as a young adult. There is better evidence for (drug-induced?) parkinsonism in Hitler's later years (Christian Moll shared with me a video clip of Hitler with clear-cut Parkinson's rest tremor). Would world history and our understanding of ourselves and humanity (*Se questo è un uomo*[11]) look different without the frightening effects of encephalitis lethargica and parkinsonism? On the other side of the spectrum, Oliver Sachs so beautifully described the life and the effect of L-dopa therapy on postencephalitic parkinsonism patients in his novel *Awakenings*. We, students of the basal ganglia and their disorders, should carefully listen to the words of the novel's protagonist Leonard Lowe, speaking for all of our patients: "I am a living candle. I am consumed that you may learn. New things will be seen in the light of my suffering."

Other movement disorders   Other common movement disorders include dystonia, essential tremor, and hemiballismus. Dystonia and essential tremor, like Parkinson's disease, are treated very well with deep brain stimulation. Hemiballismus, often caused by STN infarcts, is usually a self-limiting disease. Huntington's disease is the most common example of a basal ganglia–related hyperkinetic disorder. I will also add to this class restless leg syndrome and attention deficit hyperactivity disorder.

Psychiatric disorders   Last, but not least, are the psychiatric disorders that are at least partially treated by the modulation of dopamine and/or serotonin brain levels. Here, I will consider obsessive-compulsive disorder, major depressive disorder, bipolar disorder, and schizophrenia.

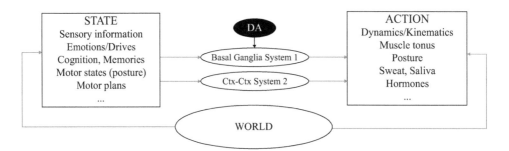

13.2 The two major functional and anatomical systems that connect state-to-action are the basal ganglia (BG) system 1 and the corticocortical (Ctx-Ctx) system 2. DA, dopamine.

WHY ARE BASAL GANGLIA DISORDERS SO POORLY COMPENSATED BY OTHER BRAIN STRUCTURES?

The above big list of basal ganglia–related disorders positions the basal ganglia as a central hub of brain and behavior. It is still surprising why other brain structures (e.g., the cortex) do not compensate for at least the abnormal negative symptoms of the basal ganglia disorders. A possible explanation is that the basal ganglia are functioning as Kahneman's system 1—the default automatic system that only when recognizing that its fast, default, and error-prone algorithms are failing "transfers the command" to the corticocortical networks system 2 (figure 13.2). I suggest that the sick basal ganglia do not let the thalamocortical treatment take control, and therefore the cortical networks cannot compensate for basal ganglia disorders.

CHAPTER SUMMARY

- The basal ganglia are part of the extrapyramidal system that parallels the pyramidal system.
- Lesions of the pyramidal system lead to paralysis, while lesions of the extrapyramidal system lead to movement disorders, altered muscle tonus, and tremor.
- Parkinson's disease is the most common basal ganglia and movement disorder. Affecting 0.3 percent of the population, it usually starts at the age of sixty.
- The motor symptoms of Parkinson's disease are due to degeneration of midbrain dopaminergic neurons and striatal dopamine depletion. However, Parkinson's disease is a broad neurodegenerative disease. Misfolding of alpha-synuclein propagates from the gut and lower brain stem neurons to midbrain dopamine neurons and to the cortex.
- Disorders of the basal ganglia can't be compensated for by other systems because the basal ganglia are the default system and continue to broadcast inhibition signals to the cortical system 2 networks.

14

ANIMAL MODELS OF PARKINSON'S DISEASE

Parkinson's disease, like many other neurodegenerative diseases such as Alzheimer's disease and schizophrenia, is a uniquely human disease. It does not exist in animals, even in monkeys and apes kept in captivity with longer life spans than those in the wild.

Early animal models of (Parkinson's) tremor were based on ablation of the substantia nigra. The midbrain is a very dense area, and ablation experiments fell short because of huge variability in the outcomes of the lesions. Interestingly, more stable results (regarding tremor; akinesia and rigidity were not observed in these midbrain ablation models) were obtained with additional lesions of the red nucleus (i.e., pointing the importance of the cerebellum in the generation of tremor). Arvind Carlsson has explored the role of dopamine in the brain by studying reserpine-induced dopamine depletion in rabbits (figure 14.1). Neither reserpine nor rabbits are commonly used today in Parkinson's disease research.

The two most common animal models of Parkinson's disease today are intracerebral injection of 6-hydroxydopamine (6-OHDA; mainly in rodents) and systemic 1-methyl-4-phenyl-1,2,3,6-tetrahydropyridine (MPTP) treatment (mainly to nonhuman primates). There are also genetic models (e.g., AT53 mutation) of Parkinson's disease in rodents and ongoing trials for the development of a genetic Parkinson's model in nonhuman primates. I am crossing my fingers for success in these trials, as they will clearly provide the future students of Parkinson's disease with better tools. Below I will elaborate on the 6-OHDA and the MPTP models of Parkinson's disease.

THE 6-OHDA RODENT MODEL OF PARKINSON'S DISEASE

Because 6-OHDA does not cross the blood-brain barrier, it is injected directly into the brain by stereotactic methods. It is usually injected to the medial forebrain bundle

14.1   Reversal of reserpine's effects by dopa. *A*, rabbits treated with IV reserpine, 5 mg/kg. *B*, the same rabbits fifteen minutes after IV D-L-dopa, 200 mg/kg.
Source: Adapted from Carlsson 2001. The original figure probably appeared in Carlsson (1960) in an article in German titled "On the Problem of the Mechanism of Action of Some Psychopharmaca." I failed to find the original article.

(MFB) and retrogradely transported to the midbrain. Midbrain dopaminergic neurons degenerate within a few days. Since 6-OHDA is also transported by the noradrenaline reuptake system, 6-OHDA is often used in conjunction with a selective noradrena-line reuptake inhibitor (such as desipramine) to selectively destroy dopaminergic neurons (Zigmond et al. 1972; Sotelo et al. 1973; Schwarting and Huston 1996). Bilateral injections of 6-OHDA are often fatal because the rodents stop self-feeding. Therefore, unilateral injections are most often performed. The most reliable clini-cal sign of 6-OHDA–induced dopamine depletion is apomorphine or amphetamine-induced rotations. Interestingly, the induced rotations are in opposite directions. This is probably because apomorphine has a stronger effect on the oversensitive dopamine receptors at the lesioned site, while amphetamine more strongly affects the nonlesioned site, where extracellular dopamine is increased due to the blocking of the reuptake of the released dopamine.

THE MPTP MODEL OF PARKINSON'S DISEASE

The MPTP neurotoxin story started with human patients. A chemistry student and then drug addicts in California used a street-lab opium substitute. They developed the full spectrum (including rest tremor) of cardinal signs of Parkinson's disease

(Davis et al. 1979; Langston et al. 1983; Langston 2017). The amazing detective work of William Langston revealed that, unfortunately, they also injected themselves with significant amounts of MPTP neurotoxin.[1] MPTP can cross the blood-brain barrier and is transformed into MPP^+, which is transported through the dopamine transporter to the cytosol of monoaminergic neurons, especially dopaminergic neurons, leading to the death of these neurons. Thus, MPTP was offered as a model for Parkinson's disease and was used to induce Parkinson's symptoms in nonhuman primates (Chiueh et al. 1985) in several ways:

- IV (intravenous) or IM (intramuscular) systemic injections. IV methods were first used, probably, following the examples of "frozen addicts." However, IM injections have the same effect. Because they are easy to use and therefore safer to the experimenters, they are more commonly used today. These systemic models can be divided into:

  □ Acute models, in which the MPTP is injected over a few days (in our hand, five IM injections of 0.35–0.4 mg/Kg over four days; two injections in the morning and evening of the first day).

  □ Chronic models, in which the MPTP is injected once every week or two weeks, for twenty to thirty weeks.

- Intracarotid injections leading to a hemiparkinsonian state, with the big advantage that the monkey can keep handling (e.g., eating and drinking) itself. This is usually done with a complex surgery that exposes the bifurcation of the carotid artery. The external carotid is temporarily blocked, and MPTP 0.5–1 mg/Kg in 20 ml saline is slowly infused into the common carotid artery. Since the external carotid is blocked, all the MPTP will go to the brain. Because of the dopamine transporters' high affinity for MPTP, the MPTP will be taken in the first passage through the brain and will not make its way through the venous return and the cardiovascular circulation to the other side.

- MPTP can be injected locally—for example, to the caudate or putamen and by axonal retrograde transportation, leading to the degeneration of the dopamine neurons innervating this area. I am aware of only one nonhuman primate study (Kato et al. 1995; Kori et al. 1995) on using this method.

MPTP-induced parkinsonism is one of the most amazing models for human natural disease. This is because

- MPTP intoxication leads to similar patterns of neural degeneration as in human idiopathic Parkinson's disease (mainly with regard to dopamine neurons but also to the neurodegeneration of other monoaminergic neurons). Figure 14.2 depicts immunohistochemical staining of tyrosine hydroxylase, the rate-limiting enzyme for dopamine biosynthesis (figure 17.1) in healthy (control) and MPTP-treated nonhuman primates.

Healthy   MPTP-treated

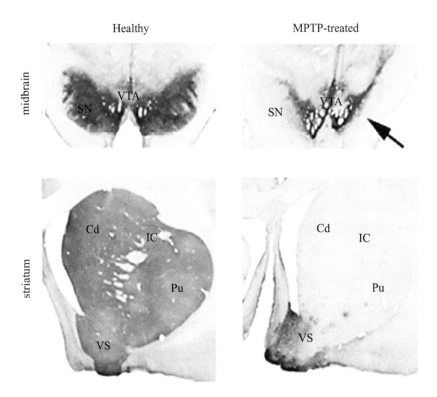

14.2 Degeneration of midbrain dopaminergic neurons and depletion of striatal dopamine by systemic MPTP treatment. Tyrosine hydroxylase staining of midbrain (*upper row*) and the striatum (*lower row*) of normal (*left column*) and MPTP-treated (*right column*) monkeys. Cd, caudate; IC, internal capsule; Pu, putamen; SN, substantia nigra; VS; ventral striatum; VTA, ventral tegmental area. Arrow indicates dopamine cell loss in the midbrain.
Source: Copied (with permission) from Deffains et al. 2016.

- MPTP-treated monkeys develop all signs of Parkinson's disease: Akinesia/bradykinesia (including a reduction in blinking frequency and drooling), flexed posture, and muscle rigidity. Macaques seldom develop tremor, but vervet monkeys (African green monkeys) often develop low-frequency (4–7 Hz) tremor (Bergman et al. 1990). Recently, we found that vervet monkeys treated with MPTP develop severe sleep insomnia, as most Parkinson's patients do.
- MPTP-treated monkeys respond to Parkinson's therapy (pharmacology, deep brain stimulation, and ablation) as do human patients. They may even develop levodopa-induced dyskinesia.
- Electrophysiological recordings in the basal ganglia of MPTP monkeys reveal a similar discharge pattern as in human patients.

Some of the similarities mentioned above (such as the beneficial effects of subthalamic nucleus (STN) ablation and deep brain stimulation and synchronized oscillations in the dopamine-depleted basal ganglia) were first described in the monkey and then extended to the human theater, revealing the predictive validity of the MPTP nonhuman primate model of Parkinson's disease.

However, "No pain, no gain." The MPTP experimental setup is not an easy one. Human safety comes first, and caution should be taken to avoid self-injection and exposure to trace amounts of MPTP that can be excreted in the urine and disseminated in the room (a negative pressure room and protective dressing are highly recommended). Non–severely affected monkeys recover quickly, and the experimental results may not be considered valid because of the nonstationarity of the biochemical and clinical effects. On the other hand, severely affected monkeys need 24/7 treatment to avoid pressure sores and must be provided with the daily intake of food, water, and therapy by nasogastric tube feedings.

ANIMAL MODELS ARE LIES THAT HELP US TO REVEAL THE TRUTH AND TO REDUCE HUMAN SUFFERING

MPTP-treated primates do not react in exactly the same way as patients with idiopathic Parkinson's disease:

- The pathogenesis is different. Following the discovery of the MPTP toxin, Langston and others have extensively searched for a similar endogenic or environmental toxin but with no real success. Probably, MPTP wipes out dopaminergic and other monoaminergic neurons in a different way than idiopathic Parkinson's. Indeed, Lewy bodies are not, or are very seldom, found in MPTP-treated monkeys.
- Parkinson's disease evolves slowly over the course of five to twenty years. Most MPTP models are acute, and the monkeys show severe Parkinson's signs within seven to ten days after starting the MPTP injections. The different time courses probably result in a lack of compensatory processes in the MPTP monkey.
- Today, most patients are on dopamine replacement therapy since diagnosis. Many years of dopamine therapy and the nonnatural fluctuation of the brain dopamine level (as induced in patients taking oral therapy three to four times a day) affect the basal ganglia.

Thus, and as always, a model is not a full replica of reality.[2] The first half of the title of this section paraphrases Pablo Picasso's quote "Art is the lie that reveals the truth." There are many similar quotes, for example, attributed to the statistician George Box's quote "All models are wrong, some are useful." They remind us that a model should be used to see the critical issues of the questions and that different models should be used for different aspects of the same problem.

Model validity (like test validity) is the extent to which a model (or a test) accurately measures what it is supposed to measure. The validity concept is heavily discussed in the psychological literature, and different definitions are given. I will use these definitions:

- Content validity is the extent to which a model represents all facets of a given construct (in psychology, face validity is the extent to which a test, or a model, looks valid).
- Criterion or predictive validity is the extent to which a model's results (e.g., modification of symptoms in an animal model) predict the actual outcome (e.g., effect on human patients).

Naturally, the content validity of animal models of Parkinson's disease is limited. Both the 6-OHDA rodent and the MPTP nonhuman primate model do not replicate the etiology and the natural history of human idiopathic Parkinson's disease. However, both models have shown very strong predictive validity. The 6-OHDA and MPTP models have been very efficient in predicating the effect of pharma and surgical therapy for human patients.

Animal models are used today under the 3Rs ethical rules of research (Replacement, Reduction, and Refinement). I am painfully aware that ethical rules are not absolute and often reflect their time period. I hope that readers of this book, even those whose ethical rules differ, accept the rights of others to have different ethics. At least, they should know that I thought about it. Most of this book is devoted to the study of the basal ganglia of nonhuman primates. We have done our best to refine and to minimize the suffering of our monkeys. I am very thankful to the generations of students, vets, and animal facility workers who have helped us in doing so. Reduction in the number of animals in experiments is very natural in monkey experiments. Most of our projects are done with $n=$ two or three. Each of our monkeys has a name, and we do our best to treat them as individuals. I try to repeat critical issues and experimental outcomes in different projects and increase the reliability of our conclusions by these repetitions. Replacement is an issue. I have decided to use the MPTP nonhuman primate as my main model because the anatomy, physiology, and pathophysiology of the basal ganglia of the monkey are much more similar to human than that of the rodent. Additionally, at the end of our experiments we do not have to kill our animals, and we usually send our monkeys to rehabilitation. Most importantly, I have made the ethical choice to conduct these experiments because they replace those on human patients and reduce the suffering of human patients who are self-aware of their brain disorders and trust us to make the correct, even if less easy, moral decisions.

CHAPTER SUMMARY

- The two most common animal models today for Parkinson's disease are based on
  - Unilateral intracerebral injection of 6-OHDA to rodents.
  - Systemic injection of MPTP to nonhuman primates.

- MPTP-treated nonhuman primates develop the cardinal motor symptoms of Parkinson's disease—akinesia, rigidity, and postural deficits. MPTP-treated vervet monkeys also develop low-frequency (3–5 Hz) tremor.

- The anatomical pattern of degeneration of dopamine and other neuromodulators in MPTP-treated nonhuman primates is very similar to the pattern observed in human patients with idiopathic Parkinson's disease.

- The clinical symptoms of MPTP-treated monkeys are ameliorated by the same treatments used in human patients (dopamine replacement therapy, ablation, and deep brain stimulation of the STN and the GPi).

15

PHYSIOLOGY OF PARKINSON'S DISEASE: ANIMAL MODELS

The MPTP nonhuman primate model's main advantage is that it enables the comparison of neural activity before and after MPTP treatment; that is, before and after development of Parkinson's clinical signs. Below we will discuss MPTP-induced changes in the discharge rate, pattern, and synchronization of basal ganglia neurons.

DISCHARGE RATE

Early studies revealed significant changes in the discharge rate of neurons in the main axis of the basal ganglia.[1] In line with the prediction of the D1/D2 direct/indirect pathways model of the basal ganglia (figures 9.1 and 9.2), the mean discharge rate of STN and GPi neurons increased, while the mean discharge rate of GPe neurons decreased (Miller and DeLong 1987; Filion and Tremblay 1991; Bergman et al. 1994). The effects of dopamine replacement treatment were also in line with the prediction of the D1/D2 direct/indirect model. Figure 15.1A shows the outcome of our pallidal recording, where we second the results given by previous publications (Miller and DeLong 1987; Filion and Tremblay 1991; Filion et al. 1991). In the normal monkey, the GPi discharge rates (*red bars*) are slightly higher than GPe (*blue bars*). Following MPTP treatment (DA, ↓↓), the GPe average discharge rate decreases and GPi increases. Finally, following dopamine replacement therapy (DRT; DA,↑↑) the GPe/GPi discharge rate ratio is reversed and the average GPe rate is very high, while the GPi rate decreases significantly. These impressive and fast changes in the activity of GPe and GPi neurons following DRT are further shown in figure 15.1B. About one hour after the administration of an oral dose of L-dopa, the discharge rate of the GPe units (*three upper subplots*) significantly increased, while the discharge rate of the GPi neurons significantly decreased (*two lower subplots*).

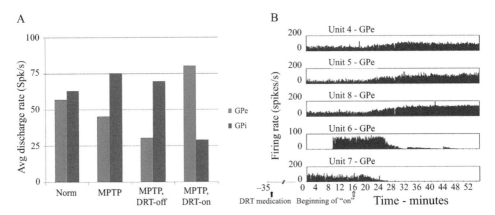

15.1 Opposite effects of dopamine tone on discharge rate in the two pallidal segments. *A*, average discharge rates of neurons in the external and internal segments of the globus pallidus (GPe and GPi, *blue and red bars*), respectively, before and after MPTP and following dopamine replacement therapy (DRT). *B*, time course of opposite changes in discharge rate of simultaneously recorded neurons of the two pallidal segments following oral DRT. *Three upper traces*, GPe neurons showing increase in discharge rate when dopamine "kicks in"; *two lower traces*, GPi with a decrease in discharge rate. *Source*: Copied (with permission) from Heimer et al. 2002.

The distributions of discharge rate of basal ganglia neurons (either before or after MPTP) are very broad, and the changes in the mean discharge after MPTP are usually smaller than the standard deviation of these broad distributions (figures 12.6 and 15.2). Changes in the GPi and SNr discharge rate were not consistently found in all studies (including some of our studies). Other descriptive parameters of neural discharge—pattern and synchronization—lead to more consistent results.

DISCHARGE PATTERN AND SYNCHRONIZATION

Figure 15.3 depicts raw traces and the correlation matrix of simultaneously recorded pallidal units in the normal (figure 15.3A) and MPTP (figure 15.3B) nonhuman primate. The normal correlation matrix was already given in figure 7.3, and we show it again for the visual contrast with the MPTP matrix. In the correlation matrices, you can see the autocorrelation histograms (*diagonal, blue*) and the cross-correlation histograms (*red*). A schematic illustration of the experimental setup is given in the inset of figure 15.3A, and the power spectrum (*blue*) and the coherence functions (*red*) are given in the inset of figure 15.3B. One picture is better than a thousand words. One can easily see that the normal pallidum units tend to fire in a Poisson (random) pattern in time with flat or slightly peaked autocorrelation functions. The cross-correlation functions are strictly flat, indicating independent activity of the pallidal units. This unexciting picture drastically changes after MPTP. The flat sea is now wavy and synchronous oscillations dominate (figure 15.3B). Figure 15.4 gives the populations results, now in the frequency domain in various nuclei of the basal ganglia.

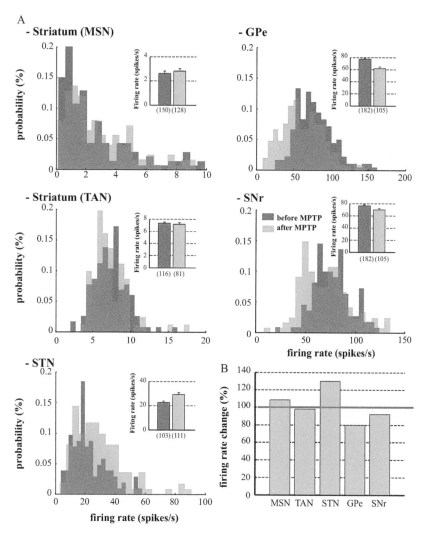

15.2 Discharge-rate distribution in the basal ganglia before and after MPTP. *A*, distribution of discharge rate in the striatum of MSNs, TANs, STN, GPe, and SNr before (control, *gray*) and after (*green*) MPTP treatment. *Inset*, mean discharge rate. *B*, relative changes to control. MSNs, medium spiny neurons; TANs, tonically active neurons; STN, subthalamic nucleus; GPe, globus pallidus external segment; SNr, substantia nigra reticulata.
Source: Copied (with permission) from Deffains et al. 2016.

The synchronous oscillations in the basal ganglia of MPTP-treated monkeys are not affected by the distance between the cells. Rea Mitelman compared the synchronous oscillations of nearby pallidal neurons (recorded by the same electrode, estimated distance <200 μm apart) and remote pallidal neurons (recorded by different electrodes, distance >500 μm apart) apart. However, he was unable to "fix" the shadowing effect (Bar-Gad et al. 2001b) in the units recorded by the same electrode because of their oscillatory discharge pattern. He therefore "broke up" the remote pairs by inducing artificial shadowing in their spike trains. The synchronous oscillations were not

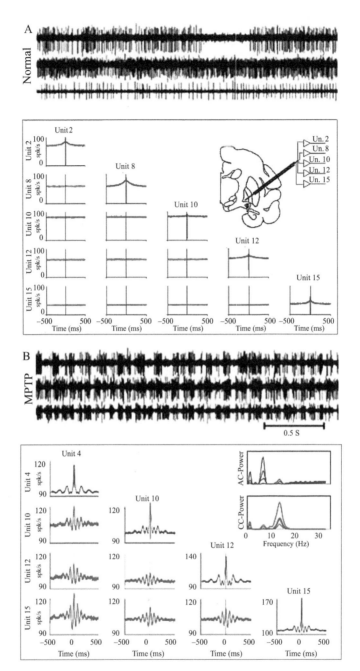

15.3 Independent random versus synchronous oscillatory spiking activity in the globus pallidus of normal and parkinsonian nonhuman primates. Examples of raw traces and correlation matrices of simultaneously recorded pallidal neurons before (A) and after (B) MPTP treatment. AC, CC, autocorrelation and cross-correlation functions.

Source: Modified (with permission) from Bergman et al. 1998.

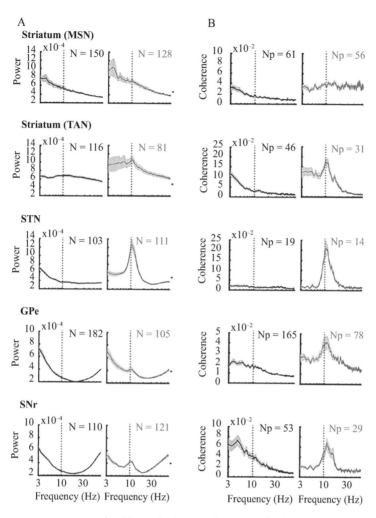

15.4 Average power spectrum densities and coherence functions of spiking activities in the basal gan-glia before and after MPTP treatment. *A*, power spectrum densities. *B*, coherence functions of spiking activity of pairs of neurons in the striatum (MSNs and TANs), STN, GPe, and SNr before (control, *gray*) and after (*green*) MPTP treatment. MSN, medium spiny neurons; TANs, tonically active neurons; STN, subthalamic nucleus; GPe, globus pallidus external segment; SNr, substantia nigra reticulata.
Source: Copied (with permission) from Deffains et al. 2016.

significantly different between the two groups (Mitelman et al. 2011). Thus, synchro-nous oscillations in the pallidum are not a local property but rather a global network phenomenon.

Most multiple-electrode physiological studies of the normal and dopamine-depleted basal ganglia are limited to neurons of the same structure. However, a few studies indicate that the synchronicity of abnormal oscillations can be detected even between different nuclei of the basal ganglia network. Aeyal Raz recorded the simultaneous activity of striatal TANs and pallidal neurons before and after MPTP treatment. Only a few pairs were weakly correlated in the normal monkey (figure

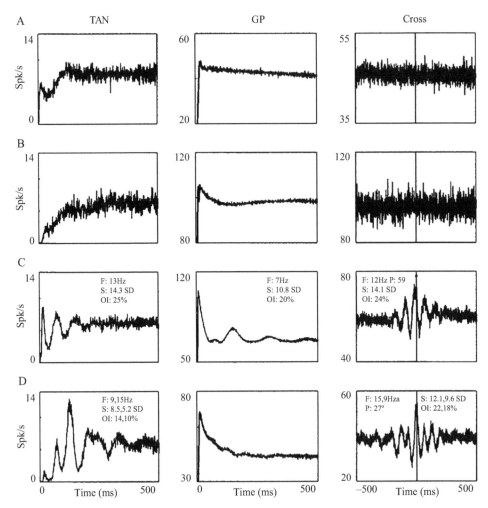

15.5 Synchronous oscillations of striatal cholinergic interneurons and pallidal neurons in the MPTP-treated nonhuman primate. Examples of autocorrelograms and cross-correlograms of striatal tonically active (TAN) and pallidal (GP) neurons. *First and second columns*, the autocorrelogram of the TAN and the pallidal cell; *third column*, the cross-correlogram. The y-axis displays the conditional firing rate. Power spectra of cross-correlograms are represented. A, B, normal monkey. C, D, MPTP-treated monkey. A, C, D, subplots of TAN-GPe pairs. B, TAN-GPi pair. F, frequency in Hertz; P, phase shift in degrees; S, signal-to-noise ratio; OI, oscillation index; CG, CT, correlation coefficient.
Source: Modified (with permission) from Raz et al. 2001.

15.5*A, B*). After MPTP treatment, significant synchronous oscillations were observed in the majority of the TAN-pallidal pairs (figure 15.5*C, D*). TANs probably represent the striatal cholinergic interneurons that do not project outside of the striatum, and the efficacy of the synaptic connections between striatal TANs and MSNs is weak (figure 5.5). The TAN-pallidum correlation probably reflects the global synchronous mode of the basal ganglia networks after MPTP treatment.

We have not conducted formal studies of the spike-to-spike correlation between simultaneously recorded cortical and pallidal neurons. Figure 15.6 shows an example

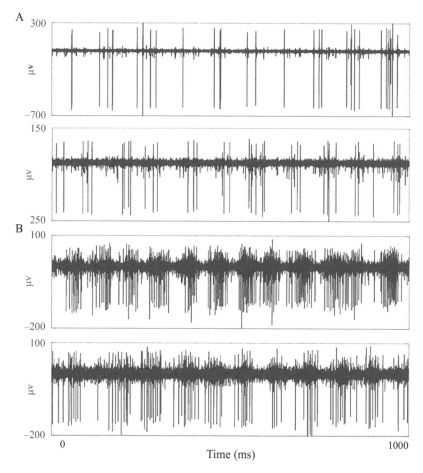

15.6 Synchronous oscillations in the cortical basal ganglia network of MPTP-treated nonhuman primates. Shown is 1 second of 300–6,000 Hz band-passed filter activity recorded from two neurons in the primary motor cortex (A) and in the globus pallidus (B) of an MPTP-treated vervet monkey. *Source*: Modified (with permission) from Rosin et al. 2011.

of simultaneous recording of two M1 and two pallidal units with synchronous 10–11 Hz oscillations. The success of closed-loop GP stimulation triggered by M1 oscillations (Rosin et al. 2011) further indicates that the spiking activity of the two structures is synchronized. Finally, the origin of field potentials recorded by single electrodes in the basal ganglia is still debated. One option is that basal ganglia field potentials represent the volume conduction of cortical activity and EEG. The second option is that as in the cortex they represent subthreshold synaptic activity. In both cases the correlation between basal ganglia local field potentials (LFP) and spiking activity (figures 15.10 and 15.11*B*) is in line with the idea of cortex-basal ganglia network synchronization in the MPTP-treated nonhuman primate.

The frequency domain of the beta oscillations in the monkey is 8–20 Hz. The range of beta oscillations in the human is defined as 13–30 Hz, often divided into low (13–20 Hz) and high (20–30 Hz) beta, with indications that the low beta oscillations are better correlated with the clinical symptoms. The lack of high beta oscillations in the monkey could be due to species differences, severity of the disease, or a lack of compensation processes. Man may work from sun to sun, but woman's work (understanding the full spectrum of beta oscillations) is never done.

ON OSCILLATIONS AND SYNCHRONIZATION

"Do two walk together unless they have agreed to do so?" (Amos 3:3). Synchronization is easily achieved in an oscillating system with weak coupling between its elements.[2] As in society, we can assume that in the brain (and basal ganglia) any random pairs of neurons will be eventually connected.[3] Thus, the brain is prone to synchronous oscillations. However, oscillations can be unsynchronized, and synchronization may be found without oscillations. Figure 15.7 shows an example of a unique nonoscillatory synchronized event that was detected by the careful eye of Gali Havetzelt-Heimer while recording with seven electrodes in the GPe of an MPTP-treated monkey (Heimer et al. 2002a). The figure depicts ten seconds of activity and a

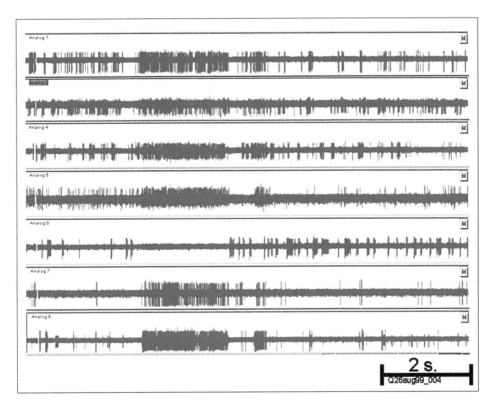

15.7 Nonoscillatory synchronized event in the basal ganglia of an MPTP-treated monkey. Simultaneous seven-electrode recording in the globus pallidus external segment of an MPTP-treated nonhuman primate.
Source: Modified (with permission) from Heimer et al. 2002b.

clear-cut nonoscillatory synchronized event of a duration of about two seconds. This exact same event is repeated more than twenty times in the recording session (Heimer et al. 2002b). While six of the recorded units were recruited with intense discharge, one of the recorded units (*fifth from the top*) did the opposite. The spontaneous activity of this unit was suppressed for the time of the event. The normal basal ganglia probably use lateral inhibition for active decorrelation and maximization of information capacity but avoid the "winner-takes-all" scenario and maintain themselves in the "winner-shares-all" state. Figure 15.7 shows that following dopamine depletion and the development of Parkinson's symptoms, the basal ganglia may fail to share.

The possible dissociation between oscillation and synchronization is often ignored by students of the brain, especially those who study EEG and LFP oscillations. The common assumption in the EEG and LFP fields is that awake low-amplitude EEG is due to the summation of many unsynchronized inputs (figure 15.8A). Large-amplitude EEG waves (e.g., delta waves) reflect a highly synchronized state (figure 15.8B). The transient reduction in beta oscillations during movement is defined as

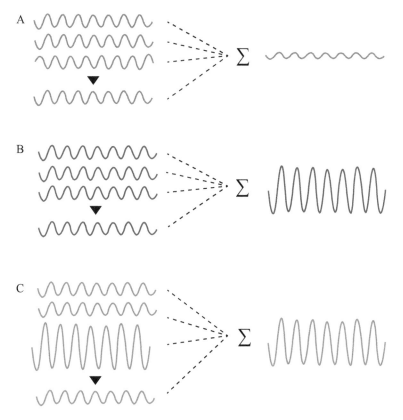

15.8 Oscillations, synchronization, and desynchronization. *A*, population of nonsynchronized oscillations are summed into small-amplitude signal. *B*, synchronized small-amplitude oscillations are summed into large-amplitude signal. *C*, unsynchronized small oscillations and one large-amplitude channel will lead to a high-amplitude average signal.

"event-related desynchronization," hinting at a change in synchronization. While not arguing that it can't be true, I would like to stress that the recording of high-amplitude oscillations of EEG and LFP can be due to an increased amplitude of one (or several) of the oscillating elements (figure 15.8C). Movement-related reduction of EEG or LFP beta oscillations can be due to reduction in the amplitude of a few major sources. Synchronization should be studied and quantified (as correlation in the time domain and coherence in the frequency domain) by multiple-electrode recording.

CHANGES IN THE SPIKING ACTIVITY OF THE MOTOR CORTEX IN MPTP-TREATED MONKEYS

Scientific thinking oscillates (unfortunately, also in a synchronized mode). Quite often, old truths are rejected, forgotten, and rediscovered decades later. The basal ganglia, as part of the classical extrapyramidal thinking, were thought to parallel the pyramidal tract and to target downstream structures, such as the spinal cord (figure 13.1). Most current models of the basal ganglia indicate that basal ganglia output targets the frontal and motor cortex (figures 9.1 and 9.2). Recent models, while not neglecting basal ganglia output to the thalamus and cortex, readmit that basal ganglia output is also directed downstream to the brain stem motor nuclei and their descending spinal tracts (figures 9.3, 9.4, 10.2, 10.7, and 11.5).

In contrast with the prediction of the D1/D2 direct/indirect model of the basal ganglia, recordings of spiking activity in the arm-related area of the primary motor cortex (M1) have not revealed any substantial decrease in discharge rate (Watts and Mandir 1992; Goldberg et al. 2002; Darbin et al. 2020). The average discharge rate was 4–5 Hz in both cases, with similar distribution of discharge rates. Before MPTP, M1 synchronized bursts were correlated with upper-arm mobility (figure 15.9A). However, following MPTP, the M1 synchronized bursts were not associated with any movement (figure 15.9B). The autocorrelation functions reveal more burst activity, and the cross-correlation functions of M1 pairs were significantly more correlated than before. These findings may suggest that the discharge rate is overrated, and other modes of information transfer in the nervous system, such as resonance, should be considered (Izhikevich 2010).

The MPTP M1 synchronized bursts failed to elicit detectable movement (figure 15.9B), indicating that even robust synchronous M1 discharge need not result in movement (Kaufman et al. 2014). The role of the primary motor cortex in movement control is probably exaggerated (Lawrence and Kuypers 1965, 1968; Kuypers 1982). Indeed, out of the twenty million fibers that descend to the spinal cord in humans, only one million are part of the corticospinal pyramidal pathway (Saliani et al. 2017). The other nineteen million constitute the brain stem-spinal pathways (figures 13.1). This is in line with the suggestion that the corticocortical system 2 network is called

A **Normal**

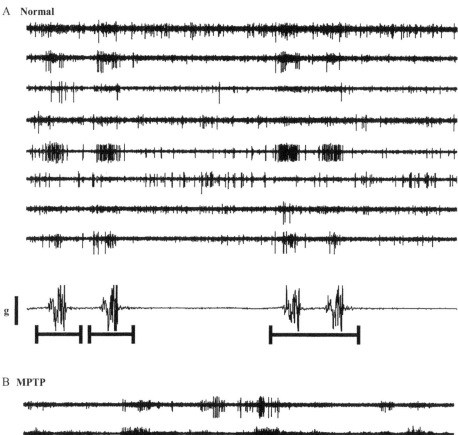

B **MPTP**

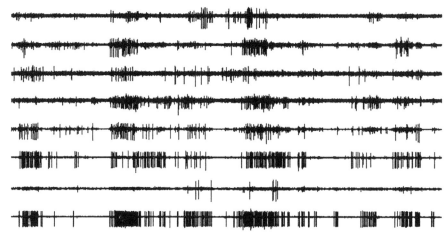

15.9 Synchronized bursts of activity in the primary motor cortex are not associated with movement in the MPTP-treated nonhuman primate. Here, eight-electrode recorded the primary motor cortex's spiking activity and accelerometer was attached to the upper limb. Mobility periods are marked below the accelerometer recording. *A, B,* before and after MPTP treatment.
Source: Copied (with permission) from Goldberg et al. 2002.

to action only for goal-directed movements (e.g., complex digit actions). The basal ganglia output to the thalamus and frontal cortex aims to inhibit the corticocortical system 2 network and to supply this network with corollary information about the actions performed by the real workers of the nervous system—the basal ganglia, the cerebellum, and the brain stem motor nuclei and their descending spinal pathways.

LOCAL FIELD POTENTIALS AND THEIR CORRELATION WITH SPIKING ACTIVITY IN THE BASAL GANGLIA OF MPTP-TREATED MONKEYS

Spiking activity represents the output of the recorded neurons and of a structure. The LFP represent the subthreshold activity and are commonly used as a proxy for the synaptic input to the recorded neurons/structure. LFP are usually recorded with 0.1–70 Hz band-pass filters. High-frequency (>70 Hz) oscillations of LFP can be used as reliable biomarkers for clinical symptoms, but they may be contaminated by leaks from spikes and will not be further discussed here. LFP of 0.1–70 Hz may be affected by movement and power supply (50/60 Hz) artifacts, and hard-core electrophysiologists, such as my mentors Moshe Abeles and Mahlon DeLong, have left them out of their recording band of interest. However, today's technology is doing better, and most research groups, both in basic and translational fields, are able and willing to record LFP.[4]

LFP are well understood in the cortex (or in other layered structures, such as the hippocampus) due to the stereotypical parallel structure of pyramidal neurons (figure 2.1A). This makes the cortex a most significant source of LFP-like signals that can be recorded as EEG outside the scalp. Because EEG was discovered first, LFP are called local EEG from time to time. However, going deeper, to less organized structures such as the striatum, the recorded field potentials can be obscured by volume conduction of the strong EEG signal. This can be overcome by differential recording of the LFP. Traditionally, spikes are recorded by microelectrodes, and LFP are recorded by macroelectrodes (e.g., DBS contacts, with a length of 1.5 mm and a diameter of 1.27 mm). However, spiking activity can also be recorded by macroelectrodes, and LFP can be recorded by microelectrodes (figure 15.10). Unluckily, the high and never uniform impedance of the microelectrodes makes impractical the differential recording of LFP, and therefore it is impossible to rule out volume conduction of EEG. Since, in any case, we use the basal ganglia LFP as a proxy for the input to the recorded structure and since the cortex is the major source of information to the basal ganglia, I will use the basal ganglia LFP as a reliable index for the synaptic input to the structure.

Figure 15.11 shows the aggregate of power spectral densities of LFPs (figure 15.11A) and the spike-triggered average (STA) of the LFP (figure 15.11B) in the various structures of the basal ganglia before and after MPTP treatment. Data were collected only while recording well-isolated and stable unit activity to ensure stability of the LFP recording. Figure 15.11A shows that, in line with the previous publications of Ann

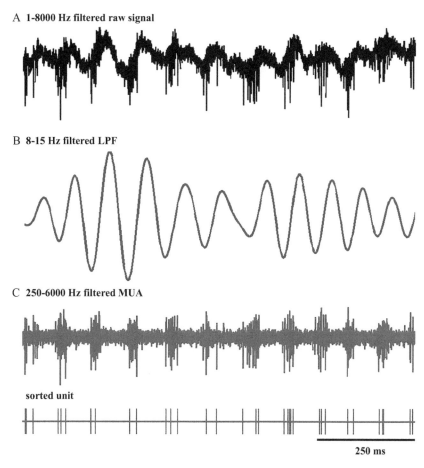

15.10 Subthalamic LFP and spiking activity after MPTP treatment. *A.* broadband microelectrode recording of both LFP and spiking activity. *B* and *C.* LFP, spiking activity and detected units. *Source:* Copied (with permission) from Deffains et al. 2016.

Graybiel, Josh Birke, and others (Courtemanche et al. 2003; Howe et al. 2011; Leventhal et al. 2012), beta (8–20 Hz) oscillations can be observed in the normal basal ganglia. However, they are augmented after MPTP treatment. The LFP recorded in the vicinity of striatal MSNs and TANs have a similar spectral signature. This may indicate that beta oscillations dominate the synaptic input to all populations of striatal neurons. However, the MSNs' discharge is not tied up to this input. This notion is further augmented by the STA of the LFP shown in figure 15.11*B*. In the cortex and in most basal ganglia structures before MPTP (figure 15.11*B, left*), one can discern a major negative LFP deflection around the time of the spike. Negative values of extracellular recorded LFP are correlated with intracellular depolarization, and it is therefore expected that action potentials will be triggered on the negative slope of the LFP wave. This picture drastically changes after MPTP treatment. Strong beta oscillations are shown around the timing of the spikes of STN, GPe, and GPi neurons but not in the vicinity of MSNs.

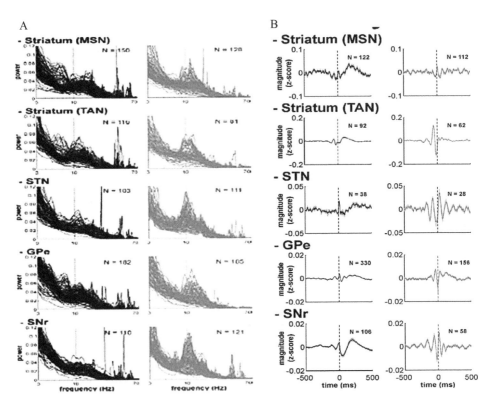

15.11 Augmented LFP oscillatory activity and LFP-spike synchronization in the MPTP-treated nonhuman primate. Population power spectral densities (*A*) and spike-triggered average of LFP (*B*) before (*black*) and after (*green*) MPTP. Striatum MSN, medium spiny neurons; striatum TAN, tonically active neurons; STN, subthalamic nucleus; GPe, globus pallidus external segment; SNr, substantia nigra reticulata. *Source*: Modified (with permission) from Deffains et al. 2016.

BETA OSCILLATIONS IN THE MPTP MONKEY DO NOT GO TO SLEEP

Parkinson's patients have severe sleep disorders. The most common sleep disorder is insomnia—that is, the difficulty of initiating and maintaining sleep overnight. There are many (and as always in biology and medicine, not mutually exclusive) reasons for Parkinson's insomnia, including comorbid diseases and effects of therapy. However, the most parsimonious explanation would be that Parkinson's insomnia is due to striatal dopamine depletion and abnormal activity in the main axis of the basal ganglia.

MPTP-treated monkeys develop severe insomnia, with increased latency for consolidated sleep and severe fragmentation of nighttime sleep (figure 15.12*A, middle hypnogram*). In our studies, the sleep architecture was partially improved by dopamine replacement therapy (figure 15.12*A, bottom hypnogram*). Synchronized beta oscillations were markedly found during the NREM and REM sleep of the MPTP-treated monkeys (see figure 15.12*B* for examples and figure 15.12*C, D* for population averages of the LFP and spiking activity). The increases in beta oscillations were correlated with the transition from sleep to an awake state on a short temporal scale

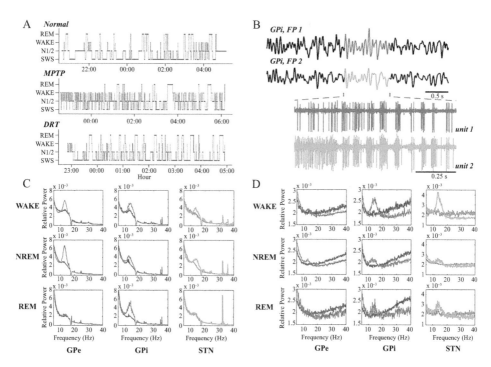

15.12 Fragmentation of sleep and basal ganglia beta oscillations in MPTP-treated nonhuman primates. *A*, hypnograms. *B*, example of LFP and spiking beta episodes. *C*, *D*, population power spectrum densities of LFP (*C*) and spiking (*D*) activity in the basal ganglia during wake, NREM sleep, and REM sleep of MPTP monkeys.
Source: Modified (with permission) from Mizrahi-Kliger et al. 2020.

and with the degree of night insomnia on the hours' timescale (Mizrahi-Kliger et al. 2020). Thus, abnormal synchronized beta oscillations in the main axis of the basal ganglia are correlated with both awake akinetic symptoms and sleep disorders in Parkinson's disease. Correlation does not imply causality. I hope that future experiments with phase-specific, closed-loop DBS methods will provide evidence for causality and, even more importantly, provide improved therapy of sleep disorders in Parkinson's disease.

CHAPTER SUMMARY

- Following MPTP treatment, changes take place in the discharge rate—mainly an increase in the discharge rate of STN and GPi and a decrease of GPe.
- Beta oscillations can be observed in the basal ganglia LFP of normal monkeys, and they are augmented after MPTP.
- Synchronized beta oscillations in the spiking activity of the basal ganglia are the hallmark of Parkinson's pathophysiology in the MPTP nonhuman primate model.
- Oscillatory systems are more prone to synchronization, but synchronization in the dopamine-depleted basal ganglia can be independent of oscillations.

16

PHYSIOLOGY OF PARKINSON'S DISEASE: HUMAN

In 1987 Alim-Louis Benabid (Ben) showed that, as for ablation, high-frequency stimulation of the Vim (the thalamic ventral intermediate nucleus, the target of the cerebellar-thalamic pathway) can be used to ameliorate tremor (Benabid et al. 1987, 2009). This suggested that deep brain stimulation (DBS) at high frequencies (80–4,000 Hz) can replace ablative surgery and provide a reversible and adjustable method for functional inactivation of deep brain targets. Unlike ablation therapy, DBS can be safely used bilaterally.

Following the demonstration of the reversal of the clinical symptoms by inactivation and ablation of the STN in the MPTP-treated nonhuman primate (Bergman et al. 1990; Aziz et al. 1991), Abdelhamid (Hamid) Benazzouz tested whether STN DBS can ameliorate parkinsonian symptoms in MPTP-treated monkeys. Luckily enough, Hamid and colleagues found DBS to be effective in the MPTP monkey (Benazzouz et al. 1993). The shift to the human theater was really fast. In 1993 Benabid and a team of gifted neurologists, including Pierre Pollak, Patricia Limousin, and Paul Krack, successfully initiated STN DBS therapy for patients with advanced Parkinson's disease and severe side effects from dopamine replacement therapy (Pollak et al. 1993; Limousin et al. 1995; Krack et al. 1997).

The STN is the only source of excitation to the GPi in the D1/D2 direct/indirect model of the basal ganglia (figures 9.1 and 9.2). Thus, it was predicted that inactivation of the STN can be replaced with inactivation of the GPi. GPi DBS was also found to be useful (it is still debated which is a better target, STN or GPi; Williams et al. 2014). Since 1993, more than two hundred thousand patients have been successfully treated by STN or GPi DBS. We will discuss the DBS therapy of Parkinson's disease and basal ganglia–related disorders more in chapter 18. In this chapter we will mainly discuss the electrophysiological findings achieved during physiological-guided navigation to DBS targets or from recording from implanted DBS leads.

THETA AND BETA OSCILLATIONS IN THE BASAL GANGLIA OF PARKINSON'S PATIENTS

Figure 16.1 depicts the best typical example of a subthalamic "tremor cell" that oscillates at 5 Hz and strongly correlates with the visible wrist tremor of patients (recorded by a goniometer). While tremor cells can be detected in tremor-dominant Parkinson's patients with overt tremor during DBS procedures, one more often encounters cells with beta oscillations (see figure 16.2 for examples of oscillatory activity in the STN of Parkinson's patients, *left*, and their power spectral density, *right*). The first trace is of a tremor cell (5 Hz oscillations), and the two lower traces of figure 16.2 depict cells that oscillate in the beta range, one at 12 Hz and the second at approximately 28 Hz.

The typical approach to the human STN is double tilted: sixty degrees from the axial plane and twenty degrees from the midsagittal plane. Starting 10–25 mm above the estimated location of the STN lower border and moving through the white matter (axons, electrophysiologically silent) of the internal capsule, one can easily detect the STN dorsolateral (upper) border from the abrupt increase in the total power of

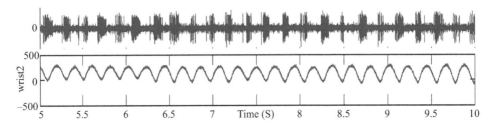

16.1 Human subthalamic tremor cell. Microelectrode recording of spiking activity (*upper trace*); wrist 5 Hz tremor (*lower trace*).

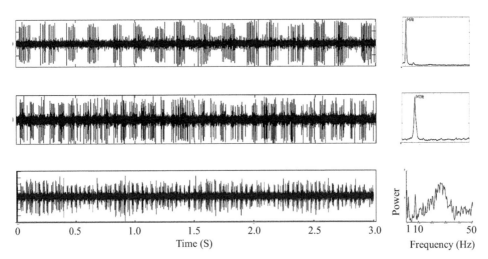

16.2 Different oscillation frequencies in the subthalamic nucleus of Parkinson's patients. *Left rows*, spiking activity; *right*, power spectrum density.

the spiking activity. Similarly, the STN exit can be detected either by transition to the white matter below the STN or directly into the substantia nigra-pars reticulata cells. While taking this anterior-dorsal-lateral to posterior-ventral-medial travel through the STN, it is very clear that there are more oscillatory cells in the first half than in the second half of the journey. Figure 16.3 displays three spectrograms of three patients (*x-axis*, distance in millimeters from the estimated target location, start and end at the entry/exit of the STN, respectively; *y-axis*, frequency in Hertz, logarithmic scale, and color scale for relative intensity of power in this frequency bin and recording location). In each recording location, the power is normalized to total power at that location, so the color scale reflects the fraction of total power per frequency and not the absolute power). Some tremor-frequency activity is observed in the third patient, but a clear-cut strip of beta oscillation is observed in each of them.

These theta (tremor-frequency) and beta oscillations can be used to divide the STN into the dorsolateral oscillatory region (shown to correspond to the subthalamic motor domain) and ventral-medial nonoscillatory region (corresponding to the nonmotor, cognitive, and limbic subdomains). The relative fraction of the two

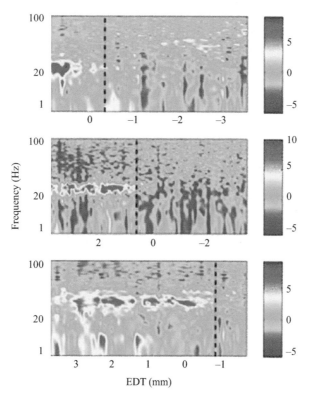

16.3 Example of subthalamic nucleus (STN) spectrograms of three patients. *X-axis*, estimated distance to STN bottom in millimeters according to imaging, starting and ending at the STN borders of each patient. *Y-axis*, logarithmic scale of frequencies. Oscillation power relative intensity is color coded. *Source*: Copied (with permission) from Zaidel et al. 2010.

subthalamic domains is a function of the particular anatomy of each patient, as well as slight modification of the trajectory angles done in order to avoid brain blood vessels and the ventricle. However, if we use the division of the STN into two domains (done by objective machine-learning tools) and normalize the x-axis to the relative length of the two domains (one to zero for the dorsolateral oscillatory region and zero to minus one for the ventral-medial nonoscillatory region) and then do the statistics for more than nine hundred trajectories recorded in our medical center, we will get the results shown in figure 16.4. It is clear the dorsolateral oscillatory domain can be characterized by low-frequency oscillations (that are composed of theta oscillations), as well as by beta 10–30 Hz oscillations. In the ventral-medial portion of the STN, this activity is replaced by gamma (40–100 Hz) oscillations.

Beta oscillations can also be observed in the GPi of Parkinson's patients (figure 16.5; Eisinger et al. 2020; Valsky, Blackwell, et al. 2020). We have seen some beta in the GPi of patients with nongenetic dystonia, but we have not observed beta in the GPi of dystonia 1 protein (DYT1) dystonia patients. Beta oscillations, as well as theta-alpha oscillation at the ventromedial segment, were also observed in the STN of obsessive-compulsive disorder (OCD) patients (Rappel et al. 2018, 2020).

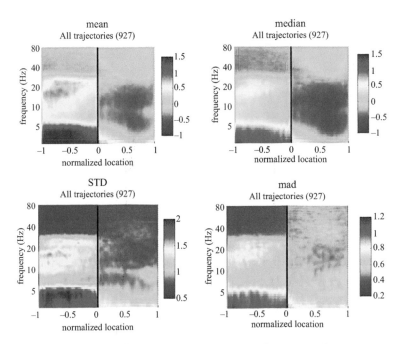

16.4 Population spectrograms of subthalamic (STN) motor and nonmotor divisions. *X-axis*, normalized location in the STN. *Left*, dorsolateral oscillatory (motor) region; *right*, ventromedial nonoscillatory (nonmotor) region. Mean, median, standard deviation (STD), and mean absolute deviation (MAD) of 927 trajectories are shown.
Source: Pnina Rappel, unpublished results.

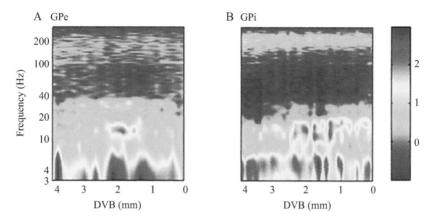

16.5 Beta oscillations in the globus pallidus of Parkinson's patients. GPe and GPi, external and internal segments. *X-axis*, DVB, distance in millimeters from ventral border of the nucleus. *Y-axis*, log (frequency); relative power intensity is color coded.
Source: Modified from Valsky, Blackwell, et al. 2020.

SUBTHALAMIC BETA OSCILLATIONS ARE SYNCHRONIZED

In the previous chapter, we showed that oscillations and synchronization can be dissociated. We therefore ask whether the beta oscillations of patients are synchronized. Physiological navigation to a DBS target is usually done with more than one electrode to provide spatial information on the DBS target. At our medical center, we routinely use two electrodes that are advanced together toward the target. When the two electrodes were located in the dorsolateral oscillatory region (DLOR; figure 16.6*A*), a significant coherence was found in the beta (10–30 Hz) domain between the spiking activities of the two electrodes (2 mm distance). A small and nonsignificant peak is observed at 5 Hz—the tremor frequency. However, when one or two of the electrodes are in the ventral-medial nonoscillatory region (VMNR; figure 16.6*B*, *C*), no coherence peak is observed.

THE FUNCTIONAL ROLES OF BETA AND THETA OSCILLATIONS

The coherence of STN DLOR activity in the beta but not at the tremor (theta) domain (figure 16.6) is in line with our previous results and further supports the notion that basal ganglia beta activity causes Parkinson's akinesia. Our working hypothesis is that the tremor is a compensatory process, probably generated by the cerebellum, and the theta (tremor-frequency) oscillations in parkinsonian patients are the result of feedback from the periphery. Since the tremor of different body parts is not coherent (Ben-Pazi et al. 2001), STN tremor cells probably affected by different muscles/body parts are not expected to be coherent. Having the tremor generator outside of the basal ganglia—for example, in the cerebellum (Bostan et al. 2010; Helmich et al.

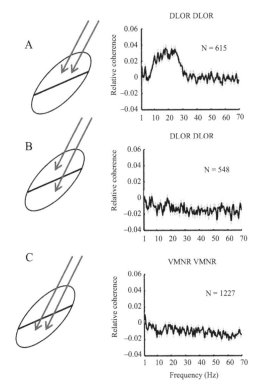

16.6 Synchronized beta oscillations in the dorsolateral oscillatory region of the subthalamic nucleus. *Left*, schematic location of the electrodes; *right*, average coherence function.
Source: Modified (with permission) from Moshel et al. 2013.

2012; Bostan and Strick 2018)—explains why the tremor is neither correlated with the other symptoms of Parkinson's disease (Stochl et al. 2008) nor correlated with the degree of dopamine depletion in the striatum (Pirker 2003).

LFP STUDIES

In many medical centers, a weeklong interval gap occurs between the lead implanta- tion and the connection of the lead to the internal pulse generator (IPG, the "brain pacemaker"). This gap can be used for studies of local field potentials (LFP) recorded from the STN and the GPi through the externalized lead. Usually, a differential (bipo- lar) configuration is used to ensure recording of local activity. In a series of studies, Peter Brown and others have shown that LFP beta oscillations can often be recorded in the STN of Parkinson's patients. The opposite phase between the beta oscilla- tions recorded from the dorsolateral and the ventromedial contacts suggests that the generator(s) of these beta oscillations is (are) located inside the STN. The oscillations are (weakly) correlated with the akinesia and the rigidity symptoms, and the ampli- tude of the oscillations is reduced by movements, dopamine-replacement treatment, and effective DBS. Thus, LFP beta oscillations are emerging as reliable markers for

Parkinson's symptoms, as well as for the contact with maximal therapeutic effects in Parkinson's patients. New technology (e.g., Medtronic ActivaPC+S or Percept PC Neuro-stimulator with BrainSense Technology) is making possible the telemetry of LFP recording from the IPG, even when the stimulation is operating. The future will tell us if this information can be used as a tool for the continuous adjustment of DBS treatment (closed-loop adaptive DBS therapy, chapter 19).

A common belief in the neuroscience community is that macroelectrodes (e.g., the contacts of DBS leads) can record only (local) field potentials. On the other hand, glass/polyamide-coated metal microelectrodes (with an exposed metal tip of 5–20 μm) should be used only for the recording of well-isolated (with high signal-to-noise ratio) action potentials (spikes). Odeya Marmor has recently tested these assumptions during a physiological navigation with two electrodes (with a microtip and one/two macrocontact(s) 3 mm above the microtip) during DBS procedures (Marmor et al. 2017). First, she demonstrated that both spiking and field potentials can be recorded by micro- and macroelectrodes. Spiking activity (band-passed filtered between 300 and 6,000 Hz) recorded by macroelectrode does not contain well-isolated spikes generated by one neuron but reflects the population spiking activity (the physiological equivalent of Olbers's bright night[1]). To record spiking activity, one needs a higher sampling rates than those needed for the recording of field potential. Luckily, this opens a completely new field of correlating field potential to spiking (input to output) activity.

By using mathematical models that compare the synchronized versus independent activity recorded by the two electrodes, Odeya showed that the recording of the field potentials by two macrocontacts is more synchronized—that is, probably more affected by volume conductance of the EEG activity—than the recording of spiking activity by microelectrodes. A simple and commonly used method to overcome the volume-conductance effect is the use of differential recording. If the two electrodes are spatially close, they will be similarly affected by the volume conductance, and therefore differential recording of their activity will leave us only with the local activity. Differential recording can be used not only to overcome the EEG volume-conductance effect but also to overcome stimulation artifacts, such as that generated by DBS. However, differential recording is a valid technique for macrocontact with low impedances. Since microelectrode impedances can range between 0.2 and 1.2 megaohms and the voltage recorded by the microelectrodes is amplified by their impedance (Ohm's law, $V = I * R$), one should be careful with regard to differential recording with microelectrodes.

Simply thinking, we can take the spiking (single or multiunit) activity as a proxy for the output message of a neuron/structure. The field potentials probably reflect the subthreshold (synaptic, input) activity. The power spectrum density of field potentials usually follows the 1/f rule, namely the power declining as the frequency goes up. This is in line with the typical EEG scenario, with high-amplitude delta

(0.1–4 Hz) and alpha (8–12 Hz) rhythms versus a much lower amplitude of beta (13–30 Hz) and gamma (31–70 Hz) rhythms. However, the technology is improving, and we can detect smaller and smaller signals, including high-frequency (150–400 Hz) oscillations in the field potentials. Several studies have shown the efficacy of high-frequency oscillations in discriminating between hypokinetic and hyperkinetic states, as well as between tremor-dominant versus akinetic-rigid Parkinson's patients (Özkur et al. 2011; Telkes et al. 2018). The methods are valid and should be used in closed-loop DBS applications (see chapter 19). However, caution should be taken regarding the biophysical meaning of these high-frequency field potential oscillations. My gut feeling is that they frequently represent leakage of the spiking activity into the field potential frequency domain. Future studies with broadband filter recording (e.g., from 0.1 to 6,000 Hz), high-rate sampling (>20 kHz), and broad-range (>24 bits) analog-to-digital converters (A/D converters, to avoid saturation of the A/D converter by the high-amplitude oscillations at the low-frequency domain) should enable us to understand better the biophysical nature of the different signals we record from the brain.

CHAPTER SUMMARY

- Theta (tremor frequency, 4–7 Hz) and beta (12–30 Hz) oscillations of spiking activity are often found in the posterior dorsolateral (motor) domain of the STN of Parkinson's patients.
- The beta oscillations are synchronized.
- LFP beta oscillations are observed in the basal ganglia of Parkinson's patients, and they are suppressed by dopamine and DBS treatments.

17

PHARMACOTHERAPY OF PARKINSON'S DISEASE

The pharmacological therapy of Parkinson's disease will be featured in this chapter, and therapy for other basal ganglia–related disorders will be left out. This is not only because of my personal biases but also because the pharmacotherapy of Parkinson's disease is a success story. To make a long story short, today we have a good symptomatic therapy (dopamine replacement therapy, DRT) for the first five to ten years after diagnosis. However, no acceptable treatment is available as a disease modification that slows the progression of the disease. No good pharmacological treatment is provided today for patients with advanced disease, on-off fluctuations, and levodopa-induced dyskinesia.

ANTICHOLINERGIC TREATMENT OF PARKINSON'S DISEASE

Some fifty years following Parkinson's essay, around 1870, Jean-Martin Charcot added muscle rigidity to the list of the symptoms of the disease he suggested be called Parkinson's disease.[1] Additionally, he (together with his intern, Leopold Ordenstein) pioneered the anticholinergic treatment of the disease (hyoscyamine, a naturally occurring tropane alkaloid and plant toxin. It is the levorotary isomer of atropine and thus sometimes known as levo-atropine). Different anticholinergic drugs can be used; however, they all have only modest therapeutic effect on Parkinson's symptoms. With the discovery of DRT, the "dopaminergic-cholinergic balance" hypothesis was quite popular (figures 11.2 and 11.6). It used to be believed that anticholinergic therapy is more effective for the tremor symptoms of Parkinson's disease. Therefore, some experts (mainly those who suspect that DRT accelerates the degeneration of midbrain dopaminergic neurons) tend to recommend anticholinergic treatment to young tremor-dominant Parkinson's patients.

DOPAMINE THERAPY FOR PARKINSON'S DISEASE

L-dopa therapy Arvind Carlson stood out from the common thinkers of his time and demonstrated that dopamine acts as a neurotransmitter in the brain and not just as a precursor for adrenaline and noradrenaline. Carlson developed a fluorescent technique to measure dopamine levels and also showed the distribution of dopamine to be highest in the striatum. The critical experiment was carried out in 1957. In a manuscript of one column with no figures, published on November 30, 1957, Carlsson and colleagues showed that reserpine depleted brain dopamine and induced a stage of severe immobility (termed "catalepsy," in line with the psychiatric terminology of a situation characterized by muscular rigidity and fixed posture regardless of external stimuli) and that dopa (3,4-dihydroxyphenylalanine) restored the normal motor vigor of reserpine dopamine-depleted rabbits (Carlsson et al. 1957; figure 14.1).

Oleh Hornykiewicz used human postmortem materials and in 1960 was the first to illustrate that Parkinson's disease patients have a dopamine deficit in the striatum and that the severity of Parkinson's is correlated with striatal dopamine depletion. Hornykiewicz persuaded Walther Birkmayer to inject dopa into patients. They reported success and continued this treatment, usually combining it with the use of a monoamine oxidase inhibitor. However, the response was limited in duration and compromised by nausea and vomiting. Subsequent trials by others failed to find any significant and consistent benefit to this treatment (Lee et al. 2015).

It was time for George Cotzias to enter the stage. Cotzias, a Greek immigrant to the US, had been looking all his life for a path to fame. He had a very special character and probably found it difficult to get along with his colleagues and patients. His biography (Patten 2012)—written by one of his students, Bernard M. Patten—mentions that when one of his patients complained, Cotzias told him to shut up because patients admitted to the Brookhaven National Laboratory Hospital were there to serve as guinea pigs for experiments. How lucky we are to live in a different world. Cotzias, like most American neurologists at that time (e.g., Derek Denny-Brown[2]), did not believe in the dopamine story and thought Parkinson's disease was caused by the depletion of melanin in the substantia nigra. To restore normal melanin in his patients, he tried several precursors of melanin. The sun started to shine when he began using dopa. I have never seen it written, but it could be that Cotzias's approach to patients as "guinea pigs" enabled him to increase the dopa dosages and, despite the nausea and vomiting side effects, attain amazing clinical effects and the amelioration of Parkinson's symptoms. Cotzias published his results in the most prestigious medical journal—the *New England Journal of Medicine*—and won fame (e.g., the Lasker prize) but died within ten years of the most important discovery of L-dopa therapy for Parkinson's disease. Even today, fifty years later, L-dopa therapy is the best, and the gold standard, treatment that we can offer to Parkinson's patients in the first five to ten years of their disease. Oliver Sachs's 1973 *Awakenings* novel (and 1990 movie directed by Penny Marshall, with great actors such as Robert DeNiro and Robin Williams) brought this story to the public's attention, focusing on surviving patients of the Spanish flu with severe post-encephalitic lethargica.

Biochemistry of dopamine, melanin, and related compounds   Dopamine is a member of the catecholamine family of neurotransmitters or neuromodulators. Its biosynthesis starts from the amino acid phenylalanine, which is converted to l-tyrosine. L-tyrosine is converted by the tyrosine hydroxylase (the rate-limiting enzyme in the chain) to L-dopa. Finally, L-dopa is converted by the aromatic amino acid decarboxylase to dopamine. In noradrenergic and adrenergic cells, dopamine is further converted to adrenaline and noradrenaline (figure 17.1).

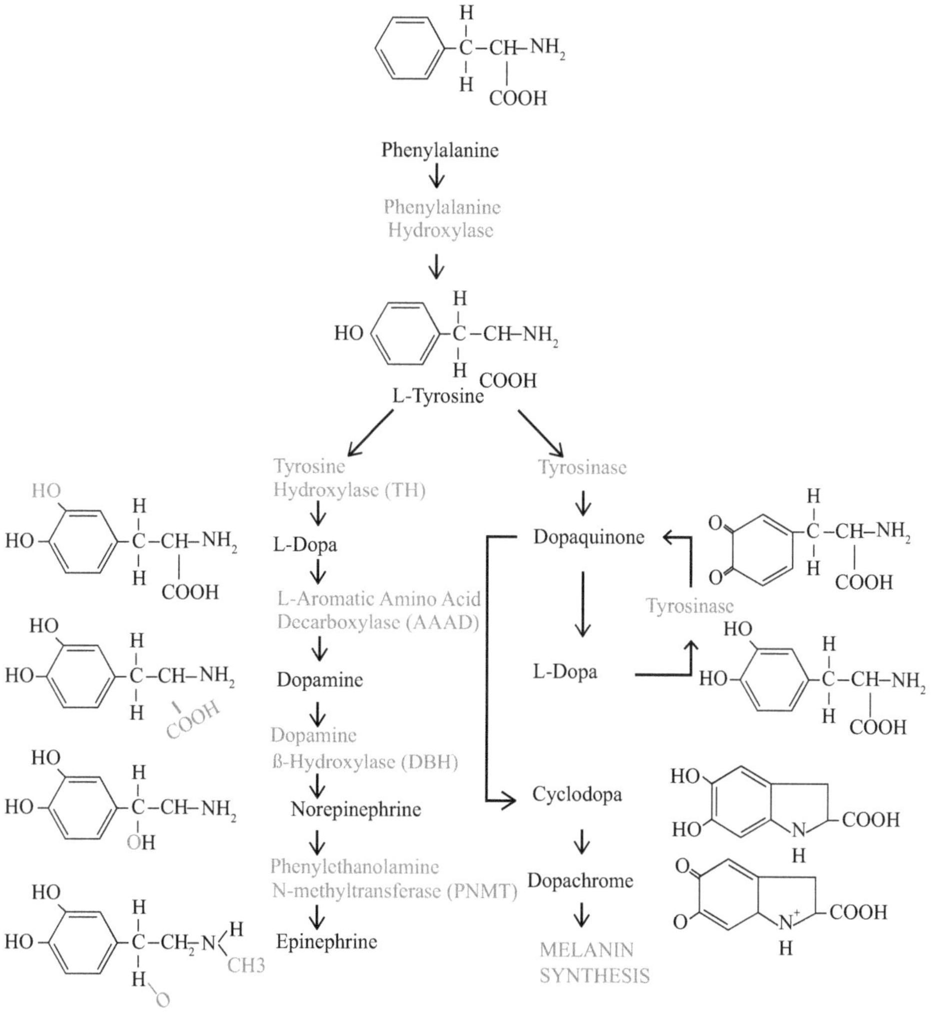

17.1   Chemical structure and biosynthesis of catecholamine and melanin. *Black*, chemical structure and names. *Orange*, change from previous molecule and enzyme names.
Source: Adapted from Végh et al. 2016 and Eisenhofer et al. 2003.

The catecholamines (dopamine, adrenaline, and noradrenaline) are part of an extended family that takes part in crucial neuronal and hormonal control. Tyrosine is also a precursor of tyramine and the thyroid hormones that, simply speaking, are built of two tyrosine and four or three iodine atoms (T4 and T3). T3 is the active form and like the catecholamine leads, on a slower time scale, to increased cardiac and respiratory functions and overactivity. Thus, the thyroid hormone levels of any patient with parkinsonism should be tested.

The catecholamines are part of the monoamine family that also includes histamine, serotonin (5-hydroxytryptamine, 5-HT), and melatonin. Serotonin and histamine are major critics in the basal ganglia network (figure 11.5). Melatonin is released by the pineal gland, under the neuronal control of the suprachiasmatic nucleus of the anterior hypothalamus. Melatonin is part of the sleep/wake control system. Figure 17.2 depicts the chemical structure and biosynthesis of serotonin and melatonin. It is nice to see the symmetry of nature and that similar (chemical) structure can lead to linked functions.

Finally, the substantia nigra ("the black substance") got its name from the melanin pigmentation of the dopaminergic neurons in the pars compacta. Similarly, the neurons of the locus coeruleus ("dark blue spot," which is the principal site for

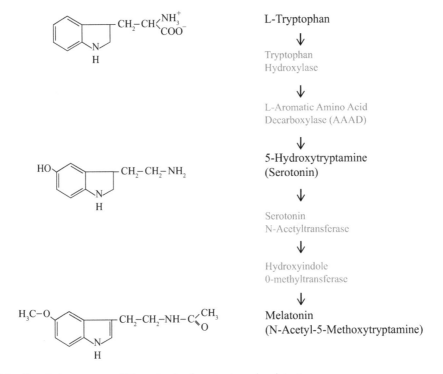

17.2 Chemical structure and biosynthesis of serotonin and melatonin.
Source: Adapted from Bertram G. Katzung, "Histamine, Seratonin, and the Ergot Alkaloids," Doctorlib, https://doctorlib.info/pharmacology/basic-clinical-pharmacology-13/16.html.

brain synthesis of noradrenaline) or the nucleus pigmentosus pontis (the pigmented nucleus of the pons) have many melanin granules. Why are the two major catechol-amine nuclei rich in melanin? Melanin is a group of polymers of tyrosine molecules (figure 17.1). There are three basic types of melanin, two of them found in the skin and hair and one (neuromelanin) in the brain. In human skin, the melanin granules are located around the nucleus. Biosynthesis of melanin is initiated by exposure to ultraviolet (UV) radiation. Melanin is an effective absorber of light and can dissipate most of the absorbed UV radiation, thus protecting the genetic material of the skin cells. Skin/hair melanin can be found in most living creatures. On the other hand, neuromelanin is found in humans and nonhuman primates but is absent in other species.

The biological function of the neuromelanin remains unknown. Neuromelanin has been shown to efficiently bind transition metals such as iron (which accumulates in the substantia nigra neurons), as well as other potentially toxic molecules. It could be that the finding of melanin in the dopamine and noradrenaline cell groups is accidental. These neurons need tyrosine for the synthesis of their neurotransmitter, and part of this tyrosine is shifted to the melanin synthesis machinery (figure 17.1). Another possibility is that the synthesis of dopamine and noradrenaline is more prone to produce toxic molecules, and therefore the melanin is protecting these important neurons. Finally, the metabolic overload of dopaminergic neurons (Pissadaki and Bolam 2013; Hernandez et al. 2019) may lead to the production of toxic molecules and the need for neuromelanin protective mechanisms.

Melanin was suspected of playing a major role in the pathophysiology of Parkinson's disease, but after the discovery of dopamine, melanin was ignored. Recent studies in rats with overexpression of human tyrosinase in the substantia nigra resulted in the age-dependent production of humanlike neuromelanin within nigral dopaminergic neurons, up to levels reached in elderly humans. The intracellular neuromelanin accumulation was associated with Parkinson's symptoms, Lewy body-like formation, and nigrostriatal neurodegeneration (Carballo-Carbajal et al. 2019). Melanin might come back to the multifaceted story of Parkinson's pathophysiology.

Augmentation of dopa treatment   Two important findings have made possible the daily use of dopa therapy with minimization of its side effects. First, both enantiomeric isomers of dopa, L-dopa, and D-dopa contribute to the side effects, but only L-dopa helps the clinical symptoms. So the switch from L-D-dopa to L-dopa has made possible the reduction of the dopa dosage by 50 percent without reducing its efficacy. Second, the dopa side effects of nausea and vomiting are due to dopamine activation of the vomiting center at the area postrema. The area postrema is located at the ventral floor of the fourth ventricle, and it is actually outside of the blood-brain barrier (BBB). Therefore, dopa therapy is enhanced by adding carbidopa, a peripheral inhibitor of the aromatic amino acid decarboxylase, the enzyme that converts L-dopa to dopamine (figure 17.1). Carbidopa does not cross the BBB and therefore inhibits the conversion of L-dopa to dopamine only in the periphery, including the area postrema. The addition of carbidopa, or other peripheral inhibitors of aromatic amino acid decarboxylase, significantly reduces the side effects of L-dopa therapy while increasing the therapeutic effect by enabling more L-dopa to cross the BBB. There are several peripheral inhibitors of aromatic amino acid decarboxylase on the market, but current Parkinson's therapy is usually based on a mixture of L-dopa and carbidopa.

Dopamine released to the brain synapses is metabolized by monoamine oxidase B (MAO-B) and the catechol-O-methyl transferase (COMT) enzymes to homovanillic acid, which is transferred to the cerebrospinal fluids, plasma, and urine. Inhibition of MAO-B and COMT leads to slow degradation of the synaptic dopamine and an increase in the amount of functional dopamine in the synapse. MAO-B and COMT inhibitors are therefore used to augment the therapeutic effect of L-dopa/carbidopa therapy.

Postsynaptic dopamine agonists   The postsynaptic dopamine agonists are the modern descendants of the ergot alkaloids. Ergot is a fungus that grows on rye and related plants. The ergot kernel contains high concentrations of ergot alkaloids. The ingestion of ergot alkaloids leads to a broad spectrum of circulatory and neurological disorders.[3]

Dopamine agonists act directly on the dopamine receptors (mainly D2 receptors) of the striatum in Parkinson's patients. Dopamine agonists are also used to treat restless legs syndrome and hyperprolactinemia (abnormally high levels of prolactin in the blood cause galactorrhea—the overproduction and spontaneous flow of breast milk and other clinical symptoms). Dopamine agonists have two subclasses: ergoline (ergot)-derived agonists (e.g., bromocriptine, cabergoline, pergolide, and lisuride) and nonergoline (e.g., pramipexole, ropinirole, rotigotine, piribedil, and apomorphine). Ergoline-like agonists are much less used nowadays because of their cardiac risk (fibrosis of the cardiac valves). Different dopamine agonists have different pharmacological profiles. Some of the agonists are said to have differential effects on the D1/D2 dopamine family of receptors, as well as on specific members of the family. My personal feeling is that the differences between the agonists matter much more to the pharma companies than to the patients.

Dopamine agonists also have different temporal profiles. Apomorphine is exceptional. Historically, it is made by boiling morphine with concentrated acid, hence the "-morphine" suffix. However, apomorphine neither contains morphine or its skeleton nor binds to opioid receptors. The "apo-" prefix relates to it being a morphine derivative ("[comes] from morphine"). Apomorphine can only be given via an IM or IV route and has ultrafast effects (figure 10.8B). Most other agonists are taken orally and have a very long (six to thirty-six hours) duration of effect. In the past there was a trend, probably driven by the finding of dopamine autotoxicity (and maybe also by the pharma companies) to start dopamine treatment of young patients with dopamine agonists rather than with L-dopa. The claim was that dopamine agonists have no toxic effects on the dopaminergic neurons, and therefore their use will delay the onset of levodopa-induced dyskinesia. The finding that toxic effects of dopamine are seen only at concentration levels never achieved in the human brain has led to the suggestion that agonists are better because of their long duration of effects and minimization of the fluctuation of dopamine level in the striatum. However, agonists produce other side effects, like hallucinations and dopamine dysregulation disorders (hypersexuality, gambling, and so on). The time from diagnosis or first motor symptoms to side effects from dopamine therapy depends more on the natural history of the disease and not on the method of treatment. Therefore, many Parkinson's disease experts today will start treatment with L-dopa and will add agonists only in the late stages of the disease.

L-dopa is taken from the extracellular medium by the remaining dopamine neurons (but also by other neural elements), so the release of dopamine to the striatum of a patient treated with L-dopa is partially controlled by physiological needs. Postsynaptic dopamine agonists continuously activate the dopamine postsynaptic receptors, and their effect cannot be modulated by the physiological needs of the patient. A fascinating question is: How can the restoration of background tone of dopamine activation by a postsynaptic agonist help the parkinsonian patient? The levels of dopamine in the striatum may exist in three states (figure 10.3). The tonic background level of dopamine encodes the "default" state in which reality is no different than the predictions. The dopamine level will rise in situations better than predicted and may decrease below the background level when reality is worse than prediction. Parkinson's disease is a chronic state of dopamine depletion in the striatum, leading to a continuous message of a reality worse than the prediction. Additionally, the lack of dopamine direct effects on excitability reduces the motor vigor and leads to Parkinson's akinesia. Restoration of the background activity of dopamine in the striatum by postsynaptic agonists restores the situation of reality equaling prediction and the normal motor vigor. Motor learning that is driven by the elevation of dopamine levels is probably not improved by dopamine agonists; however, most Parkinson's patients can live well with impaired motor learning.

Transient and continuous dopamine therapy options   In the first five years of their disease, most Parkinson's patients take their L-dopa therapy a few times a day. Although L-dopa is supposed to be buffered in the dopaminergic terminals and released only on demand, this buffering action is probably limited and declines with disease progression. The decline of the buffering action leads to leakage (in a free form and not stored in synaptic vesicles) of dopamine. Additionally, the similarity of the transporters and the biochemistry machinery of serotonin and dopamine leads to an uptake of L-dopa by serotonin terminals. The L-dopa is converted by the amino acid decarboxylase enzyme to dopamine that acts as a "false transmitter" in the serotoninergic terminals. Namely, dopamine is displacing serotonin from the terminal and is released by the action potential of the serotonergic neurons. Thus, the amount of dopamine in the striatum peaks following the ingestion of L-dopa (usually within thirty to sixty minutes, but since this depends on gastric emptying, which can be slow in Parkinson's patients, and on gut amino acid transporters, L-dopa's effects could be delayed if it is taken with an amino acid–containing food such as meat) and then subsides until the ingestion of the next pill. These fluctuations of dopamine level probably lead to abnormal sensitivity of the dopaminergic receptors. From the computational physiology point of view, the amount of dopamine in the striatum encodes the match or the signed mismatch between prediction and reality. If the amount of dopamine does not correctly encode the match between prediction and reality (e.g., because the reader of this chapter took her L-dopa an hour ago and her striatal dopamine levels are peaking now, despite being very bored by the ongoing discussion), behavioral policy will be randomly updated. Thus, after five to ten years of fluctuating dopamine treatment, most Parkinson's patients develop on-off fluctuations (alternate phases of good and bad clinical response to a single dose of dopamine therapy) and levodopa-induced dyskinesia, not only because of changes in the sensitivity of the dopamine receptors but also due to aberrant learning of motor policy.

The first attempt to overcome the fluctuating nature of dopamine replacement therapy was made with long-acting postsynaptic agonists. Then, slow-acting formulas of L-dopa/carbidopa (brand name Sinemet CR) were developed. Invasive therapies, such as duodopa (an L-dopa/Carbidopa gel directly injected into the duodenum) and apomorphine subcutaneous infusion pump, were recently introduced. Both therapies bypass the gastric emptying barrier and can provide a stable twenty-four seven level of L-dopa and dopamine levels in the brain. However, they are not easy to maintain overnight and complications may arise.

AMANTADINE THERAPY FOR PARKINSON'S DISEASE

Amantadine was developed as an anti-influenza virus therapy. It was serendipitously found to reduce Parkinson's disease motor symptoms when given prophylactically in a nursing home. It is generally accepted today that amantadine is most useful in minimizing levodopa-induced dyskinesia. The beneficial effects on the main motor symptoms of Parkinson's are modest, and therefore amantadine is not effective as a stand-alone therapy. It is usually given as a combination therapy with L-dopa. Side effects of amantadine treatment include hallucinations, dizziness; and orthostatic low blood pressure.

Amantadine affects many biological systems (a "dirty drug"). It mainly acts as a nicotinic antagonist and a noncompetitive N-methyl-D-aspartate-type (NMDA) antagonist. The classical anticholinergic treatment for Parkinson's disease is antimuscarinic. Nevertheless, nicotinic acetylcholine receptors are heavily distributed in the striatum and are responsible for some of the interactions between the cholinergic and the dopaminergic system. Nicotinic receptors are probably activated by cigarette smoking, which has a negative correlation with the prevalence and the severity of Parkinson's disease.

Noncompetitive NMDA antagonists, like ketamine, are getting to be more popular these days for the treatment of major depression and other mental disorders. The box-and-arrow models of the basal ganglia and the dominant role of the STN in controlling basal ganglia downstream structures emphasize the increased STN discharge rate and glutamate release in the pathophysiology of Parkinson's disease. NMDA antagonists, like STN inactivation, may lead to reduced glutamate excitation of the GPi/SNr and the amelioration of Parkinson's symptoms. If glutamate toxicity in the midbrain (e.g., due to increased activity in the subthalamic projections to midbrain dopaminergic neurons) contributes to the progression of Parkinson's disease, NMDA antagonists might lead to a slowing of the progression of the disease.

THE BUILT-IN PROBLEMS OF PHARMACOTHERAPY FOR PARKINSON'S DISEASE SYMPTOMS

On the one hand, dopamine therapy for Parkinson's disease is a "Cinderella story." It took 150 years of research, but at the end we now have a very good and efficient treatment. There is nothing with this level of efficacy in our current tools for the treatment of other brain disorders. On the other hand, it is far from perfect. After 5–10 years of a honeymoon of dopamine therapy, most patients develop on-off fluctuations and levodopa-induced dyskinesia. L-dopa treatment is less effective for the axial symptoms (speech, gait) of the disease, but the worst part is that it does not slow disease progression, so most patients will eventually develop Parkinson's disease dementia (PDD).

I believe pharmacotherapy can never reach the holy grail of completely treating all symptoms of any brain disorder. This is because of the opportunistic nature of evolution. Evolution does not have to care too much about systemic pharmacological effects and can use anatomical distances when employing the same transmitter-receptor pairs for multiple functions. Figure 17.3 depicts a basal ganglia–centric scheme of the state-to-action loop of the brain, with the main neuromodulators (critics) of the thalamus, cortex, basal ganglia, and spinal cord. Dopamine is used as a neuromodulator in the cortex (by the VTA mesocortical pathway), in the basal ganglia (SNc nigrostriatal pathway), and in the spinal cord (descending dopaminergic axons from the A11 dopaminergic group). Acetylcholine is used by the pedunculopontine nucleus for innervation of the intralaminar thalamic nuclei, of the whole cortex by the massive arborization of the cholinergic neurons of the nucleus basalis of Meynert, and of the striatum by the axonal arborization of striatal cholinergic interneurons. Different groups of serotonin neurons in the dorsal raphe innervate

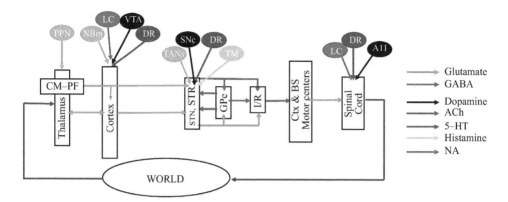

17.3 A scheme of the brain state-to-action loop with the main neuromodulator groups of the thalamus, cortex, basal ganglia, and spinal cord. The structures of the main axis are CM/PF, centromedian
and parafascicular thalamic nuclei; STN, subthalamic nucleus; STR, striatum; GPe, external segment
of the globus pallidus; I/R, internal segment of the globus pallidus and substantia nigra reticulata;
BS, brain stem; Ctx, cortex. The critics are PPN, pedunculopontine nucleus; NBm, nucleus basalis of
Meynert; LC, locus coeruleus; VTA, ventral tegmental area; DR, dorsal raphe; TANs, striatal tonically
active neurons; SNc, substantia nigra compacta; TM, tuberomammillary nucleus; A11, posterior peri-
ventricular and intermediate hypothalamic nuclei. *Red arrows*, excitatory main axis connections; *green
arrows*, inhibitory main axis connections. *Black, gray, orange, yellow,* and *blue arrows*: dopamine,
acetylcholine (ACh), serotonin (5-HT), histamine, and noradrenaline (NA).

the cortex, striatum, and spinal cord. This is of course a very partial list, but it demonstrates that evolution can dissociate function and pharmacology.

Pharmacologists can adjust therapy according to specific receptors and transporters
that might differ at various anatomical locations. This might explain the use of Ritalin
(methylphenidate, which, like amphetamine or alpha-methylphenethylamine, is a
dopamine transporter blocker that elevates the dopamine level in the synapse) and
bromocriptine (dopamine agonist) for attention deficit hyperactivity disorder and
restless legs syndrome, respectively. These hyperkinetic disorders might be caused by
the abnormal (low) activity of dopamine in the cortex or in the spinal cord. However, cross-reactivity is always expected, and therefore the pharmacological therapy
of brain neuromodulators will eventually result in side effects. In the next chapter,
we will discuss the deep brain stimulation therapy option. It is invasive, but it is very
well localized and therefore might overcome the problems of pharmacotherapy.

ALPHA-SYNUCLEIN THERAPY FOR PARKINSON'S DISEASE

Prevention is better than a cure. If abnormal misfolding and the aggregation of
alpha-synuclein are the major processes leading to the death of dopaminergic and
other brain cells in Parkinson's disease, alpha-synuclein immune therapy should
prevent or slow the progression of the disease. Unfortunately, and as in the case of
beta-amyloid in Alzheimer's disease, all clinical trials with alpha-synuclein immune

therapy have failed. It is not clear if this is due to a late start of the treatment, to the fact that alpha-synuclein is correlated with but is not the cause of Parkinson's disease, or for other reasons. The search for a treatment that will prevent or slow down the progression of the disease is just beginning.

LIFESTYLE THERAPY FOR PARKINSON'S DISEASE

Prevention is especially better than a cure when achieved without pharmacotherapy. Three lifestyle components are negatively correlated with the development and the severity of Parkinson's disease. These are smoking, drinking coffee, and exercising.

There is a negative correlation between smoking and Parkinson's disease. However, correlation does not imply causality. It could be that people with more dopamine in their brain are less aversive to risk and more prone to the unhealthy habit of smoking. On the other hand, the activation of nicotinic receptors affects dopamine transmission. Schizophrenic patients under neuroleptic (antidopaminergic) treatment are often compulsive smokers. Could it be that a reduction in dopamine (because of neuroleptic treatment or prodromal Parkinson's) encourages self-therapy with smoking? The jury is still out, but even if smoking alleviates Parkinson's symptoms, it clearly has a terrible effect on aerobic functions and the risk of lung diseases and cancer. We have better therapy for Parkinson's disease than for chronic lung disease; I therefore do not recommend smoking as a preventive treatment for Parkinson's disease.

There is also a negative correlation between coffee drinking and Parkinson's disease. This is good since I am not aware of any adverse effects of drinking coffee. Coffee is a natural antagonist of the adenosine A_{2A} receptor. Adenosine is one of four nucleoside building blocks to RNA, and its derivatives include the energy carrier adenosine triphosphate (ATP). The body's continuous use of ATP builds up a concentration of adenosine over the day. Adenosine is one of the major substances whose accumulation leads to fatigue and a desire to sleep. Adenosine levels decrease during the course of a night's sleep.[4]

Adenosine A_{2A} receptors are highly enriched in the striatum. They are predominantly expressed in D2 MSNs. The A_{2A} receptor interacts structurally and functionally with the dopamine D2 receptor. Eventually, A_{2A} and D2 receptors exhibit reciprocal antagonistic interactions. Dopamine reduces the excitability of D2 MSNs and through the indirect pathway activates the motor thalamocortical networks and motor vigor. Adenosine's antagonistic effect on D2 MSNs would lead to a reduction in motor activation, fatigue, and sleepiness. Most of us use coffee to overcome the accumulation of adenosine and to enhance our arousal level over the day. Thus, coffee is a natural medicine against the reduction of motor vigor (akinesia) associated with dopamine depletion in the striatum.

Exercise is also negatively correlated with Parkinson's disease. This is not surprising. An active lifestyle has great value for almost every human disease, including cancer, diabetes, and mental disorders. Aerobic exercise is not enough. For good health, one should add resistance (power) and flexibility exercises. Finally, I advise Parkinson's disease patients (and all of us) to add movement skills activity (e.g., dancing, martial arts). If the basal ganglia (Kahneman's system 1) are there to enable implicit (motor) learning, we will probably do better if we urge our basal ganglia to dance with us.

CHAPTER SUMMARY

- Anticholinergic therapy has a modest effect on Parkinson's disease, with significant side effects.
- Dopamine replacement therapy by L-dopa/carbidopa is the gold standard treatment for Parkinson's disease.
- L-dopa treatment can be augmented by inhibition of the enzymes that metabolize dopamine.
- Postsynaptic dopamine agonists have a longer duration effect than L-dopa.
- Amantadine (an NMDA antagonist) reduces levodopa-induced dyskinesia.
- Pharmacotherapy of brain disorders is limited because the brain is using the same transmitter-receptor pair in diverse areas with different functions.
- Alpha-synuclein immune therapy is still under investigation.
- Lifestyle (coffee drinking, exercise) therapy is highly recommended.

18

DEEP BRAIN STIMULATION THERAPY FOR PARKINSON'S DISEASE

Since 1993, more than two hundred thousand patients have been treated by deep brain stimulation (DBS) at more than one thousand centers worldwide.[1] Most of these patients underwent surgery because of advanced Parkinson's disease. A small fraction were treated for dystonia and for essential tremor. A few patients received DBS for mental disorders such as depression and obsessive-compulsive disorder (OCD). Ongoing research programs are currently experimenting with DBS for almost any brain disorder, from Alzheimer's disease to anorexia nervosa and cluster headaches.

DBS TECHNOLOGY

Anatomical targets of DBS vary depending not only on the disease treated but also on the experience and preferences of the treatment centers. There are two targets for Parkinson's disease—the STN and the GPi. Most centers prefer the STN. For dystonia, the GPi is the preferred target. For essential tremor, the Vim of the thalamus or another location along the cerebellar-rubral-thalamic tract (e.g., the posterior subthalamic area, PSA) is preferred. Other targets (e.g., subgenual cingulate gyrus, anterior limb of the internal capsule, medial forebrain bundle, fornix, and more) have been suggested and are still under trials for different diseases. As my friend Andres Lozano says, "There is no brain disease or brain target that is safe from DBS."

The insertion of the DBS leads into the target is performed by a neurosurgeon. Usually, high-quality T1 and T2 MRI images are taken before surgery. The morning of the surgery, a metal stereotactic frame is attached to the patient's head and CT is performed. Computer programs enable the registration of the three imaging sets as well as the identification of the target and the optimal path in the stereotactic frame's coordinates. After the target is pinpointed, the X, Y, and Z coordinates and the optimal (straight) trajectory are selected. Typical angles (to the STN) are sixty degrees from the axial plane and twenty degrees from the midsagittal plane. The path is further adjusted to start anterior to the motor cortices, from a gyrus (to avoid blood vessels in the sulci), to avoid any major blood vessel (seen in T1 images done with contrast media) and to avoid the ventricle (and possible mechanical electrode shift and bleeding from the ependyma and the choroid plexus). Some centers do the procedure based on imaging and under general anesthesia. Other centers prefer to verify the lead positioning by electrophysiological recording and testing for the therapeutic window of stimulation at the target. This is usually done with the patient awake or under minimal or moderate sedation.

DBS leads today usually have four contacts. The lead diameter is 1.27 mm, and each contact length is 1.5 mm. The contacts are separated by 0.5 or 1.5 mm. Segmented leads have similar dimensions, but two contacts are divided into three segments, yielding eight contacts/lead. Typical impedance values are 2–5 kilohms at 30 Hz. The leads are connected to an internal pulse generator (IPG, "brain pacemaker") that delivers continuous high-frequency stimulation to the brain. The stimulation parameters can be programmed according to the patient's needs. Current stimulation parameters include

- Geometry.
  - Monopolar cathodal (negative, attract cations) stimulation through one of the contacts. The anode (positive, attract anions) is located at the IPG case.
  - Bipolar or multipolar. One (or more) of the contacts is defined as a cathode, and another one (or more) is defined as an anode. Bipolar configuration enables more focused stimulation.
- Voltage or current stimulation mode and amplitude. The IPG stimulator can be set as a voltage or current source. Typical values are 0.25 to 6 V and 0.25 to 6 mA. The almost equal range of values would be expected for contact with impedance of 1 $k\Omega$ (kilohm); however, for contacts of more than 1 $k\Omega$, one would expect a lower range of current stimulation. The impedance is measured at 30 Hz, while the typical pulse width of DBS is 60–90 μS. Thus, the therapeutic (in the actual setting of the stimulation) impedance is closer to 1 $k\Omega$.
- Frequency. The typical value is 130 Hz; however, settings between 80 and 180 Hz are commonly found.
- Pulse width. The typical value is 60 μS. Higher pulse widths of 90 and up to 180 μS can be applied. Some of the new devices enable a reduction of the pulse width to 30 μS. The short cathodal pulse is followed by a long (3 ms) and shallow anodal pulse to achieve an electrical balance of the delivered stimulation and to prevent the deposition of metal lead particles in the brain.

Today, the programming of the stimulation parameters is an art and is usually performed by movement disorders experts. DBS programming requires multiple visits to the neurology clinics. Patients typically come to the clinic for follow-up every one to six months, according to their needs. Thus, the main advantages of DBS over ablation therapy are that it is adjustable and reversible.[2]

DBS CLINICAL AND ADVERSE EFFECTS

The clinical and adverse effects of DBS are mainly the result of the chosen location in the brain. If the electrode is well located in the STN, one should see a marked reduction in rigidity and bradykinesia with low-amplitude (0.25–1 mA) stimulation. Tremor arrest is also seen but may need a more prolonged stimulation duration to be observed. Upon increasing the stimulation amplitude, one may encounter side effects. Stimulation-induced dyskinesia (mostly affecting the leg in the operating room) can be observed, but this indicates good positioning of the lead and does not lead to problems during the programming sessions. If the lead is too lateral in the STN, it will be close to the internal capsule, and this may result in dysarthria and muscle contractions at the contralateral corner of the mouth. If the lead is too posterior, we will be too near the medial lemniscus, which conveys sensory information to the ventral caudal (Vc) nucleus of the thalamus, and the patient will report temporary paresthesia (tingling). Because the paresthesia is transient, we do not consider it an adverse side effect. Finally, if the lead is too medial, the patient may experience gaze palsy due to the stimulation of the fiber of the oculomotor (III) nerve or of the medial longitudinal fasciculus (MLF). In the GPi, we will get internal capsule effects if we are too medial and phosphenes (flashes of light) caused by current spread into the optic tract that is located ventral of the GPi.

Typically, a well-located STN DBS makes possible a reduction of the dopamine treatment to 30–50 percent of the preoperative dose. A well-located GPi DBS lead will eliminate levodopa-induced dyskinesia; however, to ameliorate Parkinson's symptoms, a similar dose of dopamine therapy should be continued.

MECHANISM OF DBS

The mechanism of DBS is still debated. The similarity between DBS effects and the effect of inactivation or ablation therapy has been the main reason for the suggestion that DBS mimics ablation therapy. Twenty-five years of extensive research in patients, animal models, in vitro (brain slices), and in silico (computer modeling) have not resulted in a consensus on this matter (Marsden and Obeso 1994).

I take the view that the main effect of DBS is the functional inactivation (information jam) of the neurons in the stimulated area. The initial hypothesis was that this is created by depolarization block. The membranes of neurons are electrically polarized. The membrane resting potential is 50 to 80 mv, with the inside more negative than the outside. The membrane can be depolarized by excitatory synaptic inputs, by changes in the ionic composition of intracellular and/or extracellular fluids, or by current delivered to the cell. When the excitable membrane is depolarized by 10–30 mv, it reaches the action-potential threshold level. Threshold crossing initiates the positive feedback all-or-none short (~1 ms in neurons) event of the action potential. Continuous depolarization of the excitable membrane can be achieved by pharmacological depolarizing agonists, an increase in extracellular potassium levels, and current injections. Such continuous depolarization will prevent the generation of future action potentials because of the inactivation of sodium channels (depolarization block).

Axons, especially those of small diameter, are more sensitive than cell bodies to artificially delivered currents (Ranck 1975). Thus, electrical stimulation at the DBS target can lead to 1) orthodromic (action potentials running in the direction from soma to the axon terminals) activation of the afferent axons to the stimulated structure, 2) orthodromic activation of the efferent axons of the neurons of the stimulated structure, and 3) antidromic activation of the afferent neurons to the structure leading to activation of the neurons in the areas that innervate the structure (e.g., primary motor cortex in the case of STN stimulation). Most studies reveal an increase in the discharge rate and complex locking of the stimulation pulses following high-frequency stimulation in the basal ganglia (figure 18.1; see also the pioneering studies of Jerry Vitek; Hashimoto et al. 2003). This leads to the "information jam" theory of the DBS mechanism. The locking of the discharge of basal ganglia neurons to DBS high-frequency stimulation replaces and masks the abnormal parkinsonian activity of these structures.

However, while axons can maintain a high-frequency discharge for a prolonged interval of time (one of the criteria for successful antidromic stimulation of a neuron is frequency following; Lemon 1984), this is not the case for most brain synapses. Most synapses would be depleted of their neurotransmitter molecules and vesicles following a few minutes of high-frequency (>80 Hz) stimulation. Thus, high-frequency stimulation probably mimics ablation by functional disconnection of the stimulated area. Both afferent and efferent axons are affected.

DBS does not have a single mechanism. While information jamming and synaptic failure are in line with the functional inactivation effects, there are other impacts of DBS. Plastic effects are important. DBS effects may be delayed and observed weeks to months after the beginning of DBS therapy (e.g., for dystonia, mental disorders). Activation of astrocytes and the reorganization of neural networks have been shown to occur after DBS. This led to the suggestion by Keyoumars Ashkan (Ash) to change the nomenclature from "deep brain stimulation" to "deep brain modulation" (Ashkan et al. 2017). I am very supportive of this suggestion since "stimulation" is misleading, especially if the major mechanism of DBS is to mimic inactivation, either owing to information jamming or to synaptic failure.

A DBS hard question is how could inactivation of a system lead to therapy? The box-and-arrow models may suggest that STN or GPi inactivation restores the normal

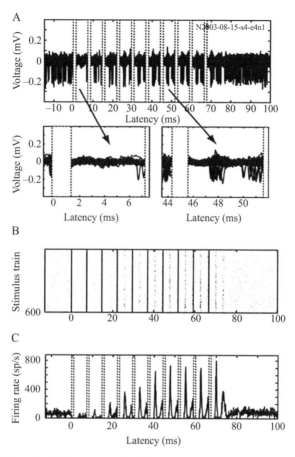

18.1 Complex locking of pallidal activity to high-frequency stimulation. *A,* one-trial row data. *B, C,* raster display and peristimulus histogram.
Source: Copied (with permission) from Bar-Gad et al. 2004.

discharge rate of the motor cortices and therefore ameliorates Parkinson's akinesia. However, we know today that these models are not accurate enough and cannot explain the effects of DBS on other Parkinson's symptoms like rigidity and tremor. If indeed the basal ganglia operate as one of many state-to-action loops and act as the default brain system (Kahneman's system 1; figure 13.2), then other brain systems can't compensate for abnormal activity (rate, pattern, and synchronization) of the basal ganglia. Inactivation of the basal ganglia enables other systems (e.g., the corticocortical network) to compensate and provide a better link between state and action.

OPTIMIZATION OF DBS THERAPY FOR PARKINSON'S DISEASE

DBS is a success story. Unlike ablation therapy it is revisable and adjustable, and the cost-benefit ratio for the patients, their caregivers, and society is positive. Below and in the next chapter, I will describe how DBS therapy can be improved. I hope that the issues discussed in these two chapters will bolster the need for a strong relationship between good medicine, academic research, and industry.

BETTER DBS NAVIGATION

For many years, two approaches have been taken to DBS surgery. The first is based on intraoperative microelectrode recording (MER) of the brain's electrical activity to verify accurate lead positioning in the DBS target. The other approach is image/anatomy based. Usually, both approaches use intraoperative stimulation of the target to verify a good therapeutic window. But the hard-core supporters of the imaging/anatomy approach would skip even the stimulation phase and perform the full DBS procedure under deep general anesthesia and muscle paralysis.

There are many advantages and disadvantages to each approach. The disadvantages of the physiological navigation approach include:

- The need for the patient to be awake during surgery. DBS is not an easy procedure. The patient is in an unfamiliar operating room with a metal stereotactic frame holding his head immobile. The patient can hear and feel the drilling into the skull and so on. This clearly increases stress and anxiety, which may lead to an inferior outcome.
- The classical physiological setting of Alim-Louis Benabid is five sharp microelectrodes spaced two millimeters apart in the "BenGun." Many authors argue that there is no more bleeding in cases performed with microelectrode-assisted navigation in comparison with cases performed without microelectrode recordings. Theoretically, however, the insertion of microelectrodes into the brain may lead to an increased risk of bleeding.
- Microelectrode recording may significantly increase surgery time. Traditionally, the physiologist examines (looks and hears) the activity of each electrode in each station on the way to the target. Early physiological exploration for DBS targets lasted several hours. The longer the surgery, the greater the risk of a poor outcome, either because of deep vein thrombosis or postoperative delirium.
- Dopamine therapy significantly reduces the physiological characteristics of DBS targets for Parkinson's disease (increased activity and beta oscillations). Therefore, most centers that perform physiological navigation and stimulation testing of the therapeutic window stop dopamine therapy twelve to seventy-two hours before surgery. Parkinson's patients are less comfortable without their dopamine therapy, and the risk for postoperative delirium increases.
- Finally, it is not easy to find a good physiologist for the DBS procedure. This person should know the physiology of the basal ganglia, but there are too few basal ganglia research groups. The DBS physiologist should have chutzpah and

be able to tell the big boss and the person in charge of the operating room—the neurosurgeon—that he or she missed the target. Unfortunately, some physiologists lack chutzpah, and after all efforts, they might declare that the "STN was found" even though they are unsure of their findings and should declare "STN yok."[3]

On the other hand, proponents of physiological navigation claim that with today's (imaging and neurosurgical) limited technology, it is too risky to skip intraoperative physiology. The STN is a remote (80–90 mm from the skull entry point) and small target. The dimensions of the human STN are about 9 mm in the long axis (on the axial plane from anterior-medial to posterior-lateral) and 6–7 mm on the largest span in the coronal and sagittal axes, with a total volume of 20 mm^3. Today, the practical resolution of a 3T clinical MRI is 1 * 0.4 * 0.4 mm. Moreover, for DBS navigation we fuse MRI T1, T2, and CT images, and the fusion might not be perfect. MRI images might look very beautiful and still be distorted (i.e., like a nice map that is not a reliable description of the real geography). Current (anatomical) MRI technology does not discriminate between the STN and the adjacent substantia nigra and cannot discriminate the STN functional domains (Wang et al. 2020). More advanced MRI techniques, like diffusion tensor imaging (DTI), are less reliable and are not routinely performed. Finally, if we keep our patient awake for assessment of the stimulation therapeutic window, we can use physiological navigation to augment the results of the examination of the stimulation effects. With today's broad distribution of DBS centers (30–50 percent of them are probably low-volume centers handling fewer than ten cases per year), about 10 percent of DBS procedures may be out of target. A recent study of two big American databases revealed a significant number of revision procedures that were carried out due to improper targeting or lack of therapeutic effect (Rolston et al. 2016). For elective surgery we should do better and achieve higher accuracy.

To improve the method of physiological navigation and outcome of DBS procedures, several generations of our research group (Anan Moran, Adam Zaidel, and Dan Valsky), together with the gifted engineers of Alpha-Omega, Nazareth, Israel, have developed computerized methods for automatic detection of DBS targets (STN and GPi) and their subdomains (Moran et al. 2006; Zaidel et al. 2010; Valsky, Blackwell, et al. 2020). These methods enable the physiological exploration of one hemisphere in less than fifteen minutes. The computer is not afraid of the neurosurgeon, so the program can declare "STN yok" or suggest the best trajectory and Z coordinates for the DBS lead (Thompson et al. 2018). Today, we develop automatic methods to assess the therapeutic window of the stimulation and to verify by physiological measurement that the lead is implanted at the optimal location as suggested by the physiological navigation.

"ASLEEP" DBS

The current estimate is that only 25 percent of the patients eligible for DBS will decide to undergo the therapy. Patients who can benefit from DBS surgery decline it for many reasons. The first is the patients' and the neurologists' fear of the neurosurgery. I share the fear, but for me and for Western medicine, to be eligible for DBS means that this medical treatment has been optimized and that our estimate of the gain/risk ratio of DBS surgery is highly positive. Second, the DBS failures are more loudly publicized than the success stories. DBS experts all over the world are trying to develop better methods for the procedure (better selection criteria, imaging, physiological navigation, and more) that will lead to fewer failures. Finally, patients fear being awake during a procedure. Patients therefore ask for the "asleep" DBS procedure.

Many research teams, including ours, have tested the effect of different sedation and anesthesia agents on the quality of physiological navigation to the DBS targets. Most of these studies were on a small number of patients and used different methods and levels of sedation/anesthesia. For all tested drugs, the physiological parameters used for navigation (discharge rate and pattern) are affected. The effect is stronger with deeper stages of sedation and anesthesia. This is in line with the clinical impression of a significant reduction in STN discharge if the patient falls asleep (in line with nonhuman primate studies; figure 8.2).

Jing Guang, Halen Baker, Orilia Ben-Yishay Nizri, and Shimon Firman tested the effects of different sedation agents on the neural activity of the basal ganglia (Guang et al. 2020). Propofol (as also remifentanil and dexmethomidine) markedly increases the power of the low-frequency (theta, alpha, beta) domains and so may mask the typical beta oscillations of Parkinson's disease used for navigation. Ketamine increases the power at the gamma range and therefore does not mask the parkinsonian activity. We have already performed few DBS procedures with an interleaved propofol-ketamine (IPK) sedation protocol. The propofol was stopped about twenty to thirty minutes before beginning physiological navigation. Then we started continuous infusion of a subanesthetic dose of ketamine (10–20 mg/hour for an adult patient). Our experience so far is that we obtained high-quality physiological recording and reliable examination of the stimulation therapeutic window. The patients' memories of the procedure were very positive. With Idit Tamir and her team at the Rabin medical center, we are carrying out a double-arm study to verify these preliminary results. Hopefully, most patients will soon be able to benefit from "asleep" interleaved ketamine-propofol DBS procedures that will minimize their stress and anxiety and still enable high-quality navigation to the DBS targets.

WHAT IS THE BEST TARGET IN SPACE FOR DBS FOR PARKINSON'S DISEASE?

Both the STN and the GPi are used as targets for the DBS treatment of Parkinson's disease (Williams et al. 2014). Most experts (at least in Europe) will agree that the

STN is a better target but more risky. On the other hand, the GPi is more forgiving and a "good-enough" DBS target. Several head-to-head studies of the efficacy of STN versus GPi DBS have been performed. The results are not consistent (US studies find a match; European studies are in favor of the STN). Most big centers will therefore use the STN as their default target and will go to the GPi in cases on the borderline of the exclusion/inclusion criteria.

Here, I will discuss a different question. Having both the STN and the striatum as the input structures of the basal ganglia that project to downstream structures in the three-layer model of the main axis of the basal ganglia (figures 9.4 and 10.2*B*), we ask ourselves: Why is the STN selected as the target for DBS and not the striatum? The striatum is the major input to the basal ganglia, has two orders of magnitude more neurons than the STN (see table 2.1), and is the major site of dopamine depletion in Parkinson's disease. In three sets of experiments (Deffains et al. 2016; Mizrahi-Kliger et al. 2018), we recorded the activity of neurons in the three layers of the basal ganglia under similar experimental conditions (during a behavioral task, during sleep, and following MPTP treatment). We found that the activity of basal ganglia downstream neurons resembles the activity of the subthalamic neurons and is different than the activity of the striatal projection neurons—the MSNs.

Stella Papa and colleagues have recently reported enhancement of LFP beta oscillations in the striatum of MPTP-treated NHPs (Singh and Papa 2020). These results are in line with our report (Deffains et al. 2016). However, their previous study (Singh et al. 2016) reported drastic changes in the discharge rate and pattern of MSNs in MPTP monkeys (compared to normal recording) and Parkinson's patients (striatal activity was compared to the MSN activity of patients with essential tremor). Their studies have reported an approximately fifteenfold increase in discharge rate (from two to thirty spikes/second) and more bursting discharge patterns in the parkinsonian subjects. Our studies, both in MPTP-treated monkeys (figures 15.2 and 15.4) and Parkinson's patients (Valsky, Heiman-Grosberg, et al. 2020), have not demonstrated such drastic modulation of striatal MSN discharge. There are several possibilities for the discrepancy in results—for example, differences in monkey species and MPTP procedures; however, this does not explain the difference in the human recording. I believe that the different results might be due to the possible inclusion of injured cells and a bias toward high-frequency discharge activity. The jury is still out, but until other groups report their results supporting one of the two sites, I will stay with our results that demonstrate no major change in MSN spiking activity in Parkinson's disease.

Thus, the STN is the winner in the different sets of experiments comparing the activity of striatal and subthalamic neurons to the activity of the basal ganglia downstream (GPe and GPi/SNr) neurons. David beats Goliath again. Steve Kitai was right, and the STN is the driving force of basal ganglia physiology, in health and disease, and the better target for DBS. This conclusion is bolstered by the small and compact structure of the STN, making possible its control by one/few DBS contacts.

Nevertheless, one should not undervalue the importance of the striatum because of the above discussion. The STN and the striatum may be compared with the pedal and the keyboard of a piano. The STN is the driving force, exercising broad and diverse, but not specific, control of the activity of basal ganglia downstream structures. The striatum provides the melody. Our limited understanding and technology only enable us to modify the less delicate parts of the system. Future generations may be able to restore the motor melody of Parkinson's patients by correcting the keyboard (the striatum).

CHAPTER SUMMARY

- DBS is the gold-standard therapy for patients with advanced Parkinson's disease, dystonia, and essential tremor.
- The most used DBS targets are STN for Parkinson's disease, GPi for dystonia, and the cerebellar-rubral-thalamic tract for tremor.
- DBS is a twenty-four seven continuous high-frequency stimulation. Typical parameters are 130 Hz, 1–6 mA (2–8 V), and cathodal pulses of 60–180 μS pulse width.
- The mechanism of DBS is still debated. Probably, it is multifaceted. The most important feature is the synaptic depletion and functional inactivation of nodes in the basal circuitry, mimicking the effect of ablation.
- DBS inactivation of critical nodes of the basal ganglia shuts down the aberrant activity of the brain default (Kahneman's system 1) network and enables the cortex (Kahneman's system 2) and other networks to compensate.
- DBS outcome is improved by better automatic physiological navigation to the DBS targets.
- New sedation algorithms will make possible physiological navigation under moderate sedation and will reduce patients' stress and anxiety.
- The STN is the driving force of the physiological activity of basal ganglia downstream structures in health and disease, and therefore the STN is the optimal target for DBS therapy for basal ganglia–related disorders.

19

CLOSING THE LOOP ON PARKINSON'S DISEASE AND BASAL GANGLIA–RELATED DISORDERS

Today, deep brain stimulation (DBS) therapy is a twenty-four seven continuous treatment. However, the symptoms of Parkinson's disease are not stable over time and fluctuate on a short timescale (for example, tremor can be very episodic, on a timescale of minutes) and long timescale (the never-ending progression of striatal dopamine depletion and severity of Parkinson's symptoms). To overcome the slow progression of the disease, the DBS patient would see his/her neurologist once every three to six months for adjusting the DBS parameters. However, the dynamics of Parkinson's might be faster than the frequency of visits to the neurologist. Moreover, the modulation of the symptoms in the doctor's office setting may make the programming less optimal. In this chapter I will describe the ongoing efforts toward developing a closed-loop (adaptive) DBS therapy.

CLOSED-LOOP DBS

Closed-loop systems have changed life in the twentieth century and are highly suitable for optimization of the performance of physical systems that face dynamic environmental states such as room temperature. Boris Rosin started in 2008 to develop a closed-loop DBS system and tested it in the nonhuman primate. We decided to use cortical or pallidal beta oscillations as a trigger for the closed-loop DBS. The results (Rosin et al. 2011) were clear and loud. Closed-loop DBS (GPi stimulation triggered by cortical beta oscillations) is superior to open-loop continuous DBS and other control conditions tested (figure 19.1). The basal ganglia system can be observed and controlled.

The nonhuman primate experiments provided the proof of the principle of closed-loop adaptive DBS. Shortly after, the research groups of Peter Brown of Oxford and Alberto Priori of the University of Milan took the lead in a series of most elegant studies in human patients and demonstrated that closed-loop adaptive DBS triggered

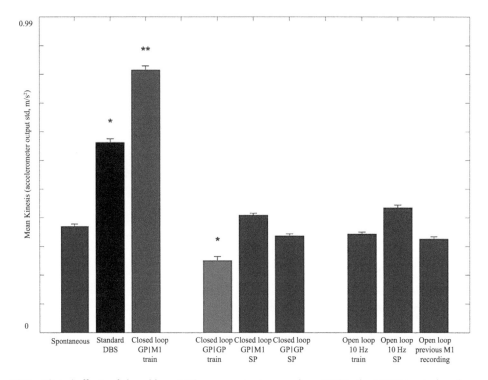

19.1 Clinical effects of closed-loop DBS are superior to open-loop DBS in the MPTP-treated nonhuman primate. Kinesia (movements/time unit). From left to right: with no stimulation, with standard open-loop DBS and with closed-loop DBS. *Right light and dark blue bars*, control stimulation patterns. *Source*: Copied (with permission) from Rosin et al. 2011.

by LFP beta oscillations is highly effective in ameliorating Parkinson's symptoms (Little et al. 2013, 2016; Priori et al. 2013; Arlotti et al. 2018).

NOT ALL BETAS ARE CREATED EQUAL: ON BAD AND GOOD BETA OSCILLATIONS

In the previous chapter we showed that

- Beta oscillations are observed in basal ganglia spiking activity after but not before MPTP (figure 15.4).
- Beta oscillations are augmented in basal ganglia LFP after MPTP (figure 15.11).
- STN beta oscillations and akinetic-rigid symptoms in Parkinson's patients are correlated.
- Beta oscillations in Parkinson's patients are reduced by movements and effective treatments (L-dopa, DBS).

Thus, one may be impelled to assume that all beta oscillations are bad and that closed-loop DBS should aim at complete suppression of basal ganglia beta oscillations. However, it is an unreasonable assumption because

- Beta oscillations are observed in normal EEG. Hans Berger in his 1924 exploration of the physiological mechanism of telepathy recorded brain electrical activity above the scalp. He first described the alpha large-amplitude oscillations observed in the occipital part of the head when the subject closed her eyes. The alpha rhythm is replaced by beta (β) smaller-amplitude oscillations in a subject awake with eyes open.
- Beta oscillations are observed in the LFP recorded in the cortex and the basal ganglia of normal monkeys.
- Beta oscillations are observed in the LFP and spiking activity of the STN and GPi of human patients with obsessive-compulsive disorder (OCD) and (nongenetic) dystonia.

The role of beta oscillations in normal physiology is still debated. One of the most popular hypotheses has been offered by Andreas Engel and Pascal Fries (2010). They suggested that beta oscillations maintain the "status quo." Thus, when a subject has to maintain the same motor set, her brain will depict beta oscillations. These oscillations will be "desynchronized," or more correctly said, reduced (figure 15.8), if the subject is instructed to perform a movement (Schmidt et al. 2019).

Marc Deffains studied the temporal profile of basal ganglia beta oscillations before and after MPTP treatment (Deffains et al. 2018). He first observed that beta oscillations are episodic—that is, even after MPTP treatment, beta oscillations are not persistent. Rather, the amplitude of the beta oscillations fluctuates with time. Figure 19.2A depicts our method for the detection of beta episodes (bursts). We divide our data into two frequency domains: the beta domain (8–15 Hz) and the beta flanking domains (6–8 and 15–20 Hz, the same width of 7 Hz). We calculate the mean and the variability (standard deviation, SD) of the beta flanking domains and set a threshold at this mean + 2.5 SD. We define beta episodes as those time segments in which the beta power crosses the threshold. There are other methods for defining episodes. For example, Peter Brown and colleagues calculated the mean and the variability of the beta domain and defined the threshold as the seventy-fifth percentile (Tinkhauser, Pogosyan, Little, et al. 2017, Tinkhauser, Pogosyan, Tan, et al. 2017). We prefer our methods since they may also apply to cases with continuous beta. To be sure, Peter Brown's method is simpler and more easily adapted for real-time applications.

Irrespective of the parameters used for the detection of the beta episodes (threshold level, bin size), the prevalence (frequency * duration) of beta episodes is significantly higher in the parkinsonian state for all layers of the basal ganglia axis. However, one can find beta episodes in the normal state as well. Thus, not all beta episodes are bad. We need beta in the basal ganglia to maintain the status quo of our motor behavior. This normal beta activity is pathologically augmented (increased frequency, magnitude, and, most importantly, duration of beta episodes; figure 19.2B; Deffains et al. 2018) during Parkinson's disease. The scientific world is a small farm, and trends are

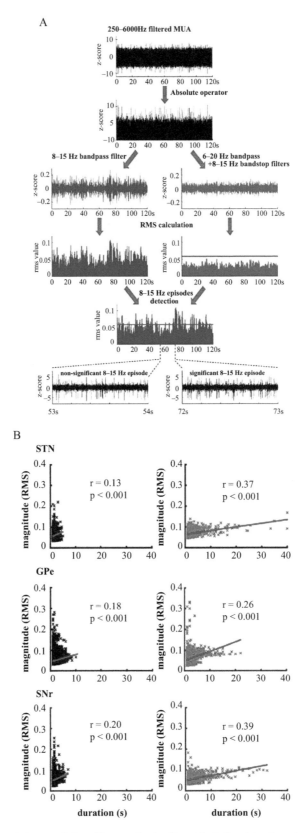

19.2 Detection and characteristics of beta episodes (bursts). *A*, the detection algorithm. Blue-beta domain, red-flanking frequencies. *B*, duration is a more reliable metric than amplitude for discrimination between healthy and pathological beta episodes. *Black*, control; *green*, after MPTP treatment. STN, subthalamic nucleus; GPe, globus pallidus external segment; SNr, substantia nigra reticulata. *Source*: Copied (with permission) from Deffains et al. 2018.

felt by many sensitive researchers all over the world. Thus, during the same time that we performed our beta episode study in normal and MPTP monkeys, Peter Brown and colleagues pursued the same searches in LFP recorded from externalized leads in Parkinson's patients, before and after dopamine replacement therapy, and reported similar results (Tinkhauser, Pogosyan, Little, et al. 2017; Tinkhauser, Pogosyan, Tan, et al. 2017).

A BASIC GANGLIA–CORTEX HOLISTIC VIEW OF BETA OSCILLATIONS

There are several theories about the functional role of brain/basal ganglia beta oscillations. Here I would like to extend the "status-quo" hypothesis of Engel and Fries (2010) and apply it to the basal ganglia/corticocortical Kahneman's system 1 and 2 networks model (table 12.1, figure 13.2).

I posit that the basal ganglia are the fast and automatic default—Kahneman's system 1—that most of the time inhibit the logical, slow, and goal-directed—Kahneman's system 2—corticocortical networks. The basal ganglia's control of automatic behavior is carried out by their brain stem projections (Redgrave et al. 2010) that descend to the muscular apparatus through the brain stem–spinal pathways (figure 13.1). The basal ganglia upward projections do not drive the cortex but rather inhibit it from taking an active part in the state-to-action loop. Here I propose that basal ganglia beta oscillations inform the corticocortical networks that they are in control, and the cortex can keep doing its own business. The basal ganglia do not have to do it continuously (beta activity is fluctuating and episodic; figure 19.2) since the beta bursts might exert a long-duration effect on the cortical network. Only when the basal ganglia automatic system faces problems does it raise the flag—a decrease in beta oscillations—that summons the cortex to take control.

The more persistent beta oscillations following dopamine depletion and the development of Parkinson's disease never let the cortex take control. However, the automatic control of behavior through the brain stem descending pathway is also affected, and the patient develops Parkinson's akinesia. External cuing (metronome, lines on the floor) can lead to recruitment of the goal-directed, cortical system 1 network. Alerting stimuli might recruit even more primitive state-to-action loops (fight or flight) and lead to "paradoxical kinesia."

THE OTHER SIDE OF THE MOON: KETAMINE, HYPERKINETIC STATES, AND GAMMA OSCILLATIONS

Most of our discussion so far has been devoted to the prototypical symptoms of hypokinetic disorders—the akinesia/bradykinesia of Parkinson's disease. After a five-to-ten-year honeymoon period on dopamine replacement therapy, most Parkinson's patients develop adverse effects from the treatment, such as levodopa-induced dyskinesia (abnormal involuntary movements, hyperkinetic symptoms). The spectrum of the hyperkinetic symptoms is broad (maybe because that even for good neurologists, it is easier to identify the different barks than to recognize the dogs that do not bark) and includes levodopa-induced and tardive (due to chronic neuroleptic/antischizophrenic treatment) dyskinesia, Huntington's chorea, hemiballismus, dystonia, tics, and hyperactivity states. While hypokinetic disorders are usually associated with depletion of striatal dopamine, hyperkinetic disorders can result from damage to the indirect basal ganglia pathway (Huntington's disease due to degeneration of D2 MSNs, hemiballismus due to STN infarct).

Levodopa-induced dyskinesia usually happens at the peak dose of brain dopamine level. Diphasic dyskinesias occur when the patient is just beginning to turn "on" and again when the patient begins to turn "off." This is also known as dyskinesia-improvement-dyskinesia (D-I-D) syndrome and may be related to the slope of the change in the level of dopamine. Although we depict levodopa-induced dyskinesia as the mirror image of Parkinson's akinesia, most Parkinson's patients prefer to be "on" with some dyskinesia rather than being "off" and akinetic.

Our early studies of the effects of dopamine therapy and levodopa-induced dyskinesia on neural activity in the basal ganglia (figure 15.1) were carried out under the assumption of the D1/D2 direct/indirect model of the basal ganglia. Moreover, we overlooked gamma oscillations at that time (low-beta 10–15 Hz oscillations were termed double tremor-frequency oscillations at that motorcentric period). LFP recordings from DBS leads of Parkinson's patients were the first to reveal that gamma (30–70 Hz) oscillations replace beta oscillations when the patient is treated with dopamine replacement therapy and do so even more when the patient develops dyskinesia.

The 1/f nature of brain activity results in a low amplitude of gamma oscillations. I remember a hot debate between me and Pascal Fries in a conference in Eilat (near the Red Sea) when I told Pascal that even if gamma oscillations exist in the brain they are too small for us to care about. I apologize, Pascal. I was wrong. I learned this when Maya Slovik recorded the effect of N-methyl-D-aspartate-type (NMDA) antagonists on neural activity in the basal ganglia (figure 19.3). Synchronous gamma oscillations were clear and loud (figure 19.3A). I can't neglect gamma oscillations anymore.

Why should we be interested in the behavioral and neuronal effects of NMDA antagonists? Because the borders between neurology and psychiatry are fuzzy. There are many similarities between Parkinson's disease and schizophrenia (see more below). Could it be that schizophrenic delusions and hallucinations are the same as dyskinesia of the limbic domains of the basal ganglia? Indeed, levodopa-induced dyskinesias are often associated with visual hallucinations. I was therefore very happy to learn that monkeys treated with NMDA antagonists (e.g., ketamine, a dissociative sedation agent) can be used as a valid animal model of schizophrenia. The treatment of schizophrenic patients with NMDA antagonists increases the severity of their schizophrenic

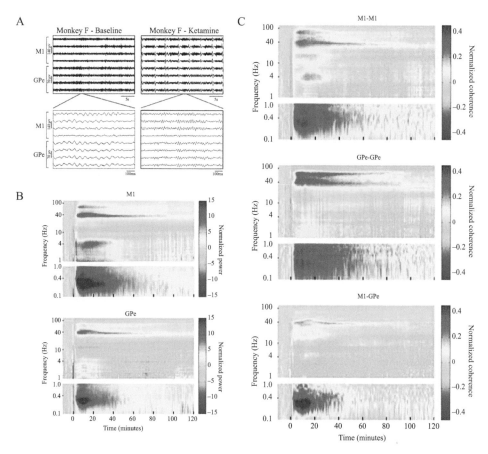

19.3 Delta and gamma oscillations at the cortex-basal ganglia network following systemic ketamine injections. *A*, example of LFP activity. *B*, average spectrograms. *C*, coherograms before and after IM ketamine injections in the primary motor cortex (MI) and globus pallidus external segment (GPe) following ketamine injection at zero time.
Source: Modified (with permission) from Slovik et al. 2017.

symptoms. Nonhuman primates develop negative schizophrenic symptoms after chronic treatment with NMDA antagonists.

The pathophysiology of schizophrenia is still debated. The dopamine hypothesis claims that the negative and positive symptoms are correlated with hypodopaminergic and hyperdopaminergic states, respectively. The glutamate hypothesis emphasizes the difference between dopamine in the frontal cortex and in the striatum and points toward modification of the frontal cortex to VTA glutamate projections as the main cause of symptoms in schizophrenia and in the NMDA antagonist animal models of this disease.

Unlike MPTP, there is little consensus on the validity of the NMDA antagonist model for schizophrenia. Classical psychologists and psychiatrists claim that because schizophrenia is defined by the presence of delusions and hallucinations, we cannot expect to find these symptoms in animals without a language. Genetics clearly plays a significant role in schizophrenia (the probability of a diagnosis of schizophrenia increases up to 50 percent if a monozygotic twin brother suffers from schizophrenia). Intrauterine factors, such as exposure to viral disease during pregnancy, may play a role (the second hit theory).

The jury is still out with regard to the construct and predictive validity of the NMDA antagonist model of schizophrenia. In rodents, developmental models such as methylazoxymethanol acetate (MAM) injection during gestation and the social isolation of pups are frequently used. These models probably mimic better the etiology of human schizophrenia (content validity). Harry Harlow demonstrated the severe effects of social isolation on young monkeys.[1] However, our moral values no longer tolerate intrauterine injections and the social isolation of monkeys. We hope that at least a predictive validity exists for the NMDA antagonist model. In any case, ketamine was very effective in bringing on excessive synchronous gamma oscillations (figure 19.3) and can be used as a tool for their study. We hope that as for beta oscillations, gamma oscillations can be divided into good and bad. Would it be the same as for the beta episodes, with short, good and long, bad episodes? The ketamine effects leading to very prolonged gamma oscillations (figure 19.3) suggest that this might be correct, but we have yet to carry out the related analysis.

DREAMS MAY COME TRUE: PHASE-SPECIFIC CLOSED-LOOP DBS

Up to now DBS was considered a better alternative than ablation therapy because it is revisable and adjustable. However, DBS mimics inactivation of the stimulated area. Closed-loop (adaptive) DBS enables us to inactivate the basal ganglia only when they misbehave (e.g., have long beta episodes); however, neither classical continuous DBS nor closed-loop DBS enables us to restore normal activity.

Recently, several research groups have pointed out that phase-specific stimulation might enable us not only to suppress "bad" oscillations but also to augment the good ones. Oren Peles recorded neural activity in the primary motor cortex of monkeys that have been trained to increase their cortical beta oscillations using brain-machine interface (BMI) methods. Oren showed that stimulation at the rising phase of the field potentials augments the oscillations, while stimulation at the falling phase of the oscillation suppresses them (figure 19.4). Similar results were reported in the basal ganglia/Parkinson's disease context by Andrew Sharott and Peter Brown (Holt et al. 2019).

Bad and good brain oscillations are context dependent. Thus, in a hypokinetic state, beta oscillations are bad, and gamma oscillations are good. The reverse will be true in the hyperkinetic state, where gamma oscillations are probably bad, and beta oscillations are good. Future phase-specific closed-loop DBS would be context

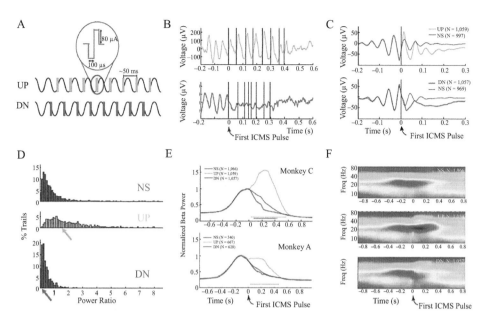

19.4 Phase-specific closed-loop brain stimulation can both suppress and augment neuronal oscillatory activity. *A, green and red*, stimulation at the up- and down-going phase augments and suppresses beta oscillations in the cortex of a nonhuman primate. *B, C*, example and population time averages. NS, no stimulation (*blue*). *D, E*, power ratio and normalized beta power. *F*, average spectrograms. *Source*: Modified (with permission) from Peles et al. 2020.

sensitive (for example, by assessing the relative fraction of long/short beta/gamma oscillations) and will be able to tailor the stimulation to the context.

Treating the sleep disorders of Parkinson's disease might be easier. Parkinson's insomnia is correlated with a reduction of basal ganglia slow oscillations (delta, 0.5–2 Hz) and with an increase of beta oscillations (figure 15.12). Stimulating at the falling phase of the beta oscillations and the rising phase of the delta oscillations may restore normal sleep to Parkinson's patients. Since delta oscillations are slow and of very different frequency than beta oscillations, delta triggering might be easier than gamma triggering. Today, open-loop, continuous high-frequency DBS improves the subjective assessment of sleep quality; however, it does not completely restore normal sleep. The treatment of sleep disorders with conventional hypnotic drugs (sleeping pills) is helping to reduce the latency to sleep but does not restore (and maybe even prevents) normal sleep architecture. We are starting now to understand the importance of normal and ample sleep for the maintenance of a healthy brain. Brain beta-amyloids accumulate in the brain during the day and become washed out during sleep by the glymphatic system. Sleep deficits (caused by any reason such as sleep apnea, sleep disorders, and lifestyle issues) may result in the insufficient elimination of beta-amyloids and the development of Alzheimer's disease. Could it be that it is the same story for alpha-synuclein and Parkinson's disease? Increased severity of

Parkinson's disease frequently leads to more severe sleep disorders. Eliminating this vicious cycle by means of phase-specific closed-loop DBS might help in restoring normal sleep architecture and be the first treatment to significantly slow the progression of Parkinson's disease.

LEARNING FROM THE BASAL GANGLIA: MACHINE-LEARNING DBS

Our discussion of closed-loop DBS therapy so far is rooted in classical control theory. We look for the physiological markers of Parkinson's disease (e.g., long episodes of beta oscillations) and use classical DBS (high-frequency) pulse stimulation to inactivate the basal ganglia. Even the phase-specific approach is naive because it divides the brain waves into two classes, good and bad, and assumes that all Parkinson's symptoms are the same.

But Parkinson's disease is multifaceted, and treatments that may alleviate one symptom may worsen other symptoms. For example, Nir Asch showed that basal ganglia beta oscillations may be correlated with the brake pedal of a car. If we release the brake pedal (by reducing beta oscillations), there is a higher probability for tremor to emerge (Asch et al. 2020). On the other hand, the spectrum of stimulation modalities is almost infinite. Today's settings of stimulation parameters enable different geometries (from single to multiple contacts), current or voltage modes, amplitudes, train frequencies, and pulse widths. However, this is only the tip of the iceberg. One can change the shape of the pulses (e.g., to sinus wave), switch between different frequencies (as has been recently shown by Luming Lee and colleagues; Jia et al. 2018), or use theta bursts and random patterns. What is the best method for optimizing DBS?

The basal ganglia use reinforcement-learning algorithms to optimize the agent behavior with the use of reward feedback. Future generations of closed-loop DBS will do the same. The DBS built-in computer will explore the world of possible stimulation paradigms and use feedback from the brain (e.g., the prevalence of long beta episodes) from telemetry of the patient's symptoms (e.g., smartphone monitoring of the patient's akinesia, speech, and other symptoms) and from the patient's subjective evaluation (e.g., by the use of analog visual scales on the smartphone) to optimize the stimulation. Like the basal ganglia, the DBS computer will be able to follow the patient's instructions toward more exploratory behavior (e.g., when the patient is in a safe environment) versus exploitation and risk-aversive behavioral policy (e.g., when the patient is driving a car). Multiobjective optimization of the different symptoms will enable machine-learning DBS to achieve an optimal balance between the subjective and objective symptoms of the patient, hopefully avoiding the common situations in today's DBS clinics in which "the neurologist is happy, but the patient is not."

ON DREAMS AND DELUSIONS IN THE BASAL GANGLIA: CLOSED-LOOP DBS FOR SCHIZOPHRENIA

For many years, the treatment of Parkinson's disease has been on the front line of treatment of neurodegenerative diseases. The battery of treatments includes L-dopa, postsynaptic dopamine agonists, conventional open-loop high-frequency DBS, and phase-specific closed-loop DBS. Following the success of DBS treatment for Parkinson's, many trials have been initiated for the treatment of mental disorders such as major depression (Mayberg et al. 2005; Döbrössy et al. 2020; Sankar et al. 2020) and obsessive-compulsive disorders (Fridgeirsson et al. 2020; Haber et al. 2020; Li et al. 2020). My dreams (admittedly, some of my less enthusiastic colleagues might define them as delusions) are DBS therapy for the negative symptoms of schizophrenia.

Schizophrenia is a very severe and common mental disorder with a global lifetime prevalence of about 0.3–0.7 percent. The clinical symptoms of schizophrenia can be divided into positive and negative symptoms. Positive symptoms are abnormal behaviors, thoughts, or feelings that are normally absent. Negative symptoms are the absence of normal behaviors, thoughts, or feelings. Schizophrenia positive symptoms include hallucinations, delusions, and disordered thoughts and speech. The negative symptoms include flat or blunted affect and emotion, poverty of speech (alogia), inability to experience pleasure (anhedonia), lack of desire to form relationships (asociality), and lack of motivation (avolition).

The onset of symptoms typically occurs in young adulthood and is earlier for males than females. The presentation of the first symptom is typically a psychotic episode with positive symptoms. However, negative symptoms can be recognized before the development of the first psychotic episode. The life course of the disease is composed of a slope of increased severity of negative symptoms, interrupted by psychotic episodes. The frequency, duration, and intensity of the psychotic episodes affect the speed of the decline of the negative symptoms (American Psychiatric Association 2013; Sadock et al. 2014).

Treatment is with antipsychotic drugs. The first generation of these drugs is the D2 dopamine antagonist family; however, they are prone to inducing extrapyramidal (Parkinson's-like) symptoms and tardive dyskinesia (involuntary movements of the orofacial muscular apparatus[2]). Atypical antipsychotic (neuroleptic) agents were introduced with the claim that they exert fewer extrapyramidal side effects. This is debated today, and in any case, atypical antipsychotic treatment is more prone to result in the development of metabolic syndrome (obesity, diabetes). Antipsychotic drugs affect the positive but not the negative symptoms. Most patients after several years of the disease are not able to conduct normal social life and live as homeless persons (in countries with less-developed social medical systems) or in hostels and mental institutes.

Medical education usually positions Parkinson's disease and schizophrenia on two extreme sides of the spectrum. Overdosing of schizophrenia patients with antipsychotics

Table 19.1 Head-to-head comparison of positive and negative symptoms of schizophrenia and Parkinson's disease.

	Schizophrenia	Parkinson's disease
Negative symptoms	Avolition, catatonia, psychomotor slowing	Akinesia, bradykinesia
	Flat affect	Masked face
	Alogia	Reduced verbal fluency, dysphonia
	Anhedonia	Apathy, depression
Positive symptoms	Delusions, hallucinations	Dystonia, levodopa-induced dyskinesia
	Disorganized thinking	Frontal signs
	Agitation	Akathisia
	Motor symptoms	Rigidity, tremor

(dopamine antagonists) will often lead to Parkinson's-like symptoms. On the other hand, overdosing of Parkinson's patients with dopamine therapy (and especially post-synaptic agonists) will lead to hallucinations and even psychotic behavior. Although the two diseases are quite common, it is still debated whether schizophrenia and Parkinson's disease are associated with decreased or increased reciprocal risk (Kuusimäki et al. 2021).

I claim that this antagonistic view of Parkinson's disease and schizophrenia is shaped by the emphasis on the positive symptoms of schizophrenia (delusions and hallucinations) and the negative symptoms of Parkinson's disease (akinesia). Anti-psychotics treatment of schizophrenic patients will cause akinesia, while dopamine treatment of Parkinson's patients will cause positive hyperkinetic symptoms, such as levodopa-induced dyskinesia and hallucinations. Schizophrenia and Parkinson's textbooks are written by psychiatrists and neurologists, respectively. They use different terminology to describe the same (or similar) symptoms. For example, both avolition and akinesia describe the patient who is not paralyzed but nevertheless does not display the internal motor vigor for spontaneous motor behavior. Table 19.1 provides a comparison of the terms used to characterize positive and negative symptoms of Parkinson's and schizophrenia.

To minimize the chance of being "lost in translation," here is a short summary of the psychiatric terminology.

- Avolition—Decrease in the motivation to initiate and perform self-directed, purposeful activities.
- Catatonia—A state of psychomotor immobility, in which the patient may hold rigid poses and will respond only to intense stimuli (stupor).
- Affective flattening—A loss or lack of emotional expressiveness.

- Alogia—Poverty of speech.
- Although less recognized and discussed, motor symptoms appear in many first-episode and medication-free schizophrenia patients (Walther and Strik 2012).

As might be hinted by the title of Sigmund Freud's 1917 essay ("Delusion and Dream in Jensen's *Gradiva*"[3]), Freund was aware of the fuzzy boundaries between delusions (I would use hallucinations, but I respect Freund's right to have a different opinion) and dreams. Dreams have better public relations. Every woman and man deserves a dream that may mature into reality. Martin Luther King Jr. had a good one. My dream is a phase-specific closed-loop DBS for schizophrenia. My working assumptions are that the negative symptoms of schizophrenia are correlated with abnormally long episodes of beta oscillations in the limbic domains of the basal ganglia. The positive symptoms are correlated with abnormal gamma oscillations. Continuous, high-frequency open-loop DBS of the limbic domains of the STN has been very helpful in treating obsessive-compulsive disorders. Volker Coenen, Thomas Schlaepfer, and colleagues have shown that high-frequency stimulation of the medial forebrain bundle is very effective for major depression (Coenen et al. 2019). However, this target is very close to the limbic domains of the STN (Li et al. 2020), and it could be that some of its therapeutic effects are achieved by STN inactivation. Thus, my first anatomical target for phase-specific closed-loop DBS treatment of schizophrenia would be the limbic domain of the STN.

IF YOUR DREAMS DON'T SCARE YOU, THEY AREN'T BIG ENOUGH

The subtitle of this section endorsing the "high-risk, high-gain" behavioral paradigm is taken from *This Child Will Be Great: Memoir of a Remarkable Life by Africa's First Woman President* by Ellen Johnson Sirleaf (2010). We started part III of this book with the mission given to us by Lev Zasettsky to understand and to treat better the disorders of the brain. Our society is much more supportive of patients wounded in wars or tackling neurological disorders than patients with psychiatric disorders. Again, the words of Clifford Whittingham Beers (1876–1943) in *A Mind That Found Itself*, written in 1907 (yes, more than 110 years ago) will speak much better than I can: "A pen rather than a lance has been my weapon of offence and defense; for with its point I have felt sure that I should one day prick the civic conscience into a compassionate activity and thus bring into a neglected field earnest men and women who should act as champions for those afflicted thousands least able to fight for themselves" (50). Beers was one of five children, all of whom suffered psychological distress and died in mental institutions, including Beers himself. Today, we have the neuroleptic treatment for the positive symptoms of schizophrenia. But there is still no good treatment for the negative symptoms of schizophrenia.

Desperate situations call for desperate measures. Indeed, the history of schizophrenia is filled with dramatic stories of failed clinical trials. Probably, the biggest failure has been the "frontal lobotomy" carried out on thousands of patients in Europe and the US in the mid-twentieth century.[4] Honestly, the history of DBS has some dark roots as well.[5] Closed-loop DBS for schizophrenia is therefore a dream that scares me. I hope that we can learn from the mistakes of António Caetano de Abreu Freire Egas Moniz and Walter Freeman—the founders of the frontal lobotomy. Our first and critical goal should be the quality of life of the patient and not the benefit of society. We should keep improving neurosurgical techniques to provide safe procedures with minimal mortality, morbidity, and stress (better imaging of brain blood vessels, better navigation tools, and "asleep" DBS procedures). We should take the advice of and work with patient associations and patient advocacy teams. Finally, we should welcome close supervision by ethics committees and slowly start our research with a few patients before we turn into big experimental programs. The life of today's schizophrenic patient is not much better than Clifford Beers's. We should not repeat the mistakes of the past but must follow Beers's request to "act for those afflicted thousands least able to fight for themselves."

CHAPTER SUMMARY

- The basal ganglia can be observed and controlled, and therefore closed-loop (adaptive) DBS has better outcomes than open-loop DBS.
- Current experimental closed-loop adaptive DBS mainly uses beta oscillations as a biomarker for Parkinson's pathophysiology.
- Not all betas are bad. Short episodes of beta oscillations are common in the normal brain and probably maintain the status quo. Too many and too lengthy beta episodes lead to Parkinson's akinesia.
- Hyperdopaminergic hyperkinetic states and treatments with an NMDA antagonist (e.g., ketamine) are correlated with basal ganglia gamma oscillations.
- Phase-specific closed-loop stimulation techniques would make possible the suppression of bad oscillations (e.g., long beta episodes) and reinforce good oscillations (e.g., those of sleep delta oscillations).
- Future machine-learning multiobjective optimization methods based on brain signals, objective telemetry assessment of clinical signs, and the patient's subjective evaluation of his or her quality of life will optimize DBS therapy.
- There are many similarities between the negative and positive symptoms of Parkinson's disease and schizophrenia. Closed-loop DBS (of STN limbic domains) may provide a cure for the negative symptoms of schizophrenia.

IV

FROM THE BASAL GANGLIA AND BRAIN TO THE MIND

20

THE BASAL GANGLIA, FREE WILL, AND HUMAN RESPONSIBILITY

In several sections of the previous chapters, I went outside of my expertise and pleasure zones. Here I am going to make a big leap to domains that my colleagues and I have rarely explored—the philosophy of mind and moral values. I apologize for the naive approach. The philosophy expert reader can skip this chapter with no regret. It is written under the luxury of attaining the first basic layers of Maslow's pyramid (figure 20.1). Therefore, I can go on to the higher levels, including humans' innate curiosity about their own nature. Over the years, we have learned that it is highly important to adjust the expectations of Parkinson's patients who consider the DBS procedure. I should also adjust the expectations of the reader. I am still in the desert searching for the promised land of understanding. Luckily, I am not alone. The influential Chinese Dau philosopher Chuang Tsu (Zhuangzi), who lived around the fourth century BC, said, "The fish trap exists because of the fish. Once you've gotten the fish you can forget the trap. The rabbit snare exists because of the rabbit. Once you've gotten the rabbit, you can forget the snare. Words exist because of meaning. Once you've gotten the meaning, you can forget the words. Where can I find a man who has forgotten words so I can talk with him?"[1]

STATE-TO-ACTION LOOPS AND FREE WILL

A recurrent theme throughout the previous chapters (figures 10.2, 10.7, 11.5, 12.1, 13.2, and 17.3) is that behavior is controlled by state-to-action loops. The subject is trying to maximize his cumulative pleasure (with regard to reinforcement-learning theories) or to reach the Pareto frontier (with regard to multiple critic, multiobjective optimization models) by interacting with the world (environment) and learning from the results of his actions. I agree with the old view that the hidden goal of our behavior is to maximize the amount of genetic information we forward to future

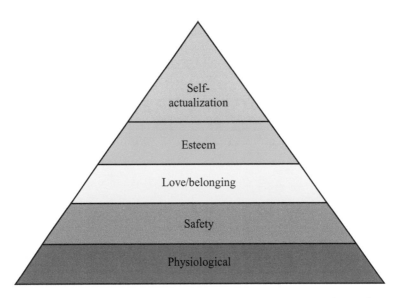

20.1 Maslow's hierarchy of needs. Human needs are represented as a pyramid, with the more basic needs at the bottom.
Source: Copied (with permission) from Wikipedia, https://en.wikipedia.org/w/index.php?title=Maslow%27s _hierarchy_of_needs&oldid=953175274.

generations. Evolution is "tricking us" with the innate rewards of food, water, and sex, on one hand, and innate risk aversion, on the other hand, to optimize behavior that will maximize the forwarding of our genetic information to the future.

However, if all our behavior is in the form of a state-to-action loop, or a very complex knee-reflex loop, then no one is in charge. How can we ask ourselves and others about moral behavior? Why do we feel that those who really misbehave should be punished if their behavior is controlled by their biological machinery?

An optional answer is Cartesian (interactionist substance) dualism. The mind and body are dissociated. Free will is a property of the mind that is somehow brought to the physical (body) world (Lowe 2000). However, I am a materialist and tend to take the side of monism versus dualism in the old debate of mind and body. I assume that the mind is created by brain/body physical activity (physical reductionism, epiphenomenalism). Dualism is not the answer for me.

The other extreme answer is that free will is a delusion. We, like all living animals, are part of a state-to-action loop. Sometimes it is a very simple loop like a knee reflex; sometimes it is a very complex loop reflecting long-term memories, emotional states, and even subliminal sensory events. However, we humans believe that we are the masters of our behavior (sense of agency, or sense of control, the subjective awareness of initiating, executing, and controlling one's own volitional actions in the world), so like the split-brain patients of Roger Sperry and Michael Gazzaniga, we fake a tale that explains our actions. We first run away from the bear and then we are afraid (James 1884; LeDoux 1996).

James's 1884 idea of "action first—emotion next" is supported by more modern findings. In 1964 Lüder Deecke and Hans Helmut Kornhuber showed that a readiness potential (*Bereitschaftspotential*) precedes spontaneous conscious decisions to perform volitional acts (Kornhuber and Deecke 2012). Benjamin Libet took an additional step and suggested an objective method of marking the subject's conscious experience of the will to perform an action in time and afterward comparing this information with the brain's electrical activity during the same interval. The subject watched a timer and was instructed to perform a spontaneous volitional movement (pressing a button) and to note the position of the timer when "he/she was first aware of the wish or urge to act." On average, two hundred milliseconds elapsed between the first appearance of the urge to press the button and the act of actually pressing it. The motor readiness potential started several hundred milliseconds before the urge to action (Libet 1985). These physiological and psychological experiments may imply that unconscious neuronal processes precede and potentially cause volitional acts that are retrospectively felt to be consciously motivated by the subject. Thus, we are left with a materialist view that denies the existence of free will. The neuroscience of free will is still debated. Libet himself regarded his experimental results to be compatible with the notion of free will, especially the ability to stop (veto) movement. Liad Mudrik and colleagues have shown that while the expected readiness potentials are found for arbitrary decisions, they are strikingly absent for more ecological, deliberate decisions (Maoz et al. 2019).

ESCAPING THE TYRANNY OF BIOLOGICAL STATE-TO-ACTION LOOPS

My answer is based on the concept that while the nervous system connects between state and action, biology gave us, animals and humans, two gifts. The first is the tool for reinforcement, not supervised, learning. Reinforcement-learning algorithms enable living creatures (including humans) to trade a negative near future outcome for a remote and cumulative positive reward. The second gift (or tragic miracle, according to John Ernst Steinbeck) has probably been given only to humans, and this is language and self-awareness. This is the lever long enough and the fulcrum on which we can place it that provide us with the pathway to escape the tyranny of the state-to-action biological loop.[2] Most of the time, we humans, like all animals, are controlled by the automatic actions of our basal ganglia system 1 neural networks connecting between state and action. However, we are aware of our behavior. We are afraid after we start to run away from the bear, and we are happy after getting rewards. Furthermore, we can plan what we will do if we meet a bear or are offered a glass of good wine. I do believe this self-awareness is a by-product of language and reflects our own internal dialogue, but this is not crucial for my hypothesis. What is more important is that the combination of our brain tools for reinforcement learning and self-awareness enables us to change our behavioral policy according to our moral values and ideas. I still hold to my deterministic thinking and claim that there is nothing more than physical rules. Nevertheless, our state-to-action deterministic system is probably chaotic and highly sensitive to initial conditions. Making small changes in our behavioral policy by reasoning might create big changes in our behavior.

The distinction between automatic, impulsive, reflex-like state-to-action behavior and the ability of adult and healthy humans to modify and control their behavioral policy accounts for the conclusions of the group wisdom of humanity. Most societies agree that a healthy adult human is responsible for her actions, especially if she planned them (i.e., adjusted her behavioral policy). However, we do not punish those who can't control their behavioral policy (the psychotic or demented patient, kids, and animals) and neither do we hold them responsible for their actions. We may have to confine or restrain them, but this is done to protect them or others and not as punishment.

IF YOU DON'T KNOW WHERE YOU'RE GOING, THEN IT DOESN'T MUCH MATTER WHICH WAY YOU GO

Like Maslow, we can speak of a hierarchy of goals. First, the evolutionary (modus operandi) goal is to maximize the genetic information we forward to next generations. To accomplish this goal, our nervous system is wired to seek primary innate rewards (food, water, sex, and escape from danger). Neuronal reinforcement-learning tools enable living subjects to establish secondary rewards—that is, cues that predict future primary rewards. Reinforcement-learning mechanisms could even generate habits, in which our search for a secondary reward is not dependent on the later acquisition of a primary reward. Finally, at the third level, we, adult and mindful humans, can reflect on our actions and modify the goals of our behavioral policy.

What should these goals be? This is a hard question for each of us. Our life goals should shape our most important decisions (Should I devote my professional life to the study of the basal ganglia or to art and philosophy?) and the daily and hourly questions such as, What should I do next?[3] My personal answer is a local, bottom-up repairing of the world (*TIKUN OLAM* in Hebrew). This goal of life lies between the active goals of Western religions to unite with God (which are outside the scope of an atheist like me) and the (passive, from my viewpoint) main purpose of Buddhism to end suffering. The *TIKUN OLAM* approach actively searches for the minimization of human suffering but does not stop there. Even a utopian world with no suffering could be further repaired and improved. While both Western and Eastern religious approaches look at nature as an enemy or as the source of human problems, the bottom-up *TIKUN OLAM* approach accepts nature's "forward your genes" command. It actively explores ways to reduce suffering and aspires to a better world in a close-to-remote approach. The close-to-remote addition respects nature, forwards your gene rules, and starts repairing our loved ones that are probably genetically closer to us. If, as stated in the small-world rule (chapter 15, note 3), all the world's people are no more than six social connections away, each of us could repair the few social and family circles around us and make the world a nicer place. Sometimes we should fight for the survival of the self and our loved ones. Humans must resist the tyranny of biology to make the hard distinction between good and evil. It is not an easy distinction, especially when it comes to actions carried out in your name by your tribe or nation.[4] Nobody promised us a rose garden.

I agonizingly admit that I cannot provide logical reasons for selecting the *TIKUN OLAM* goal of life rather than one such as Ayn Rand's ethical egoism. Immanuel Kant, in *The Critique of Practical Reason*, says, "Two things fill the mind with ever new and increasing admiration and awe, the more often and steadily we reflect upon them: the starry heavens above me and the moral law within me. I do not seek or conjecture either of them as if they were veiled obscurities or extravagances beyond the horizon of my vision; I see them before me and connect them immediately with the consciousness of my existence" (161). Like Kant, I do not seek answers to the finite/infinite (in space and time) nature of the universe. However, I am not ready to stop seeking for the foundation of human moral law.

I believe that humans have a dual nature—one of being selfish, competitive, and greedy and the other of being altruistic, cooperative, and charitable. Coming back to our biology, we are social primate species (hopefully, more a bonobo than a chimpanzee). For most people in most places most of the time, certain behaviors are either right or wrong, and the right is being prosocial, cooperative and altruistic. But even if the *TIKUN OLAM* goal aligns with the laws of nature, a major message of this chapter is that we humans can confront these rules. Thus, I can't provide a definitive life goal. Probably, being a biological free-will agent means to make the most important decisions of life with no strong logical reasons. If (as Kurt Friedrich Gödel and other great minds like Alan Turing have taught us) even mathematics is not complete, why should we be? Perhaps Rabbi Menachem Mendel of Kotzk was right and "there is nothing so whole as a broken heart." Unlike the validated dead-end (incomplete) road of mathematics, I still have hope that we will find in the future the answers to the hard questions regarding the goals of the basal ganglia, behavior, and life. In the case that we do not, I am sure we will enjoy the path.[5]

To make the long story charted in this book short, the basal ganglia are our default, automatic Kahneman's system 1. There is little free will in the knee reflex or in basal ganglia–controlled automatic behavior. Nevertheless, we humans, equipped as we are with language and self-awareness, can reflect on our practices and optimize them according to our moral values and goals of life. We can take responsibility for our behavioral policy. Maybe a good one is the repair of the world (*TIKUN OLAM*).[6]

NOTES

CHAPTER 1

1. Andreas Vesalius (1514–1564) was a sixteenth-century Flemish anatomist, physician, and author of one of the most influential books on human anatomy, *De Humani Corporis Fabrica Libri Septem* (*On the Fabric of the Human Body*). Figure 1.2 (*left*) is a production of plate 7 of volume 7 of this book. Vesalius established that nerves are nonhollow lines (i.e., disproving Galen's assertion) that transmit sensation and motion and thus refuted his contemporaries' claims that ligaments, tendons, and aponeuroses were three types of nerve units. He believed that the brain is the center of the mind and emotion, in contrast to the common Aristotelian belief that the heart is the center of the body. He correspondingly believed that nerves themselves do not originate from the heart but from the brain.

Vesalius was born in Brussels, then part of the Habsburg Netherlands. He was professor at the University of Padua and later became imperial physician at the court of Holy Roman Emperor Charles V. In 1564 Vesalius embarked on a pilgrimage to the Holy Land, probably due to the accusation of dissecting a living body and the resulting pressures imposed on him by the Inquisition. On reaching Jerusalem, he received a message from the Venetian Senate requesting him to again accept the Paduan professorship. Vesalius accepted the offer and started back. After struggling for many days with adverse winds in the Ionian Sea, he was shipwrecked on the island of Zakynthos (Zante, the "Flower of the Levant," which during World War II refused the Nazis' order to list the members of the Jewish community). Here Vesalius soon died, in such debt that a benefactor kindly paid for his funeral. At the time of his death, he was forty-nine years of age.

2. For quantitative anatomy and physiology of the cortex, see Braitenberg and Schüz (1991); Abeles (1982, 1991). The two-volume set of *The Thalamus* by Edward G. Jones (there are several editions; one of them is edited with Mircea Steriade and D. A. McCormic) is the most comprehensive source of thalamic details. For physiology (and oscillations), see Steriade et al. (1990); Sherman and Guillery (2005).

The classification and nomenclature of thalamic nuclei have been debated by many giants, like Walker, Vogt, Olzewski, Hassler, Jones, and Percheron. See Ilinsky and Kultas-Ilinsky (2001), table 1, for a comparison of the nomenclature of motor thalamic (ventral lateral tier) nuclei in relation to their major subcortical afferents. For a treasure box of basal ganglia and thalamus anatomy in the nonhuman primate, see Percheron et al. (1996).

3. The unique structure order of the gross and cellular anatomy of the cerebellum has drawn the attention of physiologists and theoreticians (Braitenberg 1977). The fundamental work (Eccles

et al. 1967) probably triggered David Marr's (1969) theoretical study of the cerebellum. For more information, see Ito (1984, 2011); Barlow (2002); Heck (2016).

4. For the history of the evolution of the limbic concept, see MacLean (1990); Lautin (2001); Heimer et al. (2008). For a discussion of the evolution of the basal ganglia and the hippocampus as (or not) the brain structures for procedural versus declarative memory, see Murray et al. (2017). For a more popular introduction to the limbic system and emotions, see LeDoux (1996); Rolls (2007).

5. From textbooks of anatomy, of which there are so many. For me, the best was *Carpenter's Human Neuroanatomy*, 9th ed. (1995) by Andre Parent, a great explorer of basal ganglia anatomy. The first editions were written by M. Carpenter, who around 1940 conducted some of the early studies of the STN and the anatomy of the basal ganglia. I also recommend *Atlas of the Human Brain* by Juergen K. Mai, Milan Majtanik, and George Paxinos, 4th ed. (2015).

CHAPTER 2

1. My anatomical gurus include:

   a. The late Gerald Percheron, Jerome Yelnik, and Chantal François in Paris, who pioneered quantitative Golgi anatomical studies of the basal ganglia. Their series of three papers on Golgi studies of the globus pallidus, published in the *Journal of Comparative Neurology* in 1984, is a masterpiece (François et al. 1984; Yelnik et al. 1984; Percheron et al. 1984).

   b. Andre Parent, another hero of basal ganglia anatomy (e.g., Parent et al. 1995; Parent and Hazrati 1995a, b). Look also for Parent's classic textbook on the comparative anatomy of the basal ganglia (Parent 1986).

   c. Suzanne Haber, who highlighted basal ganglia connectivity (e.g., Haber et al. 2011; Haynes and Haber 2013).

   d. Ann Graybiel, who introduced the concept of the striosome (patch) and matrix compartments of the striatum (Graybiel and Ragsdale 1978). For a nice review, see Brimblecombe and Cragg (2017).

2. A popular urban legend claims that normal humans utilize only 10 percent of their brains during mental activity. It might reflect the old view that there are ten times more glia than neurons, and only neurons that can generate action potentials sustain the brain's mental activity.

CHAPTER 3

1. My role model is Valentino Braitenberg (1926–2011). Valentino was a good friend of my mentor Moshe Abeles, so I was lucky to meet him quite often during my work with Moshe Abeles. I still remember his warning that "being afraid that someone will thieve your ideas is self-aggrandizement since in reality people only care about their own ideas and nobody listens to you." In this chapter I am trying to follow his cortex studies (Braitenberg and Schutz [1991] and the first two anatomy chapters of Abeles [1991]).

2. To read more on the theory and studies of brain connectivity, see Sporns (2012); Seung (2012).

3. The strong inhibition between the basal ganglia output and the thalamus reaches a record in songbirds. The basal ganglia homologue areas of the zebra finch are essential for song learning but not for song production. In the zebra finch, each thalamic neuron receives only one, or at most two, giant perisomatic calyx-like synapses from its pallidal afferent. This unique anatomy enables the recording of extracellular signals from postsynaptic cells and presynaptic terminals and reveals that pallidal neurons do not have to pause to disinhibit thalamic neurons. In line with recording in the nonhuman primate, zebra finch pallidal units tend to show moderate modulation of discharge rate, and thalamic spiking is elicited by rapid deceleration of the pallidal discharge (Person and Perkel 2007).

4. Maimonides, in Hebrew Rabi Moses ben Maimon or Rambam, in Arabic Abū ʿImrān Mūsā bin Maimūn bin ʿUbaidallāh al-Qurtabī (1138–1204). He lived in Spain, Morocco, Israel, and Egypt and worked as a rabbi, philosopher, and physician. He wrote eleven medical books, emphasizing preventive medicine, exercise, eight hours of sleep ending at sunrise, and a holistic approach to body and mind (see chapters 17, 19, and 20). The Arabic poet Ibn Sanāʾ al-Mulk wrote on Maimonides, "Galanus healed the body, and Maimonides body and soul. His knowledge earned him a reputation as the doctor of the generation. He knew how to alleviate the pain of ignorance. If the moon had come into his hand, he would have healed the spots on her face."

CHAPTER 4

1. The Blue Brain Project (https://www.epfl.ch/research/domains/bluebrain/) is a prototypical example of the bottom-up approach. For reviews, see Markram (2006); Kandel et al. (2013). For major science outcomes (so far) of the Blue Brain Project, see Gal et al. (2017); Nolte et al. (2019); Reimann et al. (2019).

2. Anderson 1972. See also Marom 2015.

3. For a college-level introduction to chaos theory, see Gleick (1987). For a more formal and still very accessible introduction, see Strogatz (2014).

4. John Hopfield published in 1982 his analysis of recurrent artificial neural networks (Hopfield 1982). In my opinion, everything published since then regarding neural networks is a footnote to this study. Still, there are good textbooks on neural network theory (Abbott and Dayan 2001; Haykin 2008; Ermentrout and Terman 2010).

   For computational models and studies of the basal ganglia, see Wickens (1993); Houk et al. (1995); Miller and Wickens (2000); Miller (2008); Chakravarthy and Moustafa (2018).

5. The seminal book by David Marr was published in 1982 (Marr 1982). In 2010, MIT Press republished the book with a foreword from Shimon Ullmann and an afterword from Tomaso Poggio.

CHAPTER 5

1. For more details regarding microelectrode recording and stimulation techniques, Roger Lemon's (1984) book is still the best reference for the methodology of extracellular recording and stimulation in behaving nonhuman primates.

2. For the history of EEG, see Borck (2018). For EEG recording techniques and insight, see Luck (2014); Nunez and Srinivasan (2006).

3. In *Song of Myself*, 48, Walt Whitman (1819–1892, Poet.org; https://poets.org/poem/song -myself-48) says, "I have said that the soul is not more than the body, And I have said that the body is not more than the soul, And nothing, not God, is greater to one than one's self is . . . I find letters from God dropt in the street, and every one is signed by God's name, And I leave them where they are." I am afraid that for most of us students of the brain, we open the letters, but we don't understand the meaning.

CHAPTER 6

1. For correlation analyses, see Abeles (1982b); Eggermont (1990). More on analog (EEG, LFP) data analysis but still a superb source is Cohen (2014). Other good books are Kass et al. (2016); Mitra and Bokil (2007).

2. Jean-Baptiste Joseph Fourier (1768–1830) was orphaned at the age of nine. Fourier was educated by the Benedictine Order of the Convent of Saint Mark and accepted a military lectureship on mathematics. He took a prominent role in his own district in promoting the French Revolution. Fourier accompanied Napoleon Bonaparte on his Egyptian expedition in 1798 as scientific

adviser and was appointed secretary of the Institut d'Égypte. In 1801, after the withdrawal of the French army from Egypt, Napoleon appointed Fourier prefect (governor) of the Department of Isère in Grenoble, where he oversaw road construction and other projects.

At Grenoble, Fourier began to experiment on the propagation of heat. He presented his paper "On the Propagation of Heat in Solid Bodies" to the Paris Institute on 1807 and published his book *The Analytical Theory of Heat* in 1822. Here he made the claim that any function can be expanded in a series of sines of different amplitudes. Together with other scholars and scientists who accompanied Bonaparte's 1798–1801 expedition to Egypt (the savants), Fourier also contributed to the monumental Description de l'Égypte. The Description of Egypt was a series of publications (1809–1829) that aimed to comprehensively catalog all known aspects of ancient and modern Egypt. The cartographic section, "Carte de l'Égypte," had approximately fifty plates of maps, was the first triangulation-based map of Egypt, Syria, and Palestine, and was used as the basis for most maps of the region for much of the nineteenth century.

As do many others of our current society, I read and learn about history from the Western/European point of view. I was very happy to "meet" Joseph Fourier from the Egyptian point of view. This meeting was through the Hebrew translation of *The Day on Which the Laws of Nature Have Changed: The Incredible Chronicle of the French Occupation in Egypt* by Abd Al-Rahmam Al-Jabarti. This book was published by Maktoob, the Arabic-Hebrew Translators' Forum that seeks to combat the segregation of Jews and Palestinians. http://maktoobooks.com/en/.

3. One of the founders of the field recently wrote a popular, provocative, and must-read book. See Pearl and MacKenzie (2019).

CHAPTER 7

1. Mahlon DeLong has provided the first description of discharge rate and pattern in the basal ganglia of awake, behaving nonhuman primates (e.g., DeLong 1971; DeLong 1972; Crutcher and DeLong 1984a, b).

CHAPTER 8

1. Poems may have their own life that is independent of the intentions of the poet. I am assuming it is the same for books, and I wish this book a good, long, and interesting life. Like many others, I used the famous lines of Robert Frost (1874–1963) to endorse nonconformist behavior (https://www.poetryfoundation.org/articles/89511/robert-frost-the-road-not-taken). I was therefore surprised to find that Robert Frost wrote "The Road Not Taken" as a gentle mocking of the indecision of his good friend the English poet Edward Thomas (1878–1917). Thomas was indecisive about which road they ought to take on their many walks together when Frost spent two years in England. Nevertheless, and despite being a mature married man who could have avoided enlisting, Thomas enlisted in the Artists Rifles in 1915 and was killed in action soon after he arrived in France at Arras on Easter Monday, April 9, 1917. Indeed, "Yet knowing how way leads on to way, I doubted if I should ever come back."

2. Thomas Kuhn's 1952 classic book *The Structure of Scientific Revolutions* is still a good reflection of the academic research society. It still shows us the differences between our romantic dreams and academic reality. For a more updated and provocative exploration, see *Academia: All the Lies: What Went Wrong in the University Model and What Will Come in Its Place* by Tamar Almog and Oz Almog (Kindle Direct, 2020). My advice to the young scientist is 1) do not forget your dreams, and 2) do not listen too much to advice.

Sometimes, an artist can better pinpoint the essence of reality. I recommend the novel *Stoner* by John Williams, *Returning to Reims* by Didier Eribon (2018; translated by Michael Lucey), and *Footnote*, a short film written and directed by Joseph Cedar. Academic human relations probably do not strongly differ between humanity and natural science and over time.

3. For an introduction to the science of sleep, start with Walker (2017); Moorcroft (2013). For sleep medicine, consult Kryger et al. (2017). More hard-core sleep physiology books include Garcia-Rill (2015); Steriade and McCarley (1990); Monti et al. (2016). For your next summer vacation, I suggest the novel *The Castle of Dreams* by Michel Jouvet (one of the first heroes of sleep physiology), written in 1992 and translated by Laurence Garey (2008).

4. Right and left brains. I found Rogers et al. (2013) to be the best introduction to the biology of right/left brain hemispheres. For human neurology and culture, see Kushner (2017; left-handed neurologist and scientist); Gazzaniga (2016).

5. The first description of situs inversus is given in a letter titled "An Account of a Remarkable Transposition of the Viscera in the Human Body," *Lond Med J*. 1789;10(Pt 2):178–197. The letter was submitted by M. Baillie to John Hunter for publication in the *Philosophical Transactions of the Royal Society*. *Philosophical Transactions* is the world's first and longest-running scientific journal, launched in March 1665 by Henry Oldenburg. John Hunter (1728–1793) was one of the most distinguished British scientists and surgeons of his day. Hunter was also an admired mentor of James Parkinson. Notably, the next account in that volume of the *Philosophical* truncations is about winemaking. I wish we could practice this diversity and open-mindedness in today high-impact journals.

6. For the sake of space, I will not deeply discuss the questions of decussation—that is, why the right hemisphere takes care of the left side of the body and vice versa ("decussation" is from the Latin *deca*, ten, written as X). This is correct for all vertebrates but not for invertebrates (96 percent of known animal species). Biological evolution starts with spherical unicellular or multicellular simple organisms. Animals then evolved to have radial to bilateral left/right symmetry along the rostroposterior axis. Finally, in the transition from invertebrate to vertebrate, brain-to-body decussation appeared, and with no exceptions.

There are, in fact, multiple theories about why tracts decussate in the vertebrate nervous system. Santiago Ramón y Cajal suggested the visual map theory in 1898. The pupils in vertebrates' eyes invert the image on the retina so that the visual periphery projects to the medial side of the retina. By the decussation of the optic tracts (the optic chiasma), the visual periphery is again on the lateral aspects of the brain's visual areas. Unfortunately, reality is that the visual tracts spiral their way from the lateral geniculate body to the visual cortex. As a result, the retinal-geniculate map depicts the visual periphery on the medial side of the visual cortex. Valentino Braitenberg (1984) suggested that decussation enables more economic control of motor approach and withdrawal behavior. Others (e.g., Shinbrot and Young 2008) suggest a significant functional topological advantage and point out that as the number of wiring connections grows, decussated arrangements become more robust against wiring errors than simpler same-sided wiring schemes.

My favorite theory (de Lussanet and Osse 2012; Kinsbourne 2013) is the evolution twist hypotheses. Bilateral invertebrates have a large collection of neurons near the front of the body (their brain) and a nerve cord that runs the length of the animal along the ventral side of the body. The (axial or somatic) twist hypotheses suggest that at some point in evolution, near the appearance of the first vertebrates, the entire body plan underwent a 180-degree twist relative to the brain. Therefore, in vertebrates the spinal cord runs along the dorsal side of the body, but with decussation.

Finally, not all brain-periphery connections decussate. The peripheral olfactory organs are connected to the ipsilateral centers of the forebrain. Unlike the basal ganglia, the connection between the cerebellum and the spinal cord are ipsilateral, with no (e.g., dorsal spinocerebellar tract, lateral vestibulospinal tract) or double decussation (e.g., ventral spinocerebellar tract). For me, the reason why the basal ganglia and cerebellum are connected with the contralateral or ipsilateral side of the body is one of the great questions waiting for an answer.

7. Probably, the most extreme example of physiological right/left-hemisphere asymmetry is the unihemispheric SWS of several aquatic and avian species (Mascetti 2016). During unihemispheric

sleep, one half of the brain is in deep sleep, while the other side of the brain is "awake" (Mukha-metov et al. 1977). This is probably enabled by unilateral cholinergic and noradrenergic innervation of the neocortex and alternation of high/low cholinergic and adrenergic tonus in the two hemispheres. The size of the corpus callosum of animals that can maintain unihemispheric sleep is diminished compared to similar-size animals, and they (at least in birds) have complete decus-sation of the optic nerve (i.e., the left hemisphere controls the right eye and not the right visual field). Notably, only SWS can be localized to one hemisphere; REM sleep is always bilateral or absent in animals with unihemispheric sleep.

Unihemispheric sleep enables animals that have to be in constant motion (like dolphins, or birds during long migratory flights) or are living in areas of high predation to have the benefits of sleep (e.g., removal of accumulated waste products from the brain). Unihemispheric sleep can play a role in social sleep behavior. In species living under the risk of predation, the animals on the edge of the group keep the lateral eye and the opposite hemisphere open and awake. On the other hand, other species follow a reverse organization, probably to maintain group formation and cohesion.

Right/left hemispheric asymmetry in sleep depth occurs during the first night for humans in a new place. The default-mode cortical network in one hemisphere is kept more vigilant to wake the sleeper upon detection of deviant stimuli. This right/left hemispheric asymmetric sleep in a novel environment may play a similar protective function as complete unihemispheric sleep (Tamaki et al. 2016). Finally, although not as drastic as the phenomenon of unihemispheric sleep, both rodents and humans have been shown to have local "off" sleeplike neuronal episodes following sleep deprivation. These local neuronal events are not synchronized with remote or even nearby neurons (Vyazovskiy et al. 2011; Nir et al. 2017) and, together with unihemispheric sleep, reveal that sleep is not a global uniform brain phenomenon.

8. National Institutes of Health, "NIH Policy on Sex as a Biological Variable," https://orwh.od .nih.gov/sex-gender/nih-policy-sex-biological-variable.

CHAPTER 9

1. Note that the definition of "direct/indirect basal ganglia pathway" has a different meaning if you are living in the dorsal or ventral striatum communities. Dorsal striatum people's (like this author) main interests are movement disorders like Parkinson's disease. Ventral striatum people care about addiction. The direct/indirect pathways of the ventral striatum community connect between the ventral striatum and the ventral tegmental area (VTA), either directly or indirectly through the ventral pallidum.

2. Modified from the quote "If the human brain were so simple, that we could understand it, we would be so simple, that we couldn't" in the 1977 book *The Biological Origin of Human Values* by George Edgin Pugh. The statement was used as a chapter epigraph with a footnote that specified an ascription to Emerson M. Pugh, who was the father of the author. Both the father and son were physicists. https://quoteinvestigator.com/2016/03/05/brain/.

CHAPTER 10

1. In my opinion there is one exceptional book relating to the topics of this chapter—Sutton and Barto (2018). The first edition of this book, published in 1998, was to me and other students the best introduction to the field of reinforcement learning. I am sure that the publisher pressured the authors for new editions. To my delight, while Sutton and Barto took their time, the second 2018 edition testifies that they used their time well. Their new chapter 14, "Psychology," and chapter 15, "Neuroscience," are masterpieces. I have never met Sutton and Barto, but I would like to thank them for teaching me reinforcement learning and that patience is also a form of action.

2. Wolfram Schultz, Peter Dayan, and Reed Montague's seminal paper (Schultz et al. 1997) constructed the first and most important bridge between reinforcement learning and dopamine

physiology. They created a new world and deserve every credit for this. By the time this book was written, their paper had been cited more than eighty-five hundred times, and in my opinion it should be considered exemplary by any student of the brain.

3. The sky is not very high for the dopaminergic neurons. In line with their wide (a few milliseconds) action potential waveforms, the maximal discharge rate reported for dopaminergic neurons is around 50 spikes/s. Nevertheless, the dynamic domain for increase in discharge rate from approximately 5 to 50 spikes/s is much larger than the negative domain, from approximately 5 to 0 spikes/s.

CHAPTER 11

1. "With four parameters I can fit an elephant and with five I can make him wiggle his trunk"— attributed to Johnny von Neumann. For proof of this statement, see Mayer et al. (2010).

CHAPTER 12

1. Sigmund Freud said, "Everywhere I go I find a poet has been there before me." Almost everywhere I go in this book, I find that cerebellum researchers have been there before me. In chapters 9–11, I described the evolution of the computational models of the basal ganglia, from the box-and-arrow models that assume the basal ganglia modulate motor vigor, to reinforcement-learning models that assume the basal ganglia modulate motor learning, to multiobjective optimization models that suggest the basal ganglia modulate both motor vigor and plasticity.

In the cerebellum literature, the motor vigor and motor plasticity models are called performance and learning theories, respectively. Valentino Braitenberg and Roger Atwood suggested in 1958 that the propagation of signals along parallel fibers allows the cerebellum to detect time relationships and to modulate ongoing behavior (delay line hypothesis). However, the conduction time along the parallel fibers is too fast compared to motor actions. David Marr (1969) and James Albus (1971) are the forerunners of the motor-learning theory of the cerebellum. They suggested that the climbing fibers and complex action potential of the Purkinje cells serve as a motor error signal that leads to enhancement (Marr) or depression (Albus) of the efficacy of the connectivity between the parallel fibers and the Purkinje cells.

2. "To move things is all that mankind can do . . . for such the sole executant is muscle, whether in whispering a syllable or felling a forest." Charles Sherrington in the Linacre lecture of 1924 (Molnár and Brown 2010).

3. The amazing behind-the-screen story of the life friendship and collaboration of Daniel Kahneman and Amos Tversky is beautifully presented in *The Undoing Project: A Friendship That Changed Our Minds* by Michael Lewis (2017).

4. The bottleneck dimensionality-reduction hypothesis (Slonim et al. 2006) is living within the framework of information theory. Tali Tishbi introduced me to bottleneck dimensionality-reduction concepts.

5. Izhikevich (2010) gives an outside-the-box view on the dynamic of neurons. It is a must-read book.

6. A unique property of thalamic relay cells is the dense distribution of calcium T channels in their soma and dendrites. This leads to two different dynamic (discharge as a function of input current, I-f curve) modes of the thalamic relay neurons (Sherman and Guillery 2005) that resemble the I-f curves of type I and type II excitability. When the thalamic relay cell is in resting potential, a depolarizing input activates the tonic mode of firing that is linearly correlated with the input current. However, if the thalamic relay cell is hyperpolarized for ≥100 ms, the calcium T channels are deinactivated, and the same depolarizing input activates the low threshold calcium spike and a burst of two to ten fast sodium action potentials. The linear tonic mode is probably

used during the awake state to reliably forward information to the cortex. The burst mode, more commonly observed during periods of relaxation and sleep, can be used as an alarm call for something that is strong or important enough (e.g., your baby's cry) to cross the thalamic gate.

7. I am not the first to admire the empty void. Lao Tzu (Laozi) wrote in the Tao Te Ching, "Thirty spokes share the hub of a wheel; yet it is its center that makes it useful. You can mould clay into a vessel; yet, it is its emptiness that makes it useful. Cut doors and windows from the walls of a house; but the ultimate use of the house will depend on that part where nothing exists. Therefore, something is shaped into what is; but its usefulness comes from what is not."

CHAPTER 13

1. The complexity of spinal cord organization should not be underestimated. There are 31 spinal segments. In the seventh lumbar segment of the dog, there are 360,000 neurons. Only 2 * 12,000 sensory fibers enter this segment through the dorsal root, and 2 * 6,000 motor neurons and motor axons leave through the ventral root (Gelfan and Tarlov 1963). The existence of more than 300,000 spinal interneurons promotes extensive spinal processing.

Extending the calculation to the whole spinal cord, we reach the total number of $31 * 360,000 = 11 * 10^6$ neurons in the spinal cord (about the same number of basal ganglia neurons in the macaque and a small fraction of the number of neurons in the cortex and cerebellum; see table 2.1). These numbers are of the same order as achieved by modern isotropic fractionator and stereology methods, $28 – 69 * 10^6$ and $197 – 222 * 10^6$ neurons in the spinal cord of the macaque and human, respectively (Bahney and von Bartheld 2018).

Extending the numbers also to the motor nuclei of the cranial nerve, we reach the conclusion that humans have about $12,000 * (31+10) = 500,000$ alpha motor neurons or motor units. Thus, our fantastic motor repertoire (including dancing and singing) is achieved with 600 named muscles and $0.5 * 10^6$ motor neurons.

It is also interesting to compare the number of $0.5 * 10^6$ motor units to the estimated number of $10 – 20 * 10^6$ afferent channels from the periphery (e.g., 10^6 photoreceptors in each eye) and to the huge expansion of $86,000 * 10^6$ neurons in between the input/output of the central nervous system.

2. There are many textbooks of Parkinson's disease and movement disorders. See Jankovic and Tolosa (2015); Donaldson et al. (2012).

3. When it comes to movement disorders, one movie is better than a thousand words. In the public domain, look for the Spring Video *Atlas of Movement Disorders* by Dr. Prodigious, https://www.youtube.com/playlist?list=PLwlvPe1bGTl9GqPMhEG5_zRVd3NVSSQ46&app=desktop. Other commercial video atlases are *Involuntary Movements: Classification and Video Atlas* by Hiroshi Shibasak, Mark Hallett, Kailash P. Bhatia, Stephen G. Reich, and Bettina Balint; *Movement Disorders: 100 Instructive Cases* by Stephen G. Reich (2008), and *Movement Disorders: A Video Atlas* by Roongroj Bhidayasiri and Daniel Tarsy (2012). The video library of the movement disorder society (MDS) houses more than eighteen hundred videos, but you must be a current MDS member to view them.

4. In most descriptions of Parkinson's disease, you will find the definition of akinesia/bradykinesia as a lack or slowness of voluntary movement. While I agree that voluntary movements are slow in patients with Parkinson's disease, I most strongly believe that automatic movements are the most absent and drastically affected. Most patients are able to overcome their akinesia when we shift their attention to the requested movement. Therefore, when the patient is asked to open/close a fist, the first movement will usually be of high and normal amplitude and speed, and the next movements will slow down and be of diminished amplitude.

5. Parkinson's disease textbooks describe Parkinson's tremor as a slow tremor. Indeed, the typical frequency of Parkinson's tremor (4–7 Hz) is below the faster cerebellar and physiological tremors

(8–12 Hz). Nevertheless, Parkinson's tremor is faster than typical human rhythmic movements, which tend to be around 2–3 Hz (Logigian et al. 1991).

The interactions between the neuronal attractors of paced voluntary movements and the involuntary tremor movements of Parkinson's disease may lead to the interesting phenomenon of movement hastening. Parkinson's patients could produce auditory-paced frequencies of 1 and 2 Hz, but at higher cue frequencies, their voluntary movements are "hastened" to the tremor frequency band (see, for example, Logigian et al. 1991; Elazary et al. 2003).

6. The amazing life of James Parkinson is described in *The Enlightened Mr. Parkinson: The Pioneering Life of a Forgotten English Surgeon* (2017) by Cherry Lewis. I was happy to read about his fights against the tyranny of the English government, child labor, and other social problems of his time. We are only two hundred years later. We should be proud of humanity for making giant steps toward a better world since that era.

7. One should speak of signs rather than symptoms, as I recently learned from a good anonymous referee. Signs are objective manifestations of the disease, such as tremor, rigidity, and so on. Symptoms are subjective manifestations, such as pain and fatigue. However, I have seen very few mentions of Parkinson's signs, so here I will hew to the old tradition of speaking on Parkinson's symptoms, with apology to the gods of correct behavior.

8. More reading on Parkinson's sleep disorders can be found in Videnovic and Högl (2015) and in Chahine (2020).

9. From the twenty-first-century Western point of view, the life story of Constantin Freiherr von Economo (1876–1931) has some positive and negative records. On the positive side of the list are his studies of encephalitis lethargica, his atlas of cytoarchitectonics of the cerebral cortex (figure 1.3B), and the identification of the "von Economo neurons," the large bipolar neurons in layer V of the frontal cortex. On the negative side, I would list von Economo's interest in "elite brains" and his hope to find microstructural characteristics in these brains that distinguish them from the average brain.

10. For more information on encephalitic lethargica and postencephalitic parkinsonism, see Vilensky (2010); Spinney (2018); Foley (2019).

11. The novel *If This Is a Man* (Survival in Auschwitz) by Primo Levi starts with a poem:

You who live safe In your warm houses, You who find warm food And friendly faces when you return home. Consider if this is a man Who works in mud, Who knows no peace, Who fights for a crust of bread, Who dies by a yes or no.

Consider if this is a woman Without hair, without name, Without the strength to remember, Empty are her eyes, cold her womb, Like a frog in winter.

Never forget that this has happened. Remember these words. Engrave them in your hearts, When at home or in the street, When lying down, when getting up. Repeat them to your children. Or may your houses be destroyed, May illness strike you down, May your offspring turn their faces from you.

A paragraph of Primo Levi's message to humanity is composed of reverberating words from the Jewish *Shema Yisrael* (Hebrew for "Hear, O Israel"). It is based on a section of the Old Testament (Torah, Deuteronomy 6:6–7: "And these words, which I command thee this day, shall be upon thy heart; and thou shalt teach them diligently unto thy children, and shalt talk of them when thou sittest in thy house, and when thou walkest by the way, and when thou liest down, and when thou risest up"). The Shema serves as a centerpiece of the morning and evening Jewish prayer services, and it is traditional for Jews to say the Shema as their last words. My mother, Ahuva Bergman Z"L, did so during the last stages of editing this book.

CHAPTER 14

1. For the detective-fiction-like (or even better) story of the discovery of MPTP, see *The Case of the Frozen Addicts* by J. William Langston and Jon Palfreman (1995).

2. "On Rigor in Science" (the original Spanish-language title is "Del rigor en la ciencia") is a one-paragraph short story written in 1946 by Jorge Luis Borges about model-reality relation. The story imagines an empire where the science of cartography becomes so exact that only a map on the same scale as the empire itself will suffice. The story continues to say that "succeeding generations came to judge a map of such magnitude cumbersome . . . In the western deserts, tattered fragments of the map are still to be found, sheltering an occasional beast or beggar." Borges's story echoes a similar theme of Lewis Carroll's *Sylvie and Bruno Concluded*, where the one-to-one scale maps were objected to by the farmers: "They said it would cover the whole country, and shut out the sunlight! So we now use the country itself, as its own map, and I assure you it does nearly as well."

CHAPTER 15

1. When I arrived at Mahlon DeLong's lab in 1987, I met a forgotten hero of the basal ganglia and Parkinson's physiology—William (Bill) Miller. Bill was an MD, PhD student of Mahlon's and had performed the first physiological recording of the globus pallidus of monkeys before and after MPTP treatment. He recorded from three monkeys, and I still keep a paper copy of his exceptional PhD thesis in my office.

Unluckily, Bill had published his results only as a book chapter in *The Basal Ganglia II*. The Basal Ganglia book series summarized the meetings of the International Basal Ganglia Society (IBAGS) and for many years was the bible of the field. In his chapter (Miller and DeLong 1987), Bill reported the changes in the discharge rate of the GPe and GPi. Mahlon's deep insight is shown in the title, which reports on altered tonic activity in the STN. Bill recorded only in the STN of one monkey and after MPTP treatment. He found that the average discharge rate in this monkey (twenty-nine spikes/second) was higher than the average (twenty-four spikes/second) reported in the previous study (Georgopoulos et al. 1983). It took Thomas Wichmann and me three years to verify the intuition of Mahlon regarding the STN discharge rate, but that is another story.

The first peer-reviewed publications of changes in pallidal activity following MPTP treatment of nonhuman primates were those of Michel Filion and his group at the Université de Montréal, Canada (Filion and Tremblay 1991). The other papers of Michel Filion on his early studies of the MPTP monkeys are a lost treasure (Filion et al. 1988; Tremblay et al. 1989; Filion et al. 1991). The MPTP primal model was the second time for Michel Filion to study the effect of dopamine tone on pallidal activity. His 1979 paper testing the effect of haloperidol (a dopamine antagonist) on pallidal discharge is a masterpiece (Filion 1979).

2. Depending on your math skills, read one of the two books by Steven Strogatz (2003, 2014). Sometimes, one movie is better than a thousand words. I suggest 1) movies on the synchronization of metronomes—for example, https://www.youtube.com/watch?v=Aaxw4zbULMs, and 2) the lateral wobbling of London Millennium Bridge on its opening, https://www.youtube.com/watch?v=gQK21572oSU.

3. The small-world rule, or six-handshakes rule, claims that all people are six, or fewer, social connections away from each other. The rule was originally set out by Frigyes Karinthy in 1929 and popularized in a 1990 play written by John Guare, *Six Degrees of Separation*, and the *Babel* (1993 and 2006) drama films. Duncan J. Watts and Steven Strogatz showed that the average path length between two nodes in a random network is equal to ln N/ln K, where N = total nodes, and K = acquaintances per node. If $N = 6 * 10^9$ and K = 30, we get a mean degrees of separation = 22.5 / 3.4 = 6.6. The math will work even better if the distances of the network connections are logarithmically (power low) distributed, with many short-distance and few long-distance connections.

4. For an excellent book on brain oscillatory activity and the origin of LFP, the reader should consult Buzsáki (2006). It is biased in favor of Gyorgy Buzsáki's pleasure zone—the hippocampus—but nobody is perfect, and this is a great book.

CHAPTER 16

1. The Olbers's paradox, named after the German astronomer Heinrich Wilhelm Olbers (1758–1840), also known as the "dark night sky paradox," is the argument that if the universe is static, homogeneous, and populated by an infinite number of stars, then the night sky should be completely illuminated and very bright. From the point of view of an electrode, the STN universe is infinite and homogeneous, and therefore the "night sky," when there is no big spike in the recording, is bright and represents the spiking activity of the population of neurons in the structure.

CHAPTER 17

1. Jean-Martin Charcot (1825–1893) was a French neurologist and professor of anatomical pathology. Charcot was interested in all fields of neurology. He made significant progress in our understanding of multiple sclerosis and other neurological disease. On the other hand, his work on hysteria is sharply criticized these days. For students of Parkinson's disease, his discovery of Parkinson's rigidity and the naming of the disease after James Parkinson are major events.

Charcot worked and taught at the Pitié-Salpêtrière Hospital in Paris for 33 years (the same place where Jean-Baptiste Pussin and Philippe Pinel ordered the removal of chains from insane women around 1800 and Gerald Percheron and his colleagues did the Golgi studies of the basal ganglia 180 years later). His reputation as an instructor drew students from all over Europe. Many of them (e.g., Sigmund Freud, Joseph Babinski, Georges Gilles de la Tourette) will be remembered forever for their psychiatric and neurological insights. *The Story of San Michel* by Axel Munthe describes the personal memories of Munthe as a student in Paris. There are question marks regarding the validity of these memories. However, a good artist does not have to be an accurate photographer of history.

2. It is striking today to see how the dopamine system was ignored by basal ganglia and movement disorders specialists. Derek Ernest Denny-Brown is one of the fathers of modern American neurology. Denny-Brown completed a research fellowship with Sir Charles Scott Sherrington and worked with John Farquhar Fulton (whose findings that lesions of the prefrontal cortex created a calming effect in the monkeys encouraged Egas Moniz to develop the frontal lobotomy). Denny-Brown was a professor of neurology at Harvard Medical School, the president of the American Neurological Society, and an active writer. In his monograph *The Basal Ganglia and Their Relation to Disorders of Movement*, in a section titled "The Enigma of Parkinsonism," Denny-Brown writes (1962), "In 1919 Tretiakoff drew attention to some loss of cells in the substantia nigra" and then said that such loss was only occasionally found in his postmortem studies. Dopamine is not mentioned in the book index.

James Purdon Martin's 1967 book titled *The Basal Ganglia and Posture* is another exemplar of careful clinical studies, but again, dopamine neurons are ignored. On the positive side, these two great clinicians remind us of the importance of axial and tonus changes in basal ganglia–related disorders.

3. Early references to ergot poisoning (ergotism) date back as far as 600 BC. The common name for ergotism is "Saint Anthony's Fire," in reference to the hospital Brothers of Saint Anthony, an order of monks established in 1095 that specialized in treating ergotism victims. There are two types of ergotism. The first is characterized by muscle spasms, fever, and hallucinations. This is believed to be caused by serotonergic stimulation of the central nervous system. The second type of ergotism is marked by violent burning, absent peripheral pulses, and shooting pain in the poorly vascularized distal organs. Other symptoms include strong uterine contractions, nausea, and vomiting. Thus, since the Middle Ages, controlled doses of ergot have been used to induce abortions and to stop maternal bleeding after childbirth. The broad medicinal uses of the ergot alkaloids have encouraged their intensive research in the first half of the twentieth century, culminating in the development of both legal drugs (e.g., postsynaptic dopamine agonists) and psychedelic agents (e.g., lysergic acid diethylamide [LSD], a serotonin and dopamine receptor agonist).

4. A popular model of sleep/wake control is the two-process model: process S and C. Process S is a process of substance accumulation, and process C is the circadian process. Usually, the two processes are synchronized over the light/dark cycle. Sleep disorders—for example, as in jet lag—can be due to a lack of synchronization of the two processes. The first candidate for S was serotonin. These days the literature is pointing toward adenosine, although it seems that adenosine is not the only S compound. In any case, serotonin has a similar day-night profile as adenosine, accumulating during the day and decreasing during sleep.

Chronic (in humans, more than seven days) sleep deprivation may be fatal, probably because of the accumulation of adenosine and its adverse effects on the heart (adenosine is used to treat supraventricular tachycardia).

The transient beneficial effects of sleep deprivation on depression symptoms might be attributed to higher levels of serotonin in the day following a night with no sleep. In the same vein, common antidepressant therapy with specific serotonin reuptake inhibitors (SSRIs) significantly impairs REM sleep. SSRI beneficial effects could be due to the effects of SSRIs on sleep architecture and the resulting increase in serotonin levels.

CHAPTER 18

1. Early DBS technology used the "know-how" of cardiac pacemaker technology. Cardiac pacemaker history is an example of how medicine and industry can benefit each other. Earl Bakken and his brother-in-law Palmer Hermundslie owned a medical equipment repair shop in Minneapolis. Bakken came to know Dr. Walton Lillehei, a heart surgeon at the University of Minnesota Medical School. Artificial heart pacemakers of the mid-twentieth century were wall powered. The shortcomings of this technology were made painfully apparent following a power outage in 1957 that affected large sections of Minnesota. A pacemaker-dependent pediatric patient of Lillehei died because of the blackout. The next day, Lillehei spoke with Bakken about developing some form of battery-powered pacemaker. Bakken modified the design for a transistorized metronome and within one month created the first battery-powered external pacemaker. In 1960 he produced an implantable pacemaker (Nelson 1993). Today, Medtronic operates worldwide, employs over one hundred thousand people, and enjoys strong competition with other companies such as Boston Scientific, Abbott, and PINS.

2. For more comprehensive and updated information regarding DBS surgery and therapy, see Marks (2015).

3. The historical expression "Malta yok" (Malta does not exist) was born, according to some history books, when the admiral of the Othman fleet, Kapudan Pasha, was ordered by Ibrahim I "The Mad," an Othman emperor of the seventeenth century, to attack Malta. Kapudan Pasha decided to avoid the order and reported back that Malta does not exist.

Kapudan Pasha deliberately lied. I do not recommend that a DBS physiologist lie. But I do recommend insisting on the truth even if it is not what is expected by your operating room emperor.

CHAPTER 19

1. Harry Harlow received the worst public relations treatment for his 1930–1940 experiments on social isolation in infant monkeys. Nevertheless, these experiments laid the groundwork for our understanding that affection is more important than a sterile environment. His findings changed our world and saved the lives of thousands of orphans. I strongly recommend reading *Love at Goon Park: Harry Harlow and the Science of Affection*, by Deborah Blum (2002), to learn again that humans are complex.

2. Why tardive and levodopa-induced dyskinesia have different body distributions (more orofacial for tardive, more upper limb for levodopa-induced) is still a mystery to me. Nor do

I understand why levodopa-induced hallucinations are mainly visual, while schizophrenia hallucinations are mainly auditory.

3. "Delusion and Dream in Jensen's Gradiva" is an essay written in 1907 by Sigmund Freud that subjects the novel *Gradiva* by Wilhelm Jensen, and especially its protagonist, to psychoanalysis.

4. Walter Freeman, who introduced the frontal lobotomy to the US, is usually depicted as a modern monster. Life is never black and white, and the complicated story of Walter Freeman is given in *The Lobotomist: A Maverick Medical Genius and His Tragic Quest to Rid the World of Mental Illness* by Jack El-Hai (2005). For the hard-core history, read *Last Resort: Psychosurgery and the Limits of Medicine* by Jack D. Pressman (1998).

5. For the dark side of DBS history, see *The Pleasure Shock: The Rise of Deep Brain Stimulation and Its Forgotten Inventor* by Lone Frank (2018).

CHAPTER 20

1. I hope that the reader still remembers our discussions of sleep, dreams, hallucinations, and the basal ganglia. The most famous of Zhuangzi stories—"Zhuang Zhou Dreams of Being a Butterfly"—reminds us of the fuzzy border between sleep and hallucinations: "Once, Zhuang Zhou dreamed he was a butterfly, a butterfly flitting and fluttering about, happy with himself and doing as he pleased. He didn't know that he was Zhuang Zhou. Suddenly he woke up and there he was, solid and unmistakable Zhuang Zhou. But he didn't know if he was Zhuang Zhou who had dreamt he was a butterfly, or a butterfly dreaming that he was Zhuang Zhou. Between Zhuang Zhou and the butterfly there must be some distinction! This is called the Transformation of Things" (18).

2. "Give me a lever long enough and a fulcrum on which to place it, and I shall move the world." Attributed to Archimedes of Syracuse (c. 287–c. 212 BC).

3. How critical it is to set a life goal is beautifully clarified by Lewis Carroll in *Alice's Adventures in Wonderland* (1865): "'Is this the right way?' said Alice to the Cheshire cat. 'That depends a lot on where you want to go,' said the cat. 'I don't know where I'm going,' said Alice. 'Then it doesn't much matter which way you go,' said the cat" (71).

4. As George Orwell suggested in the introductory sentence to his 1941 essay "The Lion and the Unicorn: Socialism and the English Genius," "As I write, highly civilized human beings are flying overhead, trying to kill me. They do not feel any enmity against me as an individual, nor I against them. They are 'only doing their duty,' as the saying goes. Most of them, I have no doubt, are kind-hearted law-abiding men who would never dream of committing murder in private life" (1).

5. Constantine Peter Cavafy (1863–1933) was an Egyptian Greek poet, journalist, and civil servant. One of his poems is "Itahaka":

As you set out for Ithaka, hope your road is a long one, full of adventure, full of discovery. Laistrygonians, Cyclops, angry Poseidon—don't be afraid of them: you'll never find things like that on your way, as long as you keep your thoughts raised high, as long as a rare excitement stirs your spirit and your body. Laistrygonians, Cyclops, wild Poseidon—you won't encounter them unless you bring them along inside your soul, unless your soul sets them up in front of you.

Hope your road is a long one. May there be many summer mornings when, with what pleasure, what joy, you enter harbors you're seeing for the first time; may you stop at Phoenician trading stations to buy fine things, mother of pearl and coral, amber and ebony, sensual perfume of every kind—as many sensual perfumes as you can; and may you visit many Egyptian cities to learn and go on learning from their scholars.

Keep Ithaka always in your mind. Arriving there is what you're destined for. But don't hurry the journey at all. Better if it lasts for years, so you're old by the time you reach the island, wealthy with all you've gained on the way, not expecting Ithaka to make you rich.

Ithaka gave you the marvelous journey, Without her you wouldn't have set out. She has nothing left to give you now. And if you find her poor, Ithaka won't have fooled you. Wise as you will have become, so full of experience, you'll have understood by then what these Ithakas mean.

C. P. Cavafy, "The City," in *C. P. Cavafy: Collected Poems*, trans. Edmund Keeley and Philip Sherrard (Princeton, NJ: Princeton University Press, 1975).

6. Marcus Tullius Cicero (106 BC–43 BC) said, "A room without books is like a body without a soul." Below are some of the books that have shaped my life and this chapter:

Motivation and Personality by Abraham H. Maslow (1954)

Man on His Nature by Charles Sherrington (1940)

Becoming Human: Evolution and Human Uniqueness by Ian Tattersall (1998)

Phi: A Voyage from the Brain to the Soul by Giulio Tononi (2012)

The Feeling of Life Itself: Why Consciousness Is Widespread but Can't Be Computed by Christof Koch (2019)

Beyond Freedom and Dignity by B. F. Skinner (1971)

Who's in Charge? Free Will and the Science of the Brain by Michael S. Gazzaniga (2012)

Man's Search for Meaning by Viktor E. Frankl (1946)

REFERENCES

Abdi A, Mallet N, Mohamed FY, et al. Prototypic and arkypallidal neurons in the dopamine-intact external globus pallidus. *J Neurosci.* 2015;35(17):6667–6688.

Abeles M. Quantification, smoothing, and confidence limits for single-units' histograms. *J Neurosci Methods.* 1982a;5(4):317–325.

Abeles M. *Local Cortical Circuits: An Electrophysiolgical Study.* Springer-Verlag; 1982b.

Abeles M. *Coticonics, Neural Circuits of the Cerebral Cortex.* Cambridge University Press; 1991.

Abitz M, Nielsen RD, Jones EG, Laursen H, Graem N, Pakkenberg B. Excess of neurons in the human newborn mediodorsal thalamus compared with that of the adult. *Cereb Cortex.* 2007;17(11):2573–2578.

Adam Y, Kim JJ, Lou S, et al. Voltage imaging and optogenetics reveal behaviour-dependent changes in hippocampal dynamics. *Nature.* 2019;569(7756):413–417.

Adler A, Finkes I, Katabi S, Prut Y, Bergman H. Encoding by synchronization in the primate striatum. *J Neurosci.* 2013;33(11):4854–4866.

Adler A, Joshua M, Rivlin-Etzion M, et al. Neurons in both pallidal segments change their firing properties similarly prior to closure of the eyes. *J Neurophysiol.* 2010;103(1):346–359.

Adler A, Katabi S, Finkes I, Israel Z, Prut Y, Bergman H. Temporal convergence of dynamic cell assemblies in the striato-pallidal network. *J Neurosci.* 2012;32(7):2473–2484.

Adler A, Katabi S, Finkes I, Prut Y, Bergman H. Different correlation patterns of cholinergic and GABAergic interneurons with striatal projection neurons. *Front Syst Neurosci.* 2013;7:47.

Aertsen AM, Gerstein GL. Evaluation of neuronal connectivity: sensitivity of cross-correlation. *Brain Res.* 1985;340(2):341–354.

Aertsen AM, Gerstein GL, Habib MK, Palm G. Dynamics of neuronal firing correlation: modulation of "effective connectivity." *J Neurophysiol.* 1989;61(5):900–917.

Ahissar E, Arieli A. Seeing via miniature eye movements: a dynamic hypothesis for vision. *Front Comput Neurosci.* 2012;6:89.

Alcacer C, Andreoli L, Sebastianutto I, Jakobsson J, Fieblinger T, Cenci MA. Chemogenetic stimulation of striatal projection neurons modulates responses to Parkinson's disease therapy. *J Clin Invest*. 2017 Feb 1;127(2):720–734.

Albin RL, Aldridge JW, Young AB, Gilman S. Feline subthalamic nucleus neurons contain glutamate-like but not GABA-like or glycine-like immunoreactivity. *Brain Res*. 1989;491(1):185–188.

Albin RL, Young AB, Penney JB. The functional anatomy of basal ganglia disorders. *Trends Neurosci*. 1989;12(10):366–375.

Albus JS. A theory of cerebellar function. *Mathematical biosciences* 1971;10:25–61.

Alexander GE, DeLong MR. Microstimulation of the primate neostriatum. I. Physiological properties of striatal microexcitable zones. *J Neurophysiol*. 1985a;53(6):1401–1416.

Alexander GE, DeLong MR. Microstimulation of the primate neostriatum. II. Somatotopic organization of striatal microexcitable zones and their relation to neuronal response properties. *J Neurophysiol*. 1985b;53(6):1417–1430.

Alexander GE, DeLong MR, Strick PL. Parallel organization of functionally segregated circuits linking basal ganglia and cortex. *Annu Rev Neurosci*. 1986;9:357–381.

American Psychiatric Association. *Diagnostic and Statistical Manual of Mental Disorders*. 5th ed. American Psychiatric Association; 2013.

An K, Zhao H, Miao Y, et al. A circadian rhythm-gated subcortical pathway for nighttime-light-induced depressive-like behaviors in mice. *Nat Neurosci*. 2020;23(7):869–880.

Anderson B. There is no such thing as attention. *Front Psychol*. 2011 Sep 23;2:246.

Anderson PW. More is different. *Science*. 1972;177(4047):393–396.

Andersson C, Hamer RM, Lawler CP, Mailman RB, Lieberman JA. Striatal volume changes in the rat following long-term administration of typical and atypical antipsychotic drugs. *Neuropsychopharmacology*. 2002;27(2):143–151.

Aosaki T, Kawaguchi Y. Actions of substance P on rat neostriatal neurons in vitro. *J Neurosci*. 1996;16(16):5141–5153.

Aragona BJ, Day JJ, Roitman MF, Cleaveland NA, Wightman RM, Carelli RM. Regional specificity in the real-time development of phasic dopamine transmission patterns during acquisition of a cue-cocaine association in rats. *Eur J Neurosci*. 2009;30(10):1889–1899.

Arecchi-Bouchhioua P, Yelnik J, François C, Percheron G, Tandé D. 3-D tracing of biocytin-labelled pallido-thalamic axons in the monkey. *Neuroreport*. 1996;7(5):981–984.

Arecchi-Bouchhioua P, Yelnik J, François C, Percheron G, Tandé D. Three-dimensional morphology and distribution of pallidal axons projecting to both the lateral region of the thalamus and the central complex in primates. *Brain Res*. 1997;754(1–2):311–314.

Arkadir D, Ben-Shaul Y, Morris G, et al. False detection of dynamic changes in pallidal neuron interactions by the joint peri-stimulus histogram. In: Nicholson LFB, Faull RLM, eds. *The Basal Ganglia VII, Advances in Behavioral Biology*. Volume 52. Kluwer Academic/Plenum; 2002:181–187.

Arlotti M, Marceglia S, Foffani G, et al. Eight-hours adaptive deep brain stimulation in patients with Parkinson disease. *Neurology*. 2018;90(11):e971–e976.

Asch N, Herschman Y, Maoz R, Aurbach-Asch C, Valsky D, Abu-Snineh M, Arkadir D, Linetsky E, Eitan R, Marmor O, Bergman H, Israel Z. Independently together: subthalamic theta and beta opposite roles in predicting Parkinson's tremor. *Brain Commun* 2020 Aug 18;2(2):fcaa128.

Ashkan K, Rogers P, Bergman H, Ughratdar I. Insights into the mechanisms of deep brain stimulation. *Nat Rev Neurol.* 2017;13(9):548–554.

Aziz TZ, Peggs D, Sambrook MA, Crossman AR. Lesion of the subthalamic nucleus for the alleviation of 1-methyl-4-phenyl-1,2,3,6-tetrahydropyridine (MPTP)-induced parkinsonism in the primate. *Mov Disord.* 1991;6(4):288–292.

Bahney J, von Bartheld CS. The cellular composition and glia-neuron ratio in the spinal cord of a human and a nonhuman primate: comparison with other species and brain regions. *Anat Rec* (Hoboken). 2018;301(4):697–710.

Barbeau A. The pathogenesis of Parkinson's disease: a new hypothesis. *Can Med Assoc J.* 1962;87:802–807.

Bar-Gad I, Elias S, Vaadia E, Bergman H. Complex locking rather than complete cessation of neuronal activity in the globus pallidus of a 1-methyl-4-phenyl-1,2,3,6-tetrahydropyridine-treated primate in response to pallidal microstimulation. *J Neurosci.* 2004;24(33):7410–7419.

Bar-Gad I, Heimer G, Ritov Y, Bergman H. Functional correlations between neighboring neurons in the primate globus pallidus are weak or nonexistent. *J Neurosci.* 2003;23(10):4012–4016.

Bar-Gad I, Morris G, Bergman H. Information processing, dimensionality reduction and reinforcement learning in the basal ganglia. *Prog Neurobiol.* 2003;71(6):439–473.

Bar-Gad I, Ritov Y, Bergman H. The neuronal refractory period causes a short-term peak in the autocorrelation function. *J Neurosci Methods.* 2001a;104(2):155–163.

Bar-Gad I, Ritov Y, Vaadia E, Bergman H. Failure in identification of overlapping spikes from multiple neuron activity causes artificial correlations. *J Neurosci Methods.* 2001b;107(1–2):1–13.

Barlow JS. *The Cerebellum and Adaptive Control.* Cambridge University Press, 2002.

Baufreton J, Kirkham E, Atherton JF, et al. Sparse but selective and potent synaptic transmission from the globus pallidus to the subthalamic nucleus. *J Neurophysiol.* 2009;102(1):532–545.

Bayer HM, Lau B, Glimcher PW. Statistics of midbrain dopamine neuron spike trains in the awake primate. *J Neurophysiol.* 2007;98(3):1428–1439.

Beatty JA, Sullivan MA, Morikawa H, Wilson CJ. Complex autonomous firing patterns of striatal low-threshold spike interneurons. *J Neurophysiol.* 2012;108(3):771–781.

Beers C. *A Mind That Found Itself.* Redclassic; 2009.

Benabid AL, Chabardes S, Torres N, et al. Functional neurosurgery for movement disorders: a historical perspective. *Prog Brain Res.* 2009;175:379–391.

Benabid AL, Pollak P, Louveau A, Henry S, de Rougemont J. Combined (thalamotomy and stimulation) stereotactic surgery of the VIM thalamic nucleus for bilateral Parkinson disease. *Appl Neurophysiol.* 1987;50(1–6):344–346.

Benazzouz A, Gross C, Féger J, Boraud T, Bioulac B. Reversal of rigidity and improvement in motor performance by subthalamic high-frequency stimulation in MPTP-treated monkeys. *Eur J Neurosci.* 1993;5(4):382–389.

Benhamou L, Bronfeld M, Bar-Gad I, Cohen D. Globus pallidus external segment neuron classification in freely moving rats: a comparison to primates. *PLoS One.* 2012;7(9):e45421.

Bennett CM, Baird AA, Miller MB, Wolford GL. Neural correlates of interspecies perspective taking in the post-mortem Atlantic salmon: an argument for proper multiple comparisons correction. *J Serendipitous Unexpected Results* 2009;1:1–5.

Ben-Pazi H, Bergman H, Goldberg JA, Giladi N, Hansel D, Reches A, Simon ES. Synchrony of rest tremor in multiple limbs in Parkinson's disease: evidence for multiple oscillators. *J Neural Transm* (Vienna). 2001;108(3):287–296.

Ben-Shaul Y, Bergman H, Ritov Y, Abeles M. Trial to trial variability in either stimulus or action causes apparent correlation and synchrony in neuronal activity. *J Neurosci Methods*. 2001;111(2):99–110.

Bergman H, Feingold A, Nini A, et al. Physiological aspects of information processing in the basal ganglia of normal and parkinsonian primates. *Trends Neurosci*. 1998;21(1):32–38.

Bergman H, Wichmann T, DeLong MR. Reversal of experimental parkinsonism by lesions of the subthalamic nucleus. *Science*. 1990;249(4975):1436–1438.

Bergman H, Wichmann T, Karmon B, DeLong MR. The primate subthalamic nucleus. II. Neuronal activity in the MPTP model of parkinsonism. *J Neurophysiol*. 1994;72(2):507–520.

Berridge KC, Kringelbach ML. Pleasure systems in the brain. *Neuron*. 2015;86(3):646–664.

Berthier NE, Singh SP, Barto AG, Houk JC. Distributed representation of limb motor programs in arrays of adjustable pattern generators. *J Cogn Neurosci*. 1993;5(1):56–78.

Bevan MD, Francis CM, Bolam JP. The glutamate-enriched cortical and thalamic input to neurons in the subthalamic nucleus of the rat: convergence with GABA-positive terminals. *J Comp Neurol*. 1995;361(3):491–511.

Bevan MD, Hallworth NE, Baufreton J. GABAergic control of the subthalamic nucleus. *Prog Brain Res*. 2007;160:173–188.

Bezard E, Boraud T, Chalon S, Brotchie JM, Guilloteau D, Gross CE. Pallidal border cells: an anatomical and electrophysiological study in the 1-methyl-4-phenyl-1,2,3,6-tetrahydropyridine-treated monkey. *Neuroscience*. 2001;103(1):117–123.

Blonder LX, Gur RE, Gur RC, Saykin AJ, Hurtig HI. Neuropsychological functioning in hemiparkinsonism. *Brain Cogn*. 1989;9(2):244–257.

Bolam JP, Ellender TJ. Histamine and the striatum. *Neuropharmacology*. 2016;106:74–84.

Borck C. *Brainwaves: A Cultural History of Electroencephalography*. Routledge; 2018.

Bostan AC, Dum RP, Strick PL. The basal ganglia communicate with the cerebellum. *Proc Natl Acad Sci USA*. 2010;107(18):8452–8456.

Bostan AC, Strick PL. The basal ganglia and the cerebellum: nodes in an integrated network. *Nat Rev Neurosci*. 2018;19(6):338–350.

Bowden DM, Martin RF. *Primate Brain Maps: Structure of the Macaque Brain: A Laboratory Guide with Original Brain Sections*. Printed Atlas and Electronic Templates for Data and Schematics. Elsevier; 2000.

Boyden ES, Zhang F, Bamberg E, Nagel G, Deisseroth K. Millisecond-timescale, genetically targeted optical control of neural activity. *Nat Neurosci*. 2005;8(9):1263–1268.

Braak H, Del Tredici K, Rüb U, de Vos RA, Jansen Steur EN, Braak E. Staging of brain pathology related to sporadic Parkinson's disease. *Neurobiol Aging*. 2003;24(2):197–211.

Brain Development Cooperative Group. Total and regional brain volumes in a population-based normative sample from 4 to 18 years: the NIH MRI Study of Normal Brain Development. *Cereb Cortex*. 2012;22(1):1–12.

Braitenberg V. *On the Texture of Brains: An Introduction to Neuroanatomy for the Cybernetically Minded*. Springer-Verlag; 1977.

Braitenberg V. *Vehicles: Experiments in Synthetic Psychology*. MIT Press; 1984.

Braitenberg V, Atwood RP. Morphological observations on the cerebellar cortex. *J Comp Neurol*. 1958;109(1):1–33.

Braitenberg V, Schüz A. *Anatomy of the Cortex: Statistics and Geometry*. Springer-Verlag; 1991.

Branch SY, Sharma R, Beckstead MJ. Aging decreases L-type calcium channel currents and pacemaker firing fidelity in substantia nigra dopamine neurons. *J Neurosci*. 2014;34:9310–9318.

Brimblecombe KR, Cragg SJ. The striosome and matrix compartments of the striatum: a path through the labyrinth from neurochemistry toward function. *ACS Chem Neurosci*. 2017;8(2):235–242.

Brimblecombe KR, Threlfell S, Dautan D, Kosillo P, Mena-Segovia J, Cragg SJ. Targeted Activation of cholinergic interneurons accounts for the modulation of dopamine by striatal nicotinic receptors. *eNeuro*. 2018;5(5):ENEURO.0397–17.2018.

Bromberg-Martin ES, Hikosaka O. Lateral habenula neurons signal errors in the prediction of reward information [published correction appears in *Nat Neurosci*. 2011 Dec;14(12):1617]. *Nat Neurosci*. 2011;14(9):1209–1216.

Bromberg-Martin ES, Matsumoto M, Hikosaka O. Distinct tonic and phasic anticipatory activity in lateral habenula and dopamine neurons. *Neuron*. 2010a Jul 15;67(1):144–155.

Bromberg-Martin ES, Matsumoto M, Nakahara H, Hikosaka O. Multiple timescales of memory in lateral habenula and dopamine neurons. *Neuron*. 2010b Aug 12;67(3):499–510.

Brooks D, Halliday GM. Intralaminar nuclei of the thalamus in Lewy body diseases. *Brain Res Bull*. 2009;78(2–3):97–104.

Brown P, Oliviero A, Mazzone P, Insola A, Tonali P, Di Lazzaro V. Dopamine dependency of oscillations between subthalamic nucleus and pallidum in Parkinson's disease. *J Neurosci*. 2001;21(3):1033–1038.

Burke DA, Rotstein HG, Alvarez VA. Striatal local circuitry: a new framework for lateral inhibition. *Neuron*. 2017;96(2):267–284.

Buzsáki G. *Rhythms of the Brain*. Oxford University Press; 2006.

Buzsáki G, Anastassiou CA, Koch C. The origin of extracellular fields and currents—EEG, ECoG, LFP and spikes. *Nat Rev Neurosci*. 2012;13(6):407–420.

Buzsáki G, Kaila K, Raichle M. Inhibition and brain work. *Neuron*. 2007;56(5):771–783.

Cabello CR, Thune JJ, Pakkenberg H, Pakkenberg B. Ageing of substantia nigra in humans: cell loss may be compensated by hypertrophy. *Neuropathol Appl Neurobiol* 2002;28:283–291.

Campbell PK, Jones KE, Normann RA. A 100 electrode intracortical array: structural variability. *Biomed Sci Instrum*. 1990;26:161–165.

Cano J, Pasik P, Pasik T. Early postnatal development of the monkey globus pallidus: a Golgi and electron microscopic study. *J Comp Neurol*. 1989 Jan 15;279(3):353–367.

Carballo-Carbajal I, Laguna A, Romero-Giménez J, et al. Brain tyrosinase overexpression implicates age-dependent neuromelanin production in Parkinson's disease pathogenesis. *Nat Commun*. 2019 Mar 7;10(1):973.

Carlsson A. On the problem of the mechanism of action of some psychopharmaca. *Psychiatr Neurol* (Basel). 1960 July;140:220–222.

Carlsson A. A paradigm shift in brain research. *Science*. 2001;294(5544):1021–1024.

Carlsson A, Lindqvist M, Magnusson T. 3,4-Dihydroxyphenylalanine and 5-hydroxytryptophan as reserpine antagonists. *Nature*. 1957;180(4596):1200.

Carrol L. *Alice's Adventures in Wonderland*. Open Books Electronic Ed.; 2007. http://www.open-bks .com/alice-71-72.html.

Casagrande M, Bertini M. Laterality of the sleep onset process: which hemisphere goes to sleep first? *Biol Psychol*. 2008a;77(1):76–80.

Casagrande M, Bertini M. Night-time right hemisphere superiority and daytime left hemisphere superiority: a repatterning of laterality across wake-sleep-wake states. *Biol Psychol*. 2008b;77(3):337–342.

Celec P, Ostatníková D, Hodosy J. On the effects of testosterone on brain behavioral functions. *Front Neurosci*. 2015;9:12.

Chahine L. *Disorders of Sleep and Wakefulness in Parkinson's Disease: A Case-Based Guide to Diagnosis and Management*. Elsevier; 2020.

Chakravarthy VS, Moustafa AA. *Computational Neuroscience Models of the Basal Ganglia*. Springer; 2018.

Charney D, Southwick S. *Resilience: The Science of Mastering Life's Greatest Challenges*. Cambridge University Press; 2018.

Chiueh CC, Burns RS, Markey SP, Jacobowitz DM, Kopin IJ. Primate model of parkinsonism: selective lesion of nigrostriatal neurons by 1-methyl-4-phenyl-1,2,3,6-tetrahydropyridine produces an extrapyramidal syndrome in rhesus monkeys. *Life Sci*. 1985;36(3):213–218.

Chu HY, Atherton JF, Wokosin D, Surmeier DJ, Bevan MD. Heterosynaptic regulation of external globus pallidus inputs to the subthalamic nucleus by the motor cortex. *Neuron*. 2015;85(2):364–376.

Cisek P. Resynthesizing behavior through phylogenetic refinement. *Atten Percept Psychophys*. 2019;81(7):2265–2287.

Coenen VA, Bewernick BH, Kayser S, Kilian H, Boström J, Greschus S, Hurlemann R, Klein ME, Spanier S, Sajonz B, Urbach H, Schlaepfer TE. Superolateral medial forebrain bundle deep brain stimulation in major depression: a gateway trial. *Neuropsychopharmacology*. 2019 Jun;44(7):1224–1232.

Cohen ME, Eichel R, Steiner-Birmanns B, Janah A, Ioshpa M, Bar-Shalom R, Paul JJ, Gaber H, Skrahina V, Bornstein NM, Yahalom G. A case of probable Parkinson's disease after SARS-CoV-2 infection. *Lancet Neurol*. 2020 Oct;19(10):804–805.

Cohen MR, Kohn A. Measuring and interpreting neuronal correlations. *Nat Neurosci*. 2011 Jun 27;14(7):811–819.

Cohen MX. *Analyzing Neural Time Series Data: Theory and Practice*. MIT Press; 2014.

Collins AGE, Cockburn J. Beyond dichotomies in reinforcement learning. *Nat Rev Neurosci*. 2020 Oct;21(10):576–586.

Coudé D, Parent A, Parent M. Single-axon tracing of the corticosubthalamic hyperdirect pathway in primates. *Brain Struct Funct*. 2018;223(9):3959–3973.

Courtemanche R, Fujii N, Graybiel AM. Synchronous, focally modulated beta-band oscillations characterize local field potential activity in the striatum of awake behaving monkeys. *J Neurosci*. 2003;23(37):11741–11752.

Crick F. *What Mad Pursuit: A Personal View of Scientific Discovery*. Basic Books; 1988.

Crutcher MD, DeLong MR. Single cell studies of the primate putamen. I. Functional organization. *Exp Brain Res*. 1984a;53(2):233–243.

Crutcher MD, DeLong MR. Single cell studies of the primate putamen. II. Relations to direction of movement and pattern of muscular activity. *Exp Brain Res*. 1984b;53(2):244–258.

Cubo E, Martín PM, Martin-Gonzalez JA, Rodríguez-Blázquez C, Kulisevsky J, ELEP Group Members. Motor laterality asymmetry and nonmotor symptoms in Parkinson's disease. *Mov Disord*. 2010;25(1):70–75.

Cui G, Jun SB, Jin X, et al. Concurrent activation of striatal direct and indirect pathways during action initiation. *Nature*. 2013;494(7436):238–242.

Czubayko U, Plenz D. Fast synaptic transmission between striatal spiny projection neurons. *Proc Natl Acad Sci USA*. 2002;99(24):15764–15769.

Dahan L, Astier B, Vautrelle N, Urbain N, Kocsis B, Chouvet G. Prominent burst firing of dopaminergic neurons in the ventral tegmental area during paradoxical sleep. *Neuropsychopharmacology*. 2007;32(6):1232–1241.

Darbin O, Hatanaka N, Takara S, et al. Parkinsonism differently affects the single neuronal activity in the primary and supplementary motor areas in monkeys: an investigation in linear and nonlinear domains. *Int J Neural Syst*. 2020;30(2):2050010.

da Silva JA, Tecuapetla F, Paixão V, Costa RM. Dopamine neuron activity before action initiation gates and invigorates future movements. *Nature*. 2018;554(7691):244–248.

Dautan D, Huerta-Ocampo I, Gut NK, et al. Cholinergic midbrain afferents modulate striatal circuits and shape encoding of action strategies. *Nat Commun*. 2020 Apr 8;11(1):1739.

Davis GC, Williams AC, Markey SP, et al. Chronic parkinsonism secondary to intravenous injection of meperidine analogues. *Psychiatry Res*. 1979;1(3):249–254.

Dayan P, Abbott LE. *Theoretical Neuroscience: Computational and Mathematical Modeling of Neural Systems*. MIT press; 2001.

De A, El-Shamayleh Y, Horwitz GD. Fast and reversible neural inactivation in macaque cortex by optogenetic stimulation of GABAergic neurons. *Elife*. 2020;9:e52658.

Deffains M, Iskhakova L, Katabi S, Haber SN, Israel Z, Bergman H. Subthalamic, not striatal, activity correlates with basal ganglia downstream activity in normal and parkinsonian monkeys. *Elife*. 2016;5:e16443.

Deffains M, Iskhakova L, Katabi S, Israel Z, Bergman H. Longer β oscillatory episodes reliably identify pathological subthalamic activity in parkinsonism. *Mov Disord*. 2018;33(10):1609–1618.

Deister CA, Dodla R, Barraza D, Kita H, Wilson CJ. Firing rate and pattern heterogeneity in the globus pallidus arise from a single neuronal population. *J Neurophysiol*. 2013;109(2):497–506.

de Jong LW, Wang Y, White LR, Yu B, van Buchem MA, Launer LJ. Ventral striatal volume is associated with cognitive decline in older people: a population based MR-study. *Neurobiol Aging*. 2012;33(2):424.e1–424.10.

DeLong MR. Activity of pallidal neurons during movement. *J Neurophysiol*. 1971;34(3):414–427.

DeLong MR. Activity of basal ganglia neurons during movement. *Brain Res*. 1972;40(1):127–135.

Del Río JP, Alliende MI, Molina N, Serrano FG, Molina S, Vigil P. Steroid hormones and their action in women's brains: the importance of hormonal balance. *Front Public Health*. 2018 May 23;6:141.

Del Tredici K, Braak H. To stage, or not to stage. *Curr Opin Neurobiol.* 2020;61:10–22.

de Lussanet MHE, Osse JWM. An ancestral axial twist explains the contralateral forebrain and the optic chiasm in vertebrates. *Animal Biology.* 2012; 62(2):193–216.

Denk W, Horstmann H. Serial block-face scanning electron microscopy to reconstruct three-dimensional tissue nanostructure. *PLoS Biol.* 2004;2(11):e329.

Denny-Brown D. *Disease of the Basal Ganglia and Subthalamic Nucleus.* Oxford University Press; 1946.

Denny-Brown D. *The Basal Ganglia and Their Relation to Disorders of Movement.* Oxford University Press; 1962.

Deschênes M, Bourassa J, Doan VD, Parent A. A single-cell study of the axonal projections arising from the posterior intralaminar thalamic nuclei in the rat. *Eur J Neurosci.* 1996;8(2):329–343.

Dewey J. The reflex arc concept in psychology. *Psychol Rev.* 1896;(3);357–370.

Difiglia M, Pasik P, Pasik T. A Golgi and ultrastructural study of the monkey globus pallidus. *J Comp Neurol.* 1982 Nov 20;212(1):53–75.

Döbrössy MD, Ramanathan C, Ashouri Vajari DA, Tong Y, Schlaepfer T, Coenen VA. Neuromodulation in psychiatric disorders: experimental and clinical evidence for reward and motivation network deep brain stimulation—focus on the medial forebrain bundle. *Eur J Neurosci.* 2020 Sep 15. https://doi.org/10.1111/ejn.14975.

Dodson PD, Dreyer JK, Jennings KA, et al. Representation of spontaneous movement by dopaminergic neurons is cell-type selective and disrupted in parkinsonism. *Proc Natl Acad Sci USA.* 2016;113(15):E2180–E2188.

Donaldson I, Marsden D, Schneider SA, Bhatia KP. *Marsden's Book of Movement Disorders.* Oxford University Press; 2012.

Dorph-Petersen KA, Pierri JN, Sun Z, Sampson AR, Lewis DA. Stereological analysis of the mediodorsal thalamic nucleus in schizophrenia: volume, neuron number, and cell types. *J Comp Neurol.* 2004;472(4):449–462.

Doupe AJ, Solis MM, Kimpo R, Boettiger CA. Cellular, circuit, and synaptic mechanisms in song learning. *Ann NY Acad Sci.* 2004;1016:495–523.

Eban-Rothschild A, Rothschild G, Giardino WJ, Jones JR, de Lecea L. VTA dopaminergic neurons regulate ethologically relevant sleep-wake behaviors. *Nat Neurosci.* 2016;19(10):1356–1366.

Eccles JC. *The Neurophysiological Basis of Mind.* Clarendon Press; 1953.

Eccles JC, Ito M, Szentagothai J. *The Cerebellum as a Neuronal Machine.* Springer-Verlag; 1967.

Edlow, BL, Mareyam, A, Horn, A, Polimeni, JR, Witzel, T, Tisdall, MD, et al. 7 Tesla MRI of the ex vivo human brain at 100 micron resolution. *Scientific Data.* 2019;6(1):244.

Eggermont JS. *The Correlative Brain: Theory and Experiment in Neural Computation.* Springer-Verlag; 1990.

Eid L, Parent M. Cholinergic neurons intrinsic to the primate external pallidum. *Synapse.* 2015;69(8):416–419.

Eid L, Parent M. Chemical anatomy of pallidal afferents in primates. *Brain Struct Funct.* 2016 Dec;221(9):4291–4317.

Eisenhofer G, Tian H, Holmes C, Matsunaga J, Roffler-Tarlov S, Hearing VJ. Tyrosinase: a developmentally specific major determinant of peripheral dopamine. *FASEB J.* 2003;17(10):1248–1255.

Eisinger RS, Cagle JN, Opri E, et al. Parkinsonian beta dynamics during rest and movement in the dorsal pallidum and subthalamic nucleus. *J Neurosci.* 2020;40(14):2859–2867.

Eitan R, Shamir RR, Linetsky E, et al. Asymmetric right/left encoding of emotions in the human subthalamic nucleus. *Front Syst Neurosci.* 2013 Oct 29;7:69.

Elazary AS, Attia R, Bergman H, Ben-Pazi H. Age-related accelerated tapping response in healthy population. *Percept Mot Skills.* 2003;96(1):227–235.

Elias S, Joshua M, Goldberg JA, et al. Statistical properties of pauses of the high-frequency discharge neurons in the external segment of the globus pallidus. *J Neurosci.* 2007;27(10):2525–2538.

Emmerich MTM, Deutz AH. A tutorial on multiobjective optimization: fundamentals and evolutionary methods. *Nat Comput.* 2018;17:585–609.

Emmi A, Antonini A, Macchi V, Porzionato A, De Caro R. Anatomy and connectivity of the subthalamic nucleus in humans and non-human primates. *Front Neuroanat.* 2020;14:13.

Engel AK, Fries P. Beta-band oscillations—signalling the status quo? *Curr Opin Neurobiol.* 2010;20(2):156–165.

Engelhard B, Finkelstein J, Cox J, et al. Specialized coding of sensory, motor and cognitive variables in VTA dopamine neurons. *Nature.* 2019;570(7762):509–513.

English DF, Ibanez-Sandoval O, Stark E, et al. GABAergic circuits mediate the reinforcement-related signals of striatal cholinergic interneurons. *Nat Neurosci.* 2011 Dec 11;15(1):123–130.

Ermentrout GB, Terman DH. *Mathematical Foundations of Neuroscience.* Springer; 2010.

Evarts EV. Temporal patterns of discharge of pyramidal tract neurons during sleep and waking in the monkey. *J Neurophysiol.* 1964;27:152–171.

Ewert S, Plettig P, Li N, et al. Toward defining deep brain stimulation targets in MNI space: a subcortical atlas based on multimodal MRI, histology and structural connectivity. *Neuroimage.* 2018;170:271–282.

Fee MS, Goldberg JH. A hypothesis for basal ganglia-dependent reinforcement learning in the songbird. *Neuroscience.* 2011;198:152–170.

Féger J, Bevan M, Crossman AR. The projections from the parafascicular thalamic nucleus to the subthalamic nucleus and the striatum arise from separate neuronal populations: a comparison with the corticostriatal and corticosubthalamic efferents in a retrograde fluorescent double-labelling study. *Neuroscience.* 1994;60(1):125–132.

Fifel K, Meijer JH, Deboer T. Circadian and homeostatic modulation of multi-unit activity in midbrain dopaminergic structures. *Sci Rep.* 2018;8(1):7765.

Filion M. Effects of interruption of the nigrostriatal pathway and of dopaminergic agents on the spontaneous activity of globus pallidus neurons in the awake monkey. *Brain Res.* 1979; 178(2–3):425–441.

Filion M, Tremblay L. Abnormal spontaneous activity of globus pallidus neurons in monkeys with MPTP-induced parkinsonism. *Brain Res.* 1991;547(1):142–151.

Filion M, Tremblay L, Bédard PJ. Abnormal influences of passive limb movement on the activity of globus pallidus neurons in parkinsonian monkeys. *Brain Res.* 1988;444(1):165–176.

Filion M, Tremblay L, Bédard PJ. Effects of dopamine agonists on the spontaneous activity of globus pallidus neurons in monkeys with MPTP-induced parkinsonism. *Brain Res.* 1991; 547(1):152–161.

Fiorillo CD, Tobler PN, Schultz W. Discrete coding of reward probability and uncertainty by dopamine neurons. *Science*. 2003 Mar 21;299(5614):1898–1902.

Floresco SB, West AR, Ash B, Moore H, Grace AA. Afferent modulation of dopamine neuron firing differentially regulates tonic and phasic dopamine transmission. *Nat Neurosci*. 2003 Sep;6(9):968–973.

Foley PB. *Encephalitis Lethargica: The Mind and Brain Virus*. Springer; 2019.

Fox CA, Andrade AN, Lu Qui IJ, Rafols JA. The primate globus pallidus: a Golgi and electron microscopic study. *J Hirnforsch*. 1974;15(1):75–93. PMID: 4135902.

François C, Percheron G, Yelnik J, Heyner S. A Golgi analysis of the primate globus pallidus. I. Inconstant processes of large neurons, other neuronal types and afferent axons. *J Comp Neurol*. 1984;227(2):182–199.

Freund TF, Powell JF, Smith AD. Tyrosine hydroxylase-immunoreactive boutons in synaptic contact with identified striatonigral neurons, with particular reference to dendritic spines. *Neuroscience*. 1984;13(4):1189–1215.

Frey S, Pandya DN, Chakravarty MM, Bailey L, Petrides M, Collins DL. An MRI based average macaque monkey stereotaxic atlas and space (MNI monkey space). *Neuroimage*. 2011;55(4): 1435–1442.

Fridgeirsson EA, Figee M, Luigjes J, van den Munckhof P, Schuurman PR, van Wingen G, Denys D. Deep brain stimulation modulates directional limbic connectivity in obsessive-compulsive disorder. *Brain*. 2020 May 1;143(5):1603–1612.

Friedman A, Homma D, Gibb LG, et al. A corticostriatal path targeting striosomes controls decision-making under conflict. *Cell*. 2015;161(6):1320–1333.

Fujiyama F, Nakano T, Matsuda W, Furuta T, Udagawa J, Kaneko T. A single-neuron tracing study of arkypallidal and prototypic neurons in healthy rats. *Brain Struct Funct*. 2016;221(9):4733–4740.

Fukai T, Tanaka S. A simple neural network exhibiting selective activation of neuronal ensembles: from winner-take-all to winners-share-all. *Neural Comput*. 1997;9(1):77–97.

Gal, E, London, M, Globerson, A, et al. Rich cell-type-specific network topology in neocortical microcircuitry. *Nat Neurosci*. 2017;20:1004–1013.

Gale SD, Perkel DJ. Anatomy of a songbird basal ganglia circuit essential for vocal learning and plasticity. *J Chem Neuroanat*. 2010;39(2):124–131.

Galvan A, Hu X, Rommelfanger KS, et al. Localization and function of dopamine receptors in the subthalamic nucleus of normal and parkinsonian monkeys. *J Neurophysiol*. 2014;112(2): 467–479.

Garcia-Rill E. *Waking and the Reticular Activating System in Health and Disease*. Academic Press; 2015.

Gazzaniga MS. *Tales from Both Sides of the Brain*. HarperCollins; 2016.

GBD 2016 Neurology Collaborators. Global, regional, and national burden of neurological disorders, 1990–2016: a systematic analysis for the Global Burden of Disease Study 2016. *Lancet Neurol*. 2019;18(5):459–480.

Gelfan S, Tarlov IM. Altered neuron population in l7 segment of dogs with experimental hindlimb rigidity. *Am J Physiol*. 1963;205:606–616.

Georgopoulos AP, DeLong MR, Crutcher MD. Relations between parameters of step-tracking movements and single cell discharge in the globus pallidus and subthalamic nucleus of the behaving monkey. *J Neurosci*. 1983;3(8):1586–1598.

Gerfen CR, Engber TM, Mahan LC, et al. D1 and D2 dopamine receptor-regulated gene expression of striatonigral and striatopallidal neurons. *Science*. 1990;250(4986):1429–1432.

Gerrits R, Verhelst H, Vingerhoets G. Mirrored brain organization: statistical anomaly or reversal of hemispheric functional segregation bias? *Proc Natl Acad Sci USA*. 2020;117(25):14057–14065.

Gershman SJ, Uchida N. Believing in dopamine. *Nat Rev Neurosci*. 2019;20(11):703–714.

Gerstein GL, Aertsen AM. Representation of cooperative firing activity among simultaneously recorded neurons. *J Neurophysiol*. 1985;54(6):1513–1528.

Gerstein GL, Perkel DH. Simultaneously recorded trains of action potentials: analysis and functional interpretation. *Science*. 1969;164(3881):828–830.

Giguere N, Burke Nanni S, Trudeau LE. On cell loss and selective vulnerability of neuronal populations in Parkinson's disease. *Front Neurol*. 2018;9:455.

Gittis AH, Kreitzer AC. Striatal microcircuitry and movement disorders. *Trends Neurosci*. 2012;35(9):557–564.

Glajch KE, Kelver DA, Hegeman DJ, et al. Npas1+ pallidal neurons target striatal projection neurons. *J Neurosci*. 2016;36(20):5472–5488.

Gleick J. *Chaos: Making a New Science*. Penguin; 1987.

Goldberg JA, Bergman H. Computational physiology of the neural networks of the primate globus pallidus: function and dysfunction. *Neuroscience*. 2011;198:171–192.

Goldberg JA, Boraud T, Maraton S, Haber SN, Vaadia E, Bergman H. Enhanced synchrony among primary motor cortex neurons in the 1-methyl-4-phenyl-1,2,3,6-tetrahydropyridine primate model of Parkinson's disease. *J Neurosci*. 2002;22(11):4639–4653.

Goldberg JH, Adler A, Bergman H, Fee MS. Singing-related neural activity distinguishes two putative pallidal cell types in the songbird basal ganglia: comparison to the primate internal and external pallidal segments. *J Neurosci*. 2010;30(20):7088–7098.

Gourévitch B, Eggermont JJ. A simple indicator of nonstationarity of firing rate in spike trains. *J Neurosci Methods*. 2007;163(1):181–187.

Grace AA. Phasic versus tonic dopamine release and the modulation of dopamine system responsivity: a hypothesis for the etiology of schizophrenia. *Neuroscience*. 1991;41(1):1–24.

Grace AA. Dysregulation of the dopamine system in the pathophysiology of schizophrenia and depression. *Nat Rev Neurosci*. 2016 Aug;17(8):524–532.

Gradinaru V, Mogri M, Thompson KR, Henderson JM, Deisseroth K. Optical deconstruction of parkinsonian neural circuitry. *Science*. 2009;324(5925):354–359.

Graybiel AM, Aosaki T, Flaherty AW, Kimura M. The basal ganglia and adaptive motor control. *Science*. 1994;265(5180):1826–1831.

Graybiel AM, Ragsdale CW Jr. Histochemically distinct compartments in the striatum of human, monkeys and cat demonstrated by acetylcholinesterase staining. *Proc Natl Acad Sci USA*. 1978;75(11):5723–5726.

Graziano MSA. *The Intelligent Movement Machine: An Ethological Perspective of the Primate Motor System*. Oxford University Press; 2009.

Grillner S, Robertson B. The basal ganglia over 500 million years. *Curr Biol*. 2016;26(20): R1088–R1100.

Grinvald A. Real-time optical mapping of neuronal activity: from single growth cones to the intact mammalian brain. *Annu Rev Neurosci.* 1985;8:263–305.

Groves PM, Linder JC, Young SJ. 5-hydroxydopamine-labeled dopaminergic axons: three-dimensional reconstructions of axons, synapses and postsynaptic targets in rat neostriatum. *Neuroscience.* 1994;58(3):593–604.

Guang J, Baker H, Ben-Yishay Nizri O, Firman S, Werner-Reiss U, Kapuller V, Israel Z, Bergman H. Toward asleep DBS: cortico basal-ganglia neural activity during interleaved propofol/ketamine sedation mimics NREM/REM sleep activity. bioRxiv. 2020 Dec 15. https://doi.org/10.1101/2020 .12.14.422637.

Gulban OF, De Martino F, Vu AT, Yacoub E, Uğurbil K, Lenglet C. Cortical fibers orientation mapping using in-vivo whole brain 7 T diffusion MRI. *Neuroimage.* 2018;178:104–118.

Gunantara N. A review of multi-objective optimization: methods and its applications. *Cogent Eng.* 2018;5(1):1502242.

Guo CN, Machado NL, Zhan SQ, Yang XF, Yang WJ, Lu J. Identification of cholinergic pallidocortical neurons. *CNS Neurosci Ther.* 2016;22(10):863–865.

Haber S. Perspective on basal ganglia connections as described by Nauta and Mehler in 1966: where we were and how this paper effected where we are now. *Brain Res.* 2016;1645:4–7.

Haber SN, Adler A, Bergman H. The basal ganglia. In: Paxinos G, Mai J, eds. *The Human Nervous System.* 2nd ed. Academic Press; 2011.

Haber SN, Yendiki A, Jbabdi S. Four deep brain stimulation targets for obsessive-compulsive disorder: are they different? *Biol Psychiatry.* 2020 Jul 25:S0006–3223(20)31773-X.

Halliday GM, Macdonald V, Henderson JM. A comparison of degeneration in motor thalamus and cortex between progressive supranuclear palsy and Parkinson's disease. *Brain.* 2005 Oct;128(Pt 10): 2272–2280.

Hammond C, Deniau JM, Rizk A, Feger J. Electrophysiological demonstration of an excitatory subthalamonigral pathway in the rat. *Brain Res.* 1978;151(2):235–244.

Hammond C, Shibazaki T, Rouzaire-Dubois B. Branched output neurons of the rat subthalamic nucleus: electrophysiological study of the synaptic effects on identified cells in the two main target nuclei, the entopeduncular nucleus and the substantia nigra. *Neuroscience.* 1983;9(3):511–520.

Hardman CD, Henderson JM, Finkelstein DI, Horne MK, Paxinos G, Halliday GM. Comparison of the basal ganglia in rats, marmosets, macaques, baboons, and humans: volume and neuronal number for the output, internal relay, and striatal modulating nuclei. *J Comp Neurol.* 2002; 445:238–255.

Hariz GM, Nakajima T, Limousin P, et al. Gender distribution of patients with Parkinson's disease treated with subthalamic deep brain stimulation: a review of the 2000–2009 literature. *Parkinsonism Relat Disord.* 2011;17(3):146–149.

Harman PJ, Carpenter MB. Volumetric comparisons of the basal ganglia of various primates including man. *J Comp Neurol.* 1950;93(1):125–137.

Hasenstaub A, Otte S, Callaway E, Sejnowski TJ. Metabolic cost as a unifying principle governing neuronal biophysics. *Proc Natl Acad Sci USA.* 2010;107(27):12329–12334.

Hashimoto T, Elder CM, Okun MS, Patrick SK, Vitek JL. Stimulation of the subthalamic nucleus changes the firing pattern of pallidal neurons. *J Neurosci.* 2003;23(5):1916–1923.

Haykin SO. *Neural Networks and Learning Machines*. 3rd ed. Pearson; 2008.

Haynes WI, Haber SN. The organization of prefrontal-subthalamic inputs in primates provides an anatomical substrate for both functional specificity and integration: implications for basal ganglia models and deep brain stimulation. *J Neurosci*. 2013;33(11):4804–4814.

Hazrati LN, Parent A. Projection from the external pallidum to the reticular thalamic nucleus in the squirrel monkey. *Brain Res*. 1991;550(1):142–146.

Hazrati LN, Parent A. The striatopallidal projection displays a high degree of anatomical specificity in the primate. *Brain Res*. 1992;592(1–2):213–227.

Hazrati LN, Parent A, Mitchell S, Haber SN. Evidence for interconnections between the two segments of the globus pallidus in primates: a PHA-L anterograde tracing study. *Brain Res*. 1990; 533(1):171–175.

Heck DH. *The Neuronal Codes of the Cerebellum*. Elsevier; 2016.

Heimer G, Bar-Gad I, Goldberg JA, Bergman H. Dopamine replacement therapy reverses abnormal synchronization of pallidal neurons in the MPTP primate model of parkinsonism. *J Neurosci*. 2002a;22(18):7850–7855.

Heimer G, Bar-Gad I, Goldberg JA, Bergman H. Synchronization of pallidal activity in the MPTP primate model of parkinsonism is not limited to oscillatory activity. In: Nicholson LFB, Faull RLM, eds. *The Basal Ganglia VII, Advances in Behavioral Biology*. Volume 52. Kluwer Academic/ Plenum; 2002b:29–34.

Heimer L, Van Hoesen GW, Trimble M, Zahm DS. *Anatomy of Neuropsychiatry: The New Anatomy of the Basal Forebrain and Its Implications for Neuropsychiatric Illness*. Elsevier; 2008.

Helmich RC, Hallett M, Deuschl G, Toni I, Bloem BR. Cerebral causes and consequences of parkinsonian resting tremor: a tale of two circuits? *Brain*. 2012;135(Pt 11):3206–3226.

Herculano-Houzel S. The human brain in numbers: a linearly scaled-up primate brain. *Front Hum Neurosci*. 2009;3:31.

Hernandez LF, Obeso I, Costa RM, Redgrave P, Obeso JA. Dopaminergic vulnerability in Parkinson disease: the cost of humans' habitual performance. *Trends Neurosci*. 2019;42(6):375–383.

Hernández VM, Hegeman DJ, Cui Q, et al. Parvalbumin+ neurons and Npas1+ neurons are distinct neuron classes in the mouse external globus pallidus. *J Neurosci*. 2015;35(34):11830–11847.

Hikosaka O, Wurtz RH. Visual and oculomotor functions of monkey substantia nigra pars reticulata. I. Relation of visual and auditory responses to saccades. *J Neurophysiol*. 1983a;49(5):1230–1253.

Hikosaka O, Wurtz RH. Visual and oculomotor functions of monkey substantia nigra pars reticulata. II. Visual responses related to fixation of gaze. *J Neurophysiol*. 1983b;49(5):1254–1267.

Hikosaka O, Wurtz RH. Visual and oculomotor functions of monkey substantia nigra pars reticulata. III. Memory-contingent visual and saccade responses. *J Neurophysiol*. 1983c;49(5):1268–1284.

Hikosaka O, Wurtz RH. Visual and oculomotor functions of monkey substantia nigra pars reticulata. IV. Relation of substantia nigra to superior colliculus. *J Neurophysiol*. 1983d;49(5):1285–1301.

Hill DN, Mehta SB, Kleinfeld D. Quality metrics to accompany spike sorting of extracellular signals. *J Neurosci*. 2011;31(24):8699–8705.

Hjorth JJJ, Kozlov A, Carannante I, et al. The microcircuits of striatum in silico. *Proc Natl Acad Sci USA*. 2020;117(17):9554–9565.

Hodgkin AL. The local electric changes associated with repetitive action in a non-medullated axon. *J Physiol.* 1948;107(2):165–181.

Holt AB, Kormann E, Gulberti A, et al. Phase-dependent suppression of beta oscillations in Parkinson's disease patients. *J Neurosci.* 2019;39(6):1119–1134.

Hong S, Hikosaka O. The globus pallidus sends reward-related signals to the lateral habenula. *Neuron.* 2008;60(4):720–729.

Hong S, Jhou TC, Smith M, Saleem KS, Hikosaka O. Negative reward signals from the lateral habenula to dopamine neurons are mediated by rostromedial tegmental nucleus in primates. *J Neurosci.* 2011;31(32):11457–11471.

Hopfield JJ. Neural networks and physical systems with emergent collective computational abilities. *Proc Natl Acad Sci USA.* 1982 Apr;79(8):2554–2558.

Horn A, Kühn AA, Merkl A, Shih L, Alterman R, Fox M. Probabilistic conversion of neurosurgical DBS electrode coordinates into MNI space. *Neuroimage.* 2017;150:395–404.

Horsager J, Andersen KB, Knudsen K, Skjærbæk C, Fedorova TD, Okkels N, Schaeffer E, Bonkat SK, Geday J, Otto M, Sommerauer M, Danielsen EH, Bech E, Kraft J, Munk OL, Hansen SD, Pavese N, Göder R, Brooks DJ, Berg D, Borghammer P. Brain-first versus body-first Parkinson's disease: a multimodal imaging case-control study. *Brain.* 2020 Aug 24:awaa238.

Houk J, Davis JL, Beiser DG. *Models of Information Processing in the Basal Ganglia.* MIT Press; 1995.

Howe M, Ridouh I, Allegra Mascaro AL, Larios A, Azcorra M, Dombeck DA. Coordination of rapid cholinergic and dopaminergic signaling in striatum during spontaneous movement. *Elife.* 2019;8:e44903.

Howe MW, Atallah HE, McCool A, Gibson DJ, Graybiel AM. Habit learning is associated with major shifts in frequencies of oscillatory activity and synchronized spike firing in striatum. *Proc Natl Acad Sci U S A.* 2011;108(40):16801–16806.

Hudon Thibeault AA, Sanderson JT, Vaillancourt C. Serotonin-estrogen interactions: what can we learn from pregnancy? *Biochimie.* 2019;161:88–108.

Ilinsky IA, Kultas-Ilinsky K. Neuroanatomical organization and connections of the motor thalamus in primates. In: Ilinsky IA, Kultas-Ilinsky K, eds. *Basal Ganglia and Thalamus in Health and Movement Disorders.* Kluwer Academic/Plenum; 2001:77–91.

Ilinsky IA, Yi H, Kultas-Ilinsky K. Mode of termination of pallidal afferents to the thalamus: a light and electron microscopic study with anterograde tracers and immunocytochemistry in Macaca mulatta. *J Comp Neurol.* 1997;386(4):601–612.

Ingham CA, Hood SH, Taggart P, Arbuthnott GW. Plasticity of synapses in the rat neostriatum after unilateral lesion of the nigrostriatal dopaminergic pathway. *J Neurosci.* 1998;18(12):4732–4743.

Ito M. *Cerebellum and Neural Control.* Lippincott, Williams, and Wilkins; 1984.

Ito M. *The Cerebellum: Brain for an Implicit Self.* Pearson; 2011.

Iwahori N. A Golgi study on the subthalamic nucleus of the cat. *J Comp Neurol.* 1978;182(3):383–397.

Izhikevich EM. *Dynamical Systems in Neuroscience: The Geometry of Excitability and Bursting.* Computational Neuroscience Series. MIT Press; 2010.

Jacobs E, D'Esposito M. Estrogen shapes dopamine-dependent cognitive processes: implications for women's health. *J Neurosci.* 2011;31(14):5286–5293.

Jaeger D, Gilman S, Aldridge JW. Neuronal activity in the striatum and pallidum of primates related to the execution of externally cued reaching movements. *Brain Res.* 1995 Oct 2; 694(1–2):111–127.

Jaeger D, Kita H, Wilson CJ. Surround inhibition among projection neurons is weak or nonexistent in the rat neostriatum. *J Neurophysiol.* 1994;72(5):2555–2558.

Jahnsen H, Llinás R. Electrophysiological properties of guinea-pig thalamic neurones: an in vitro study. *J Physiol.* 1984;349:205–226.

James W. What is an emotion? *Mind.* 1884;9(34):188–205.

Jan C, Pessiglione M, Tremblay L, Tandé D, Hirsch EC, François C. Quantitative analysis of dopaminergic loss in relation to functional territories in MPTP-treated monkeys. *Eur J Neurosci.* 2003;18(7):2082–2086.

Jankovic J, Tolosa E. *Parkinson's Disease and Movement Disorders.* 6th ed. Urban and Schwarzenberg; 2015.

Jeanne JM, Wilson RI. Convergence, divergence, and reconvergence in a feedforward network improves neural speed and accuracy. *Neuron.* 2015;88(5):1014–1026.

Jia F, Wagle Shukla A, Hu W, Almeida L, Holanda V, Zhang J, Meng F, Okun MS, Li L. Deep brain stimulation at variable frequency to improve motor outcomes in Parkinson's disease. *Mov Disord Clin Pract.* 2018 Oct 1;5(5):538–541.

Joel D, Vikhanski L. *Gender Mosaic: Beyond the Myth of the Male and Female Brain.* Endeavour; 2019.

Johansson Y, Silberberg G. The functional organization of cortical and thalamic inputs onto five types of striatal neurons is determined by source and target cell identities. *Cell Rep.* 2020;30(4):1178–1194.e3.

Joshua M, Adler A, Bergman H. The dynamics of dopamine in control of motor behavior. *Curr Opin Neurobiol.* 2009;19(6):615–620.

Joshua M, Adler A, Mitelman R, Vaadia E, Bergman H. Midbrain dopaminergic neurons and striatal cholinergic interneurons encode the difference between reward and aversive events at different epochs of probabilistic classical conditioning trials. *J Neurosci.* 2008;28(45):11673–11684.

Joshua M, Adler A, Prut Y, Vaadia E, Wickens JR, Bergman H. Synchronization of midbrain dopaminergic neurons is enhanced by rewarding events. *Neuron.* 2009;62(5):695–704.

Joshua M, Adler A, Rosin B, Vaadia E, Bergman H. Encoding of probabilistic rewarding and aversive events by pallidal and nigral neurons. *J Neurophysiol.* 2009;101(2):758–772.

Joshua M, Elias S, Levine O, Bergman H. Quantifying the isolation quality of extracellularly recorded action potentials. *J Neurosci Methods.* 2007;163(2):267–282.

Jouvet M. *The Castle of Dreams.* MIT Press; 2008.

Juavinett AL, Bekheet G, Churchland AK. Chronically implanted Neuropixels probes enable high-yield recordings in freely moving mice. *Elife.* 2019 Aug 14;8:e47188.

Jun JJ, Steinmetz NA, Siegle JH, et al. Fully integrated silicon probes for high-density recording of neural activity. *Nature.* 2017;551(7679):232–236.

Kac M. Can one hear the shape of a drum? *Am Math Mon.* 1966;73:1–23.

Kahneman D. *Thinking, Fast and Slow.* Farrar, Straus and Giroux; 2011.

Kajikawa Y, Smiley JF, Schroeder CE. Primary generators of visually evoked field potentials recorded in the macaque auditory cortex. *J Neurosci.* 2017;37(42):10139–10153.

Kalanithi PS, Zheng W, Kataoka Y, et al. Altered parvalbumin-positive neuron distribution in basal ganglia of individuals with Tourette syndrome. *Proc Natl Acad Sci USA.* 2005;102(37):13307–13312.

Kandel ER, Markram H, Matthews PM, Yuste R, Koch C. Neuroscience thinks big (and collaboratively). *Nat Rev Neurosci.* 2013;14(9):659–664.

Kant I. *The Critique of Practical Reason.* Digireads.com; 2006.

Kaplan A, Mizrahi-Kliger AD, Israel Z, Adler A, Bergman H. Dissociable roles of ventral pallidum neurons in the basal ganglia reinforcement learning network. *Nat Neurosci.* 2020;23(4):556–564.

Karachi C, François C, Parain K, et al. Three-dimensional cartography of functional territories in the human striatopallidal complex by using calbindin immunoreactivity. *J Comp Neurol.* 2002;450(2):122–134.

Karlsen AS, Korbo S, Uylings HB, Pakkenberg B. A stereological study of the mediodorsal thalamic nucleus in Down syndrome. *Neuroscience.* 2014;279:253–259.

Karmos G, Molnár M, Csépe V. A new multielectrode for chronic recording of intracortical field potentials in cats. *Physiol Behav.* 1982;29(3):567–571.

Karpinski M, Mattina GF, Steiner M. Effect of gonadal hormones on neurotransmitters implicated in the pathophysiology of obsessive-compulsive disorder: a critical review. *Neuroendocrinology.* 2017;105(1):1–16.

Karrer TM, Josef AK, Mata R, Morris ED, Samanez-Larkin GR. Reduced dopamine receptors and transporters but not synthesis capacity in normal aging adults: a meta-analysis. *Neurobiol Aging.* 2017;57:36–46.

Kass RE, Eden UT, Brown EN. *Analysis of Neural Data.* Springer; 2016.

Kato M, Miyashita N, Hikosaka O, Matsumura M, Usui S, Kori A. Eye movements in monkeys with local dopamine depletion in the caudate nucleus. I. Deficits in spontaneous saccades. *J Neurosci.* 1995;15(1 Pt 2):912–927.

Kaufman MT, Churchland MM, Ryu SI, Shenoy KV. Cortical activity in the null space: permitting preparation without movement. *Nat Neurosci.* 2014 Mar;17(3):440–448.

Kawaguchi Y, Karube F, Kubota Y. Dendritic branch typing and spine expression patterns in cortical nonpyramidal cells. *Cereb Cortex.* 2006;16(5):696–711.

Kawaguchi Y, Wilson CJ, Emson PC. Projection subtypes of rat neostriatal matrix cells revealed by intracellular injection of biocytin. *J Neurosci.* 1990;10(10):3421–3438.

Kemp JM, Powell TP. The connexions of the striatum and globus pallidus: synthesis and speculation. *Philos Trans R Soc Lond B Biol Sci.* 1971;262(845):441–457.

Kha HT, Finkelstein DI, Pow DV, Lawrence AJ, Horne MK. Study of projections from the entopeduncular nucleus to the thalamus of the rat. *J Comp Neurol.* 2000;426(3):366–377.

Kha HT, Finkelstein DI, Tomas D, Drago J, Pow DV, Horne MK. Projections from the substantia nigra pars reticulata to the motor thalamus of the rat: single axon reconstructions and immunohistochemical study. *J Comp Neurol.* 2001;440(1):20–30.

Kimura M, Rajkowski J, Evarts E. Tonically discharging putamen neurons exhibit set-dependent responses. *Proc Natl Acad Sci USA.* 1984;81(15):4998–5001.

Kincaid AE, Zheng T, Wilson CJ. Connectivity and convergence of single corticostriatal axons. *J Neurosci*. 1998;18(12):4722–4731.

Kinsbourne M. Somatic twist: a model for the evolution of decussation. *Neuropsychology*. 2013 Sep;27(5):511–515.

Kita H. Neostriatal and globus pallidus stimulation induced inhibitory postsynaptic potentials in entopeduncular neurons in rat brain slice preparations. *Neuroscience*. 2001;105(4):871–879.

Kita H. Globus pallidus external segment. *Prog Brain Res*. 2007;160:111–133.

Kita H, Chang HT, Kitai ST. The morphology of intracellularly labeled rat subthalamic neurons: a light microscopic analysis. *J Comp Neurol*. 1983;215(3):245–257.

Kita H, Kitai ST. The morphology of globus pallidus projection neurons in the rat: an intracellular staining study. *Brain Res*. 1994;636(2):308–319.

Kita T, Kita H. The subthalamic nucleus is one of multiple innervation sites for long-range corticofugal axons: a single-axon tracing study in the rat. *J Neurosci*. 2012;32(17):5990–5999.

Kita T, Shigematsu N, Kita H. Intralaminar and tectal projections to the subthalamus in the rat. *Eur J Neurosci*. 2016 Dec;44(11):2899–2908.

Kitai ST, Kita H. Anatomy and physiology of the subthalamic nucleus: a driving force of the basal ganglia. In: Carpenter MB, Jayaraman A, eds. *The Basal Ganglia II*. Plenum Press; 1987: 357–373.

Kokane SS, Perrotti LI. Sex differences and the role of estradiol in mesolimbic reward circuits and vulnerability to cocaine and opiate addiction. *Front Behav Neurosci*. 2020;14:74.

Koós T, Tepper JM, Wilson CJ. Comparison of IPSCs evoked by spiny and fast-spiking neurons in the neostriatum. *J Neurosci*. 2004;24(36):7916–7922.

Kori A, Miyashita N, Kato M, Hikosaka O, Usui S, Matsumura M. Eye movements in monkeys with local dopamine depletion in the caudate nucleus. II. Deficits in voluntary saccades. *J Neurosci*. 1995;15(1 Pt 2):928–941.

Kornhuber HK and Deecke L. *The Will and Its Brain: An Appraisal of Reasoned Free Will*. University Press of America; 2012.

Koshimizu Y, Fujiyama F, Nakamura KC, Furuta T, Kaneko T. Quantitative analysis of axon bouton distribution of subthalamic nucleus neurons in the rat by single neuron visualization with a viral vector. *J Comp Neurol*. 2013;521(9):2125–2146.

Krack P, Limousin P, Benabid AL, Pollak P. Chronic stimulation of subthalamic nucleus improves levodopa-induced dyskinesias in Parkinson's disease. *Lancet*. 1997;350(9092):1676.

Kravitz AV, Freeze BS, Parker PR, Kay K, Thwin MT, Deisseroth K, Kreitzer AC. Regulation of parkinsonian motor behaviours by optogenetic control of basal ganglia circuitry. *Nature*. 2010 Jul 29;466(7306):622–626.

Kravitz AV, Tye LD, Kreitzer AC. Distinct roles for direct and indirect pathway striatal neurons in reinforcement. *Nat Neurosci*. 2012;15(6):816–818.

Kryger MH, Roth T, Dement WC. *Principles and Practice of Sleep Medicine*. 6th ed. Elsevier; 2017.

Kühn AA, Williams D, Kupsch A, Limousin P, Hariz M, Schneider GH, Yarrow K, Brown P. Event-related beta desynchronization in human subthalamic nucleus correlates with motor performance. *Brain*. 2004 Apr;127(Pt 4):735–746.

Kupchik YM, Brown RM, Heinsbroek JA, Lobo MK, Schwartz DJ, Kalivas PW. Coding the direct/ indirect pathways by D1 and D2 receptors is not valid for accumbens projections. *Nat Neurosci.* 2015;18(9):1230–1232.

Kushner HI. *On the Other Hand: Left Hand, Right Brain, Mental Disorder, and History.* Johns Hopkins University Press; 2017.

Kuusimäki T, Al-Abdulrasul H, Kurki S, Hietala J, Hartikainen S, Koponen M, Tolppanen AM, Kaasinen V. Increased risk of parkinson's disease in patients with schizophrenia spectrum disorders. *Mov Disord.* 2021 Jan 6. https://doi.org/10.1002/mds.28484 (Epub ahead of print).

Kuypers HG. A new look at the organization of the motor system. *Prog Brain Res.* 1982;57:381–403.

Lanciego JL, Gonzalo N, Castle M, Sanchez-Escobar C, Aymerich MS, Obeso JA. Thalamic innervation of striatal and subthalamic neurons projecting to the rat entopeduncular nucleus. *Eur J Neurosci.* 2004;19(5):1267–1277.

Lange H, Thörner G, Hopf A, Schröder KF. Morphometric studies of the neuropathological changes in choreatic diseases. *J Neurol Sci.* 1976;28(4):401–425.

Langston JW. The MPTP story. *J Parkinsons Dis.* 2017;7(s1):S11–S19.

Langston JW, Ballard P, Tetrud JW, Irwin I. Chronic parkinsonism in humans due to a product of meperidine-analog synthesis. *Science.* 1983;219(4587):979–980.

Lansing A. *Endurance: Shackleton's Incredible Voyage to the Antarctic.* Basic Books 2000.

Larsen B, Luna B. Adolescence as a neurobiological critical period for the development of higher-order cognition. *Neurosci Biobehav Rev.* 2018;94:179–195.

Larsen KD, Sutin J. Output organization of the feline entopeduncular and subthalamic nuclei. *Brain Res.* 1978;157(1):21–31.

Lautin A. *The Limbic Brain.* Kluer Academic/Plenum; 2001.

Lawrence DG, Kuypers HG. Pyramidal and non-pyramidal pathways in monkeys: anatomical and functional correlation. *Science.* 1965 May 14;148(3672):973–975.

Lawrence DG, Kuypers HG. The functional organization of the motor system in the monkey; I. The effects of bilateral pyramidal lesions. *Brain.* 1968 Mar;91(1):1–14.

Ledergerber D, Larkum ME. Properties of layer 6 pyramidal neuron apical dendrites. *J Neurosci.* 2010;30(39):13031–13044.

Ledoux L. *The Emotional Brain: The Mysterious Underpinnings of Emotional Life.* Simon and Schuster; 1996.

Lee HJ, Weitz AJ, Bernal-Casas D, et al. Activation of direct and indirect pathway medium spiny neurons drives distinct brain-wide responses. *Neuron.* 2016;91(2):412–424.

Lee J, Wang W, Sabatini BL. Anatomically segregated basal ganglia pathways allow parallel behavioral modulation. *Nat Neurosci.* 2020;23(11):1388–1398.

Lees AJ, Tolosa E, Olanow CW. Four pioneers of L-dopa treatment: Arvid Carlsson, Oleh Hornykiewicz, George Cotzias, and Melvin Yahr. *Mov Disord.* 2015;30(1):19–36.

Legéndy CR, Salcman M. Bursts and recurrences of bursts in the spike trains of spontaneously active striate cortex neurons. *J Neurophysiol.* 1985 Apr;53(4):926–939.

Legaria AA, Licholai JA, Kravitz AV. Fiber photometry does not reflect spiking activity in the striatum. bioRxiv. 2021 Jan 20. https://doi.org/10.1101/2021.01.20.427525.

Lema Tomé CM, Tyson T, Rey NL, et al. Inflammation and αsynuclein's prionlike behavior in Parkinson's disease—is there a link? *Mol Neurobiol*. 2013;47:561–574.

Lemon R. *Methods for Neuronal Recording in Conscious Animals*. IBRO Handbook Series. Wiley; 1984.

Levakova M. Efficiency of rate and latency coding with respect to metabolic cost and time. *Biosystems*. 2017;161:31–40.

Leventhal DK, Gage GJ, Schmidt R, Pettibone JR, Case AC, Berke JD. Basal ganglia beta oscillations accompany cue utilization. *Neuron*. 2012;73(3):523–536.

Lévesque M, Parent A. The striatofugal fiber system in primates: a reevaluation of its organization based on single-axon tracing studies. *Proc Natl Acad Sci USA*. 2005;102(33):11888–11893.

Lewicki MS. A review of methods for spike sorting: the detection and classification of neural action potentials. *Network*. 1998;9(4):R53–R78.

Li N, Baldermann JC, Kibleur A, et al. A unified connectomic target for deep brain stimulation in obsessive-compulsive disorder. *Nat Commun*. 2020 Jul;11(1):3364.

Libet B. Unconscious cerebral initiative and the role of conscious will in voluntary action. *Behav Brain Sci*. 1985;8:529–566

Limousin P, Pollak P, Benazzouz A, et al. Effect of parkinsonian signs and symptoms of bilateral subthalamic nucleus stimulation. *Lancet*. 1995;345(8942):91–95.

Little S, Beudel M, Zrinzo L, et al. Bilateral adaptive deep brain stimulation is effective in Parkinson's disease. *J Neurol Neurosurg Psychiatry*. 2016;87(7):717–721.

Little S, Pogosyan A, Neal S, et al. Adaptive deep brain stimulation in advanced Parkinson disease. *Ann Neurol*. 2013;74(3):449–457.

Logigian E, Hefter H, Reiners K, Freund HJ. Does tremor pace repetitive voluntary motor behavior in Parkinson's disease? *Ann Neurol*. 1991;30(2):172–179.

Logothetis NK. What we can do and what we cannot do with fMRI. *Nature*. 2008;453(7197):869–878.

López-Azcárate J, Tainta M, Rodríguez-Oroz MC, et al. Coupling between beta and high-frequency activity in the human subthalamic nucleus may be a pathophysiological mechanism in Parkinson's disease. *J Neurosci*. 2010;30(19):6667–6677.

Lowe EJ. *An Introduction to the Philosophy of Mind*. Cambridge University Press; 2000.

Luck SJ. *An Introduction to the Event-Related Potential Technique*. 2nd ed. MIT Press; 2014.

Lüscher C, Pascoli V, Creed M. Optogenetic dissection of neural circuitry: from synaptic causalities to blue prints for novel treatments of behavioral diseases. *Curr Opin Neurobiol*. 2015;35:95–100.

Lyon L. Dead salmon and voodoo correlations: should we be sceptical about functional MRI? *Brain*. 2017;140(8):e53.

Ma SY, Roytt M, Collan Y, Rinne JO. Unbiased morphometrical measurements show loss of pigmented nigral neurones with ageing. *Neuropathol Appl Neurobiol*. 1999;25:394–399.

MacLean PD. *The Triune Brain in Evolution: Role in Paleocerebral Functions*. Kluwer Academic/ Plenum Publishers; 1990.

Mai JK, Majtanik M. Toward a common terminology for the thalamus. *Front Neuroanat*. 2019;12:114.

Mallet N, Micklem BR, Henny P, et al. Dichotomous organization of the external globus pallidus. *Neuron*. 2012;74(6):1075–1086.

Manera AL, Dadar M, Fonov V, Collins DL. CerebrA, registration and manual label correction of Mindboggle-101 atlas for MNI-ICBM152 template. *Sci Data*. 2020;7(1):237.

Mannella F, Gurney K, Baldassarre G. The nucleus accumbens as a nexus between values and goals in goal-directed behavior: a review and a new hypothesis. *Front Behav Neurosci*. 2013;7:135.

Maoz U, Yaffe G, Koch C, Mudrik L. Neural precursors of decisions that matter—an ERP study of deliberate and arbitrary choice. *Elife*. 2019;8:e39787.

Marani E, Heida T, Egbert EAJF, Usunoff KG. *The Subthalamic Nucleus: Part I: Development, Cytology, Topography and Connections*. Springer; 2008.

Markov NT, Ercsey-Ravasz M, Van Essen DC, Knoblauch K, Toroczkai Z, Kennedy H. Cortical high-density counterstream architectures. *Science*. 2013;342(6158):1238406.

Markram, H. The Blue Brain Project. *Nat Rev Neurosci*. 2006;7:153–160.

Marks WJ. *Deep Brain Stimulation Management*. 2nd ed. Cambridge University Press; 2015.

Marmor O, Valsky D, Joshua M, et al. Local vs. volume conductance activity of field potentials in the human subthalamic nucleus. *J Neurophysiol*. 2017;117(6):2140–2151.

Marom S. *Science, Psychoanalysis, and the Brain: Space for Dialogue*. Cambridge University Press; 2015.

Marr D. A theory of cerebellar cortex. *J Physiol*. 1969;202(2):437–470.

Marr D. *Vision: A Computational Investigation into the Human Representation and Processing of Visual Information*. WH Freeman; 1982.

Marsden CD, Obeso JA. The functions of the basal ganglia and the paradox of stereotaxic surgery in Parkinson's disease. *Brain*. 1994;117 (Pt 4):877–897.

Mascetti GG. Unihemispheric sleep and asymmetrical sleep: behavioral, neurophysiological, and functional perspectives. *Nat Sci Sleep*. 2016;8:221–238.

Mastro KJ, Bouchard RS, Holt HA, Gittis AH. Transgenic mouse lines subdivide external segment of the globus pallidus (GPe) neurons and reveal distinct GPe output pathways. *J Neurosci*. 2014;34(6):2087–2099.

Mathai A, Ma Y, Paré JF, Villalba RM, Wichmann T, Smith Y. Reduced cortical innervation of the subthalamic nucleus in MPTP-treated parkinsonian monkeys. *Brain*. 2015;138(Pt 4):946–962.

Mathai A, Smith Y. The corticostriatal and corticosubthalamic pathways: two entries, one target. So what? *Front Syst Neurosci*. 2011;5:64.

Matsuda W, Furuta T, Nakamura KC, et al. Single nigrostriatal dopaminergic neurons form widely spread and highly dense axonal arborizations in the neostriatum. *J Neurosci*. 2009;29(2):444–453.

Matsumoto M, Hikosaka O. Lateral habenula as a source of negative reward signals in dopamine neurons. *Nature*. 2007;447(7148):1111–1115.

Matsumoto M, Hikosaka O. Two types of dopamine neuron distinctly convey positive and negative motivational signals. *Nature*. 2009;459(7248):837–841.

Matsumoto N, Minamimoto T, Graybiel AM, Kimura M. Neurons in the thalamic CM-Pf complex supply striatal neurons with information about behaviorally significant sensory events. *J Neurophysiol*. 2001;85(2):960–976.

Max DT. *The Family That Couldn't Sleep: A Medical Mystery*. Random House; 2007.

Mayberg HS, Lozano AM, Voon V, McNeely HE, Seminowicz D, Hamani C, Schwalb JM, Kennedy SH. Deep brain stimulation for treatment-resistant depression. *Neuron*. 2005 Mar 3;45(5):651–660.

Mayer J. Drawing an elephant with four complex parameters. *Am J Phys* 2010;78:648.

Medina L, Jiao Y, Reiner A. The functional anatomy of the basal ganglia of birds. *Eur J Morphol.* 1999;37(2–3):160–165.

Mengler L, Khmelinskii A, Diedenhofen M, et al. Brain maturation of the adolescent rat cortex and striatum: changes in volume and myelination. *Neuroimage.* 2014;84:35–44.

Meoni S, Macerollo A, Moro E. Sex differences in movement disorders. *Nat Rev Neurol.* 2020;16(2):84–96.

Mesulam MM, Mash D, Hersh L, Bothwell M, Geula C. Cholinergic innervation of the human striatum, globus pallidus, subthalamic nucleus, substantia nigra, and red nucleus. *J Comp Neurol.* 1992;323(2):252–268.

Miller JD, Farber J, Gatz P, Roffwarg H, German DC. Activity of mesencephalic dopamine and non-dopamine neurons across stages of sleep and walking in the rat. *Brain Res.* 1983;273(1):133–141.

Miller R. *A Theory of the Basal Ganglia and Their Disorders.* CRC Press; 2008.

Miller R, Wickens J. *Brain Dynamics and the Striatal Complex.* Harwood Academic; 2000.

Miller WC, DeLong MR. Altered tonic activity of neurons in the globus pallidus and subthalamic nucleus in the primate MPTP model of parkinsonism. In: Carpenter MB, Jayaraman A, eds. *The Basal Ganglia II. Advances in Behavioral Biology.* Volume 32. Springer; 1987.

Minamimoto T, Hori Y, Kimura M. Complementary process to response bias in the centromedian nucleus of the thalamus. *Science.* 2005;308(5729):1798–1801.

Mink JW. The basal ganglia: focused selection and inhibition of competing motor programs. *Prog Neurobiol.* 1996 Nov; 50(4):381–425.

Mink JW, Thach WT. Basal ganglia motor control. II. Late pallidal timing relative to movement onset and inconsistent pallidal coding of movement parameters. *J Neurophysiol.* 1991 Feb;65(2):301–329.

Mishkin M. Memory in monkeys severely impaired by combined but not by separate removal of amygdala and hippocampus. *Nature.* 1978;273(5660):297–298.

Mitchell SJ, Richardson RT, Baker FH, DeLong MR. The primate nucleus basalis of Meynert: neuronal activity related to a visuomotor tracking task. *Exp Brain Res.* 1987;68(3):506–515.

Mitelman R, Rosin B, Zadka H, et al. Neighboring pallidal neurons do not exhibit more synchronous oscillations than remote ones in the MPTP primate model of Parkinson's disease. *Front Syst Neurosci.* 2011;5:54.

Mitra P, Bokil H. *Observed Brain Dynamics.* Oxford University Press; 2007.

Mitz AR, Bartolo R, Saunders RC, Browning PG, Talbot T, Averbeck BB. High channel count single-unit recordings from nonhuman primate frontal cortex. *J Neurosci Methods.* 2017;289:39–47.

Mizrahi-Kliger AD, Kaplan A, Israel Z, Bergman H. Desynchronization of slow oscillations in the basal ganglia during natural sleep. *Proc Natl Acad Sci USA.* 2018;115(18):E4274–E4283.

Mizrahi-Kliger AD, Kaplan A, Israel Z, Deffains M, Bergman H. Basal ganglia beta oscillations during sleep underlie parkinsonian insomnia. *Proc Natl Acad Sci USA.* 2020;117(29):17359–17368.

Mohebi A, Pettibone JR, Hamid AA, Wong JT, Vinson LT, Patriarchi T, Tian L, Kennedy RT, Berke JD. Dissociable dopamine dynamics for learning and motivation. *Nature.* 2019;570(7759):65–70.

Molnár Z, Brown RE. Insights into the life and work of Sir Charles Sherrington. *Nat Rev Neurosci.* 2010 Jun;11(6):429–436.

Monakow KH, Akert K, Künzle H. Projections of the precentral motor cortex and other cortical areas of the frontal lobe to the subthalamic nucleus in the monkey. *Exp Brain Res.* 1978;33(3–4):395–403.

Monti JM, Pandi-Perumal SR, Chokroverty S, eds. *Dopamine and Sleep: Molecular, Functional and Clinical Aspects.* Springer; 2016.

Moorcroft WH. *Understanding Sleep and Dreaming.* 2nd ed. Springer; 2013.

Moore GP, Segundo JP, Perkel DH, Levitan H. Statistical signs of synaptic interaction in neurons. *Biophys J.* 1970;10(9):876–900.

Moran A, Bar-Gad I, Bergman H, Israel Z. Real-time refinement of subthalamic nucleus targeting using Bayesian decision-making on the root mean square measure. *Mov Disord.* 2006 Sep;21(9):1425–1431.

Morris G, Arkadir D, Nevet A, Vaadia E, Bergman H. Coincident but distinct messages of midbrain dopamine and striatal tonically active neurons. *Neuron.* 2004;43(1):133–143.

Mosconi MW, Wang Z, Schmitt LM, Tsai P, Sweeney JA. The role of cerebellar circuitry alterations in the pathophysiology of autism spectrum disorders. *Front Neurosci.* 2015;9:296.

Moshel S, Shamir RR, Raz A, et al. Subthalamic nucleus long-range synchronization—an independent hallmark of human Parkinson's disease. *Front Syst Neurosci.* 2013;7:79.

Moss J, Bolam JP. A dopaminergic axon lattice in the striatum and its relationship with cortical and thalamic terminals. *J Neurosci.* 2008;28(44):11221–11230.

Moujahid A, d'Anjou A, Torrealdea FJ, Torrealdea F. Energy and information in Hodgkin-Huxley neurons. *Phys Rev E Stat Nonlin Soft Matter Phys.* 2011;83(3 Pt 1):031912.

Mukhametov LM, Supin AY, Polyakova IG. Interhemispheric asymmetry of the electroencephalographic sleep patterns in dolphins. *Brain Res.* 1977;134(3):581–584.

Murray EA, Wise SP, Graham KS. *The Evolution of Memory Systems: Ancestors, Anatomy and Adaptions.* Oxford University Press; 2017.

Nakanishi H, Kita H, Kitai ST. Intracellular study of rat substantia nigra pars reticulata neurons in an in vitro slice preparation: electrical membrane properties and response characteristics to subthalamic stimulation. *Brain Res.* 1987;437(1):45–55.

Nakanishi H, Kita H, Kitai ST. Intracellular study of rat entopeduncular nucleus neurons in an in vitro slice preparation: response to subthalamic stimulation. *Brain Res.* 1991;549(2):285–291.

Nambu A, Tokuno H, Takada M. Functional significance of the cortico-subthalamo-pallidal "hyperdirect" pathway. *Neurosci Res.* 2002;43(2):111–117.

Nauta WJ, Mehler WR. Projections of the lentiform nucleus in the monkey. *Brain Res.* 1966;1(1):3–42.

Neher E, Sakmann B. Single-channel currents recorded from membrane of denervated frog muscle fibres. *Nature.* 1976;260(5554):799–802.

Nelson GD. A brief history of cardiac pacing. *Tex Heart Inst J.* 1993;20(1):12–18.

Nevet A, Morris G, Saban G, Fainstein N, Bergman H. Discharge rate of substantia nigra pars reticulata neurons is reduced in non-parkinsonian monkeys with apomorphine-induced orofacial dyskinesia. *J Neurophysiol.* 2004;92(4):1973–1981.

Nicola SM, Surmeier J, Malenka RC. Dopaminergic modulation of neuronal excitability in the striatum and nucleus accumbens. *Annu Rev Neurosci.* 2000;23:185–215.

Nini A, Feingold A, Slovin H, Bergman H. Neurons in the globus pallidus do not show correlated activity in the normal monkey, but phase-locked oscillations appear in the MPTP model of parkinsonism. *J Neurophysiol.* 1995;74(4):1800–1805.

Nir Y, Andrillon T, Marmelshtein A, et al. Selective neuronal lapses precede human cognitive lapses following sleep deprivation. *Nat Med.* 2017;23(12):1474–1480.

Niv Y, Daw ND, Joel D, Dayan P. Tonic dopamine: opportunity costs and the control of response vigor. *Psychopharmacology* (Berlin). 2007;191(3):507–520.

Niv Y, Duff MO, Dayan P. Dopamine, uncertainty and TD learning. *Behav Brain Funct.* 2005 May 4;1:6.

Nolte G, Bai O, Wheaton L, Mari Z, Vorbach S, Hallett M. Identifying true brain interaction from EEG data using the imaginary part of coherency. *Clin Neurophysiol.* 2004 Oct;115(10):2292–2307.

Nolte M, Reimann MW, King JG, et al. Cortical reliability amid noise and chaos. *Nat Commun.* 2019;10:3792.

Nunez PL, Srinivasan R. *Electric Fields of the Brain: The Neurophysics of EEG.* 2nd ed. Oxford University Press; 2006.

Obeso JA, Rodriguez-Oroz MC, Stamelou M, Bhatia KP, Burn DJ. The expanding universe of disorders of the basal ganglia. *Lancet.* 2014;384(9942):523–531.

O'Doherty J, Dayan P, Schultz J, Deichmann R, Friston K, Dolan RJ. Dissociable roles of ventral and dorsal striatum in instrumental conditioning. *Science.* 2004;304(5669):452–454.

Okada N, Fukunaga M, Yamashita F, et al. Abnormal asymmetries in subcortical brain volume in schizophrenia. *Mol Psychiatry.* 2016;21:1460–1466.

O'Kusky J, Colonnier M. A laminar analysis of the number of neurons, glia, and synapses in the adult cortex (area 17) of adult macaque monkeys. *J Comp Neurol.* 1982;210(3):278–290.

Oorschot DE. Total number of neurons in the neostriatal, pallidal, subthalamic, and substantia nigral nuclei of the rat basal ganglia: a stereological study using the cavalieri and optical disector methods. *J Comp Neurol.* 1996 Mar 18;366(4):580–599.

Ortuño-Lizarán I, Sánchez-Sáez X, Lax P, et al. Dopaminergic retinal cell loss and visual dysfunction in Parkinson's disease. *Ann Neurol.* 2020;10.1002/ana.25897.

Orwell G. *The lion and the unicorn: socialism and the English genius.* Penguin Books; 2018.

Ottenheimer DJ, Bari BA, Sutlief E, Fraser KM, Kim TH, Richard JM, Cohen JY, Janak PH. A quantitative reward prediction error signal in the ventral pallidum. *Nat Neurosci.* 2020 Oct;23(10):1267–1276.

Özkurt TE, Butz M, Homburger M, Elben S, Vesper J, Wojtecki L, Schnitzler A. High frequency oscillations in the subthalamic nucleus: a neurophysiological marker of the motor state in Parkinson's disease. *Exp Neurol.* 2011 Jun;229(2):324–331.

Pakkenberg B, Moller A, Gundersen HJ, Mouritzen Dam A, Pakkenberg H. The absolute number of nerve cells in substantia nigra in normal subjects and in patients with Parkinson's disease estimated with an unbiased stereological method. *J Neurol Neurosurg Psychiatry* 1991;54:30–33.

Pandya DN, Yeterian EH. Prefrontal cortex in relation to other cortical areas in rhesus monkey: architecture and connections. *Prog Brain Res.* 1990;85:63–94.

Parent A. *Comparative Neurobiology of the Basal Ganglia.* Wiley-Interscience; 1986.

Parent A. The history of the basal ganglia: the contribution of Karl Friedrich Burdach. *Neuro Med.* 2012;3:374–379.

Parent A, Côté PY, Lavoie B. Chemical anatomy of primate basal ganglia. *Prog Neurobiol.* 1995; 46(2–3):131–197.

Parent A, Hazrati LN. Anatomical aspects of information processing in primate basal ganglia. *Trends Neurosci.* 1993;16(3):111–116.

Parent A, Hazrati LN. Functional anatomy of the basal ganglia. I. The cortico-basal ganglia-thalamo-cortical loop. *Brain Res Brain Res Rev.* 1995a;20(1):91–127.

Parent A, Hazrati LN. Functional anatomy of the basal ganglia. II. The place of subthalamic nucleus and external pallidum in basal ganglia circuitry. *Brain Res Brain Res Rev.* 1995b;20(1):128–154.

Parent A, Sato F, Wu Y, Gauthier J, Lévesque M, Parent M. Organization of the basal ganglia: the importance of axonal collateralization. *Trends Neurosci.* 2000;23(10 Suppl):S20–S27.

Parent M, Lévesque M, Parent A. Two types of projection neurons in the internal pallidum of primates: single-axon tracing and three-dimensional reconstruction. *J Comp Neurol.* 2001;439(2):162–175.

Parent M, Parent A. Single-axon tracing and three-dimensional reconstruction of centre median-parafascicular thalamic neurons in primates. *J Comp Neurol.* 2005;481(1):127–144.

Parent M, Parent A. Single-axon tracing study of corticostriatal projections arising from primary motor cortex in primates. *J Comp Neurol.* 2006;496(2):202–213.

Parent M, Parent A. The microcircuitry of primate subthalamic nucleus. *Parkinsonism Relat Disord.* 2007;13 Suppl 3:S292–S295.

Parent M, Wallman MJ, Gagnon D, Parent A. Serotonin innervation of basal ganglia in monkeys and humans. *J Chem Neuroanat.* 2011;41(4):256–265.

Parker JG, Marshall JD, Ahanonu B, et al. Diametric neural ensemble dynamics in parkinsonian and dyskinetic states. *Nature.* 2018;557(7704):177–182.

Parush N, Tishby N, Bergman H. Dopaminergic balance between reward maximization and policy complexity. *Front Syst Neurosci.* 2011;5:22.

Paterson RW, Brown RL, Benjamin L, et al. The emerging spectrum of COVID-19 neurology: clinical, radiological and laboratory findings. *Brain.* 2020;awaa240.

Patriarchi T, Cho JR, Merten K, Howe MW, Marley A, Xiong WH, Folk RW, Broussard GJ, Liang R, Jang MJ, Zhong H, Dombeck D, von Zastrow M, Nimmerjahn A, Gradinaru V, Williams JT, Tian L. Ultrafast neuronal imaging of dopamine dynamics with designed genetically encoded sensors. *Science* 2018;360(6396). https://doi.org/10.1126/science.aat4422.

Patten MB. *The Great Cotzias: Discoverer of the Treatment for Parkinson's Disease.* Peak Achievement Publishing; 2012.

Pauli WM, Nili AN, Tyszka JM. A high-resolution probabilistic in vivo atlas of human subcortical brain nuclei. *Scientific Data.* 2018;5:180063.

Pearl J, MacKenzie D. *The Book of Why: The New Science of Cause and Effect.* Penguin Books; 2019.

Pedreira C, Martinez J, Ison MJ, Quian Quiroga R. How many neurons can we see with current spike sorting algorithms? *J Neurosci Methods.* 2012;211(1):58–65.

Peles O, Werner-Reiss U, Bergman H, Israel Z, Vaadia E. Phase-specific microstimulation differentially modulates beta oscillations and affects behavior. *Cell Rep.* 2020;30(8):2555–2566.e3.

Pelled G, Bergman H, Ben-Hur T, Goelman G. Manganese-enhanced MRI in a rat model of Parkinson's disease. *J Magn Reson Imaging*. 2007;26(4):863–870.

Percheron G, Filion M. Parallel processing in the basal ganglia: up to a point. *Trends Neurosci*. 1991;14(2):55–59.

Percheron G, François C, Talbi B, Yelnik J, Fénelon G. The primate motor thalamus. *Brain Res Brain Res Rev*. 1996;22(2):93–181.

Percheron G, François C, Yelnik J, Fénelon G. The primate nigro striato-pallidal-nigral system. Not a mere loop. In: Crossman AR, Sambrook MA, eds. *Neural Mechanisms in Disorders of Movement*. John Libbey; 1989:103–109.

Percheron G, François C, Yelnik J, Fénelon G, Talbi B. The basal ganglia related system of primates: definition, description and informational analysis. In: Percheron G, McKenzie JS, Féger J, eds. *The Basal Ganglia IV. Advances in Behavioral Biology*. Volume 41. Springer; 1994.

Percheron G, Yelnik J, François C. A Golgi analysis of the primate globus pallidus. III. Spatial organization of the striato-pallidal complex. *J Comp Neurol*. 1984;227(2):214–227.

Perkins MN, Stone TW. Subthalamic projections to the globus pallidus: an electrophysiological study in the rat. *Exp Neurol*. 1980;68(3):500–511.

Perreault ML, Hasbi A, O'Dowd BF, George SR. The dopamine d1-d2 receptor heteromer in striatal medium spiny neurons: evidence for a third distinct neuronal pathway in basal ganglia. *Front Neuroanat*. 2011;5:31.

Person AL, Perkel DJ. Pallidal neuron activity increases during sensory relay through thalamus in a songbird circuit essential for learning. *J Neurosci*. 2007;27(32):8687–8698.

Petersen MV, Mlakar J, Haber SN, et al. Holographic reconstruction of axonal pathways in the human brain. *Neuron*. 2019;104(6):1056–1064.e3.

Phillips EA, Hasenstaub AR. Asymmetric effects of activating and inactivating cortical interneurons. *Elife*. 2016 Oct 10;5:e18383. https://doi.org/10.7554/eLife.18383.

Phillips PE, Stuber GD, Heien ML, Wightman RM, Carelli RM. Subsecond dopamine release promotes cocaine seeking. *Nature*. 2003;422(6932):614–618.

Pirker W. Correlation of dopamine transporter imaging with parkinsonian motor handicap: how close is it? *Mov Disord*. 2003 Oct;18(Suppl 7):S43–S51.

Pissadaki EK, Bolam JP. The energy cost of action potential propagation in dopamine neurons: clues to susceptibility in Parkinson's disease. *Front Comput Neurosci*. 2013 Mar 18;7:13.

Planert H, Szydlowski SN, Hjorth JJ, Grillner S, Silberberg G. Dynamics of synaptic transmission between fast-spiking interneurons and striatal projection neurons of the direct and indirect pathways. *J Neurosci*. 2010;30(9):3499–3507.

Pollak P, Benabid AL, Gross C, et al. Effets de la stimulation du noyau sous-thalamique dans la maladie de Parkinson [Effects of the stimulation of the subthalamic nucleus in Parkinson disease]. *Rev Neurol* (Paris). 1993;149(3):175–176.

Porat O, Hassin-Baer S, Cohen OS, Markus A, Tomer R. Asymmetric dopamine loss differentially affects effort to maximize gain or minimize loss. *Cortex*. 2014;51:82–91.

Priori A, Foffani G, Rossi L, Marceglia S. Adaptive deep brain stimulation (aDBS) controlled by local field potential oscillations. *Exp Neurol*. 2013;245:77–86.

Ranck JB Jr. Which elements are excited in electrical stimulation of mammalian central nervous system: a review. *Brain Res*. 1975;98(3):417–440.

Rappel P, Grosberg S, Arkadir D, et al. Theta-alpha oscillations characterize emotional subregion in the human ventral subthalamic nucleus. *Mov Disord*. 2020;35(2):337–343.

Rappel P, Marmor O, Bick AS, et al. Subthalamic theta activity: a novel human subcortical biomarker for obsessive compulsive disorder. *Transl Psychiatry*. 2018;8(1):118.

Ravel S, Sardo P, Legallet E, Apicella P. Reward unpredictability inside and outside of a task context as a determinant of the responses of tonically active neurons in the monkey striatum. *J Neurosci*. 2001;21(15):5730–5739.

Raz A, Feingold A, Zelanskaya V, Vaadia E, Bergman H. Neuronal synchronization of tonically active neurons in the striatum of normal and parkinsonian primates. *J Neurophysiol*. 1996;76(3):2083–2088.

Raz A, Frechter-Mazar V, Feingold A, Abeles M, Vaadia E, Bergman H. Activity of pallidal and striatal tonically active neurons is correlated in MPTP-treated monkeys but not in normal monkeys. *J Neurosci*. 2001;21(3):RC128.

Raz A, Vaadia E, Bergman H. Firing patterns and correlations of spontaneous discharge of pallidal neurons in the normal and the tremulous 1-methyl-4-phenyl-1,2,3,6-tetrahydropyridine vervet model of parkinsonism. *J Neurosci*. 2000;20(22):8559–8571.

Redgrave P, Coizet V, Comoli E, et al. Interactions between the midbrain superior colliculus and the basal ganglia. *Front Neuroanat*. 2010;4:132.

Redgrave P, Rodriguez M, Smith Y, et al. Goal-directed and habitual control in the basal ganglia: implications for Parkinson's disease. *Nat Rev Neurosci*. 2010;11(11):760–772.

Rehani R, Atamna Y, Tiroshi L, et al. Activity patterns in the neuropil of striatal cholinergic interneurons in freely moving mice represent their collective spiking dynamics. *eNeuro*. 2019; 6(1):ENEURO.0351-18.2018.

Reimann MW, Gevaert M, Shi Y, et al. A null model of the mouse whole-neocortex microconnectome. *Nat Commun*. 2019;10:3903.

Reiner A, Medina L, Veenman CL. Structural and functional evolution of the basal ganglia in vertebrates. *Brain Res Brain Res Rev*. 1998;28(3):235–285.

Renshaw B, Forbes A, Morrison BR. Activity of isocortex and hippocampus: electrical studies with micro-electrodes. *J Neurophysiol*. 1940;3(1):74–105.

Rey HG, Pedreira C, Quian Quiroga R. Past, present and future of spike sorting techniques. *Brain Res Bull*. 2015;119(Pt B):106–117.

Reynolds JN, Hyland BI, Wickens JR. A cellular mechanism of reward-related learning. *Nature*. 2001;413(6851):67–70.

Rice ME, Cragg SJ. Dopamine spillover after quantal release: rethinking dopamine transmission in the nigrostriatal pathway. *Brain Res Rev*. 2008;58(2):303–313.

Rice ME, Patel JC. Somatodendritic dopamine release: recent mechanistic insights. *Philos Trans R Soc Lond B Biol Sci*. 2015 Jul 5;370(1672):20140185.

Richardson RT, DeLong MR. Context-dependent responses of primate nucleus basalis neurons in a go/no-go task. *J Neurosci*. 1990;10(8):2528–2540.

Rigoli F, Chew B, Dayan P, Dolan RJ. The dopaminergic midbrain mediates an effect of average reward on Pavlovian vigor. *J Cogn Neurosci*. 2016;28(9):1303–1317.

Rijpkema M, Everaerd D, van der Pol C, Franke B, Tendolkar I, Fernández G. Normal sexual dimorphism in the human basal ganglia. *Hum Brain Mapp*. 2012;33(5):1246–1252.

Rippon G. *The Gendered Brain: The New Neuroscience That Shatters the Myth of the Female Brain.* Vintage; 2019.

Rivera-Garcia MT, McCane AM, Chowdhury TG, Wallin-Miller KG, Moghaddam B. Sex and strain differences in dynamic and static properties of the mesolimbic dopamine system. *Neuropsychopharmacology.* 2020;10.1038/s41386-020-0765-1.

Rogers LJ, Vallortigara G, Andrew RJ. *Divided Brains: The Biology and Behaviour of Brain Asymmetries.* Cambridge University Press; 2013.

Rolls ET. *Emotion Explained.* Oxford University Press; 2007.

Rolston JD, Englot DJ, Starr PA, Larson PS. An unexpectedly high rate of revisions and removals in deep brain stimulation surgery: analysis of multiple databases. *Parkinsonism Relat Disord.* 2016;33:72–77.

Romo R, Schultz W. Dopamine neurons of the monkey midbrain: contingencies of responses to active touch during self-initiated arm movements. *J Neurophysiol.* 1990;63(3):592–606.

Rosenbaum R, Smith MA, Kohn A, Rubin JE, Doiron B. The spatial structure of correlated neuronal variability. *Nat Neurosci.* 2017;20(1):107–114.

Rosin B, Slovik M, Mitelman R, et al. Closed-loop deep brain stimulation is superior in ameliorating parkinsonism. *Neuron.* 2011;72(2):370–384.

Rouzaire-Dubois B, Scarnati E, Hammond C, Crossman AR, Shibazaki T. Microiontophoretic studies on the nature of the neurotransmitter in the subthalamo-entopeduncular pathway of the rat. *Brain Res.* 1983;271(1):11–20.

Royce GJ, Mourey RJ. Efferent connections of the centromedian and parafascicular thalamic nuclei: an autoradiographic investigation in the cat. *J Comp Neurol.* 1985;235(3):277–300.

Rudow G, O'Brien R, Savonenko A, Resnick SM, Zonderman AB, Pletnikova O, Marsh L, Dawson TM, Crain BJ, West MJ, Troncoso JC. Morphometry of the human substantia nigra in ageing and Parkinson's disease. *Acta Neuropathol.* 2008;115:461–470.

Sadek AR, Magill PJ, Bolam JP. A single-cell analysis of intrinsic connectivity in the rat globus pallidus. *J Neurosci.* 2007;27(24):6352–6362.

Sadikot AF, Parent A, Smith Y, Bolam JP. Efferent connections of the centromedian and parafascicular thalamic nuclei in the squirrel monkey: a light and electron microscopic study of the thalamostriatal projection in relation to striatal heterogeneity. *J Comp Neurol.* 1992;320(2):228–242.

Sadock BJ, Sadock VA, Ruiz P. *Kaplan and Sadock's Synopsis of Psychiatry: Behavioral Sciences/Clinical Psychiatry.* 11th ed. Wolters Kluwer; 2014.

Saliani A, Perraud B, Duval T, Stikov N, Rossignol S, Cohen-Adad J. Axon and myelin morphology in animal and human spinal cord. *Front Neuroanat.* 2017 Dec 22;11:129.

Salvesen L, Ullerup BH, Sunay FB, et al. Changes in total cell numbers of the basal ganglia in patients with multiple system atrophy—a stereological study. *Neurobiol Dis.* 2015;74:104–113.

Sankar T, Chakravarty MM, Jawa N, Li SX, Giacobbe P, Kennedy SH, Rizvi SJ, Mayberg HS, Hamani C, Lozano AM. Neuroanatomical predictors of response to subcallosal cingulate deep brain stimulation for treatment-resistant depression. *J Psychiatry Neurosci.* 2020 Jan 1;45(1): 45–54.

Sato F, Lavallée P, Lévesque M, Parent A. Single-axon tracing study of neurons of the external segment of the globus pallidus in primate. *J Comp Neurol.* 2000;417(1):17–31

Saunders A, Oldenburg IA, Berezovskii VK, et al. A direct GABAergic output from the basal ganglia to frontal cortex. *Nature*. 2015;521(7550):85–89.

Schechtman E, Adler A, Deffains M, et al. Coinciding decreases in discharge rate suggest that spontaneous pauses in firing of external pallidum neurons are network driven. *J Neurosci*. 2015; 35(17):6744–6751.

Schmidt R, Herrojo Ruiz M, Kilavik BE, Lundqvist M, Starr PA, Aron AR. Beta oscillations in working memory, executive control of movement and thought, and sensorimotor function. *J Neurosci*. 2019;39(42):8231–8238.

Schneidman E, Berry MJ 2nd, Segev R, Bialek W. Weak pairwise correlations imply strongly correlated network states in a neural population. *Nature*. 2006;440(7087):1007–1012.

Schultz W. Predictive reward signal of dopamine neurons. *J Neurophysiol*. 1998 Jul;80(1):1–27.

Schultz W, Dayan P, Montague PR. A neural substrate of prediction and reward. *Science*. 1997;275(5306):1593–1599.

Schultz W, Romo R. Dopamine neurons of the monkey midbrain: contingencies of responses to stimuli eliciting immediate behavioral reactions. *J Neurophysiol*. 1990;63(3):607–624.

Schüz A, Palm G. Density of neurons and synapses in the cerebral cortex of the mouse. *J Comp Neurol*. 1989;286(4):442–455.

Schwarting RK, Huston JP. The unilateral 6-hydroxydopamine lesion model in behavioral brain research: analysis of functional deficits, recovery and treatments. *Prog Neurobiol*. 1996; 50(2–3):275–331.

Seabold GK, Daunais JB, Rau A, Grant KA, Alvarez VA. DiOLISTIC labeling of neurons from rodent and non-human primate brain slices. *J Vis Exp*. 2010;(41):2081.

Sellal F, Hirsch E, Lisovoski F, Mutschler V, Collard M, Marescaux C. Contralateral disappearance of parkinsonian signs after subthalamic hematoma. *Neurology*. 1992;42(1):255–256.

Seung S. *Connectome: How the Brain's Wiring Makes Us Who We Are*. Houghton Mifflin Harcourt; 2012.

Shadmehr R, Alaa AA. *Vigor: Neuroeconomics of Movement Control*. MIT Press; 2020.

Shen W, Flajolet M, Greengard P, Surmeier DJ. Dichotomous dopaminergic control of striatal synaptic plasticity. *Science*. 2008 Aug 8;321(5890):848–851.

Shepherd, G. Corticostriatal connectivity and its role in disease. *Nat Rev Neurosci* 2013;14:278–291.

Shepherd GM. *Foundations of the Neuron Doctrine: 25th Anniversary Edition*. Oxford University Press; 2015.

Sherman SM. Thalamus. *Scholarpedia*. 2006;1(9):1583.

Sherman SM, Guillery RW. *Exploring the Thalamus and Its Role in the Cortical Function*. Academic Press; 2005.

Shinbrot T, Young W. Why decussate? Topological constraints on 3D wiring. *Anat Rec* (Hoboken). 2008;291(10):1278–1292.

Shink E, Bevan MD, Bolam JP, Smith Y. The subthalamic nucleus and the external pallidum: two tightly interconnected structures that control the output of the basal ganglia in the monkey. *Neuroscience*. 1996;73(2):335–357.

Shink E, Smith Y. Differential synaptic innervation of neurons in the internal and external segments of the globus pallidus by the GABA- and glutamate-containing terminals in the squirrel monkey. *J Comp Neurol.* 1995;358(1):119–141.

Singh A, Mewes K, Gross RE, DeLong MR, Obeso JA, Papa SM. Human striatal recordings reveal abnormal discharge of projection neurons in Parkinson's disease. *Proc Natl Acad Sci USA.* 2016;113(34):9629–9634.

Singh A, Papa SM. Striatal oscillations in parkinsonian non-human primates. *Neuroscience.* 2020 Nov 21;449:116–122.

Sirleaf EJ. *This Child Will Be Great: Memoir of a Remarkable Life by Africa's First Woman President.* HarperCollins; 2010.

Sizemore RJ, Zhang R, Lin N, et al. Marked differences in the number and type of synapses innervating the somata and primary dendrites of midbrain dopaminergic neurons, striatal cholinergic interneurons, and striatal spiny projection neurons in the rat. *J Comp Neurol.* 2016;524(5):1062–1080.

Slonim N, Friedman N, Tishby N. Multivariate information bottleneck. *Neural Comput.* 2006;18(8):1739–1789.

Slovik M, Rosin B, Moshel S, et al. Ketamine induced converged synchronous gamma oscillations in the cortico-basal ganglia network of nonhuman primates. *J Neurophysiol.* 2017;118(2):917–931.

Smith MA, Kohn A. Spatial and temporal scales of neuronal correlation in primary visual cortex. *J Neurosci.* 2008;28:12591–12603.

Smith Y, Bevan MD, Shink E, Bolam JP. Microcircuitry of the direct and indirect pathways of the basal ganglia. *Neuroscience.* 1998;86(2):353–387.

Smith Y, Parent A. Neurons of the subthalamic nucleus in primates display glutamate but not GABA immunoreactivity. *Brain Res.* 1988;453(1–2):353–356.

Smith Y, Raju DV, Pare JF, Sidibe M. The thalamostriatal system: a highly specific network of the basal ganglia circuitry. *Trends Neurosci.* 2004;27(9):520–527.

Smith Y, Wichmann T, DeLong MR. Synaptic innervation of neurones in the internal pallidal segment by the subthalamic nucleus and the external pallidum in monkeys. *J Comp Neurol.* 1994;343(2):297–318.

Song P, Mabrouk OS, Hershey ND, Kennedy RT. In vivo neurochemical monitoring using benzoyl chloride derivatization and liquid chromatography-mass spectrometry. *Anal Chem.* 2012;84:412–419.

Sotelo C, Javoy F, Agid Y, Glowinski J. Injection of 6-hydroxydopamine in the substantia nigra of the rat. I. Morphological study. *Brain Res.* 1973;58(2):269–290.

Spinney L. *Pale Rider: The Spanish Flu of 1918 and How It Changed the World.* Jonathan Cape; 2018.

Sporns O. *Networks of the Brain.* MIT Press; 2012.

Steinfels GF, Heym J, Strecker RE, Jacobs BL. Behavioral correlates of dopaminergic unit activity in freely moving cats. *Brain Res.* 1983;258(2):217–228.

Steinmetz NA, Koch C, Harris KD, Carandini M. Challenges and opportunities for large-scale electrophysiology with Neuropixels probes. *Curr Opin Neurobiol.* 2018;50:92–100.

Stephenson AR, Edler MK, Erwin JM, et al. Cholinergic innervation of the basal ganglia in humans and other anthropoid primates. *J Comp Neurol.* 2017;525(2):319–332.

Steriade M, Jones EG, Llinás RR. *Thalamic Oscillations and Signaling*. John Wiley and Sons; 1990.

Steriade M, McCarley RW. *Brainstem Control of Wakefulness and Sleep*. Plenum; 1990.

Stochl J, Boomsma A, Ruzicka E, Brozova H, Blahus P. On the structure of motor symptoms of Parkinson's disease. *Mov Disord*. 2008 Jul 15;23(9):1307–1312.

Straub C, Tritsch NX, Hagan NA, Gu C, Sabatini BL. Multiphasic modulation of cholinergic interneurons by nigrostriatal afferents. *J Neurosci*. 2014;34(25):8557–8569.

Strausfeld NJ, Hirth F. Deep homology of arthropod central complex and vertebrate basal ganglia. *Science*. 2013;340(6129):157–161.

Strogatz SH. *Sync: How Order Emerges from Chaos in the Universe, Nature, and Daily Life*. Hyperion; 2003.

Strogatz SH. *Nonlinear Dynamics and Chaos*. Perseus Books; 2014.

Surmeier DJ, Reiner A, Levine MS, Ariano MA. Are neostriatal dopamine receptors co-localized? *Trends Neurosci*. 1993;16(8):299–305.

Sutton RS, Barto AG. *Reinforcement Learning: An Introduction*. 2nd ed. MIT Press; 2018.

Tamaki M, Bang JW, Watanabe T, Sasaki Y. Night watch in one brain hemisphere during sleep associated with the first-night effect in humans. *Curr Biol*. 2016;26(9):1190–1194.

Taverna S, Ilijic E, Surmeier DJ. Recurrent collateral connections of striatal medium spiny neurons are disrupted in models of Parkinson's disease. *J Neurosci*. 2008;28(21):5504–5512.

Tecuapetla F, Jin X, Lima SQ, Costa RM. Complementary contributions of striatal projection pathways to action initiation and execution. *Cell*. 2016;166(3):703–715.

Telkes I, Viswanathan A, Jimenez-Shahed J, Abosch A, Ozturk M, Gupte A, Jankovic J, Ince NF. Local field potentials of subthalamic nucleus contain electrophysiological footprints of motor subtypes of Parkinson's disease. *Proc Natl Acad Sci USA*. 2018 Sep 4;115(36):E8567–E8576.

Tepper JM, Koós T, Wilson CJ. GABAergic microcircuits in the neostriatum. *Trends Neurosci*. 2004;27(11):662–669.

Thompson JA, Oukal S, Bergman H, Ojemann S, Hebb AO, Hanrahan S, Israel Z, Abosch A. Semi-automated application for estimating subthalamic nucleus boundaries and optimal target selection for deep brain stimulation implantation surgery. *J Neurosurg*. 2018 May 18:1–10.

Thomson W. *Popular Lectures and Addresses*. Volume 1. MacMillan; 1891.

Threlfell S, Lalic T, Platt NJ, Jennings KA, Deisseroth K, Cragg SJ. Striatal dopamine release is triggered by synchronized activity in cholinergic interneurons. *Neuron*. 2012;75(1):58–64.

Tinkhauser G, Pogosyan A, Little S, et al. The modulatory effect of adaptive deep brain stimulation on beta bursts in Parkinson's disease. *Brain*. 2017;140(4):1053–1067.

Tinkhauser G, Pogosyan A, Tan H, Herz DM, Kühn AA, Brown P. Beta burst dynamics in Parkinson's disease OFF and ON dopaminergic medication. *Brain*. 2017;140(11):2968–2981.

Tkačik G, Marre O, Amodei D, Schneidman E, Bialek W, Berry MJ 2nd. Searching for collective behavior in a large network of sensory neurons. *PLoS Comput Biol*. 2014;10(1):e1003408.

Tomer R, Aharon-Peretz J. Novelty seeking and harm avoidance in Parkinson's disease: effects of asymmetric dopamine deficiency. *J Neurol Neurosurg Psychiatry*. 2004;75(7):972–975.

Trautmann EM, Stavisky SD, Lahiri S, et al. Accurate estimation of neural population dynamics without spike sorting. *Neuron*. 2019;103(2):292–308.e4.

Tremblay L, Filion M, Bédard PJ. Responses of pallidal neurons to striatal stimulation in monkeys with MPTP-induced parkinsonism. *Brain Res.* 1989;498(1):17–33.

Triarhou LC. The cytoarchitectonic map of Constantin von Economo and Georg N. Koskinas. In: Geyer S, Turner R, eds. *Microstructural Parcellation of the Human Cerebral Cortex.* Heidelberg; 2013:33–53.

Trulson ME, Preussler DW, Howell GA. Activity of substantia nigra units across the sleep-waking cycle in freely moving cats. *Neurosci Lett.* 1981;26(2):183–188.

Tunstall MJ, Oorschot DE, Kean A, Wickens JR. Inhibitory interactions between spiny projection neurons in the rat striatum. *J Neurophysiol.* 2002;88(3):1263–1269.

Turner RS, DeLong MR. Corticostriatal activity in primary motor cortex of the macaque. *J Neurosci.* 2000;20(18):7096–7108.

Valsky D, Blackwell KT, Tamir I, Eitan R, Bergman H, Israel Z. Real-time machine learning classification of pallidal borders during deep brain stimulation surgery. *J Neural Eng.* 2020 Jan 6;17(1):016021.

Valsky D, Heiman-Grosberg S, Israel Z, Boraud T, Bergman H, Deffains M. What is the true discharge rate and pattern of the striatal projection neurons in Parkinson's disease and dystonia? *Elife.* 2020 Aug 19;9:e57445.

van der Hoorn A, Burger H, Leenders KL, de Jong BM. Handedness correlates with the dominant Parkinson side: a systematic review and meta-analysis. *Mov Disord.* 2012;27(2):206–210.

Van Der Kooy D, Hattori T. Single subthalamic nucleus neurons project to both the globus pallidus and substantia nigra in rat. *J Comp Neurol.* 1980;192(4):751–768.

van Vreeswijk C, Sompolinsky H. Chaos in neuronal networks with balanced excitatory and inhibitory activity. *Science.* 1996;274(5293):1724–1726.

Végh AMD, Duim SN, Smits AM, et al. Part and parcel of the cardiac autonomic nerve system: unravelling its cellular building blocks during development. *J Cardiovasc Dev Dis.* 2016 Sep 12;3(3):28.

Videnovic A, Högl B. *Disorders of Sleep and Circadian Rhythms in Parkinson's Disease.* Springer; 2015.

Vigneswaran G, Kraskov A, Lemon RN. Large identified pyramidal cells in macaque motor and premotor cortex exhibit "thin spikes": implications for cell type classification. *J Neurosci.* 2011;31(40):14235–14242.

Vilensky JA. *Encephalitis Lethargica: During and after the Epidemic.* Oxford University Press; 2010.

Villalba RM, Wichmann T, Smith Y. Neuronal loss in the caudal intralaminar thalamic nuclei in a primate model of Parkinson's disease. *Brain Struct Funct.* 2014;219(1):381–394.

von Bartheld CS, Bahney J, Herculano-Houzel S. The search for true numbers of neurons and glial cells in the human brain: a review of 150 years of cell counting. *J Comp Neurol.* 2016; 524(18):3865–3895.

Vyazovskiy VV, Olcese U, Hanlon EC, Nir Y, Cirelli C, Tononi G. Local sleep in awake rats. *Nature.* 2011;472(7344):443–447.

Waldert S, Lemon RN, Kraskov A. Influence of spiking activity on cortical local field potentials. *J Physiol.* 2013;591(21):5291–5303.

Walker M. *Why We Sleep.* Penguin Books; 2017.

Walther S, Strik W. Motor symptoms and schizophrenia. *Neuropsychobiology*. 2012;66(2):77–92.

Wang Q, Akram H, Muthuraman M, Gonzalez-Escamilla G, Sheth SA, Oxenford S, Yeh FC, Groppa S, Vanegas-Arroyave N, Zrinzo L, Li N, Kühn A, Horn A. Normative vs. patient-specific brain connectivity in deep brain stimulation. *Neuroimage*. 2020 Aug 27:117307.

Watabe-Uchida M, Zhu L, Ogawa SK, Vamanrao A, Uchida N. Whole-brain mapping of direct inputs to midbrain dopamine neurons. *Neuron*. 2012;74(5):858–873.

Watts RL, Mandir AS. The role of motor cortex in the pathophysiology of voluntary movement deficits associated with parkinsonism. *Neurol Clin*. 1992;10(2):451–469.

Wei Z, Lin BJ, Chen TW, Daie K, Svoboda K, Druckmann S. A comparison of neuronal population dynamics measured with calcium imaging and electrophysiology. *PLoS Comput Biol*. 2020;16(9):e1008198.

Welniak-Kaminska M, Fiedorowicz M, Orzel J, et al. Volumes of brain structures in captive wild-type and laboratory rats: 7T magnetic resonance in vivo automatic atlas-based study. *PLoS One*. 2019;14(4):e0215348.

Westly E. Different shades of blue. *Scientific American Special Editions*. 2012 May;21:2s:34–41.

Wichmann T, Bergman H, DeLong MR. The primate subthalamic nucleus. III. Changes in motor behavior and neuronal activity in the internal pallidum induced by subthalamic inactivation in the MPTP model of parkinsonism. *J Neurophysiol*. 1994;72(2):521–530.

Wichmann T, Devergnas A. A novel device to suppress electrical stimulus artifacts in electrophysiological experiments. *J Neurosci Methods*. 2011;201(1):1–8.

Wickens J. *A Theory of the Striatum*. Pergamon Press; 1993.

Wightman RM, Heien ML, Wassum KM, Sombers LA, Aragona BJ, Khan AS, Ariansen JL, Cheer JF, Phillips PE, Carelli RM. Dopamine release is heterogeneous within microenvironments of the rat nucleus accumbens. *Eur J Neurosci*. 2007;26(7):2046–2054.

Wijesinghe R, Protti DA, Camp AJ. Vestibular interactions in the thalamus. *Front Neural Circuits*. 2015;9:79.

Williams CC, Kappen M, Hassall CD, Wright B, Krigolson OE. Thinking theta and alpha: mechanisms of intuitive and analytical reasoning. *Neuroimage*. 2019;189:574–580.

Williams NR, Foote KD, Okun MS. STN vs. GPi deep brain stimulation: translating the rematch into clinical practice. *Mov Disord Clin Pract*. 2014;1(1):24–35.

Williams RW, Rakic P. Elimination of neurons from the rhesus monkey's lateral geniculate nucleus during development. *J Comp Neurol*. 1988;272(3):424–436.

Wilson CJ, Chang HT, Kitai ST. Firing patterns and synaptic potentials of identified giant aspiny interneurons in the rat neostriatum. *J Neurosci*. 1990;10(2):508–519.

Wong JE, Cao J, Dorris DM, Meitzen J. Genetic sex and the volumes of the caudate-putamen, nucleus accumbens core and shell: original data and a review. *Brain Struct Funct*. 2016; 221(8):4257–4267.

Wongmassang W, Hasegawa T, Chiken S, Nambu A. Weakly correlated activity of pallidal neurons in behaving monkeys [published online ahead of print, 2020 Jul 10]. *Eur J Neurosci*. 2020;10.1111/ejn.14903.

Woolley SC, Rajan R, Joshua M, Doupe AJ. Emergence of context-dependent variability across a basal ganglia network. *Neuron*. 2014;82(1):208–223.

Wu Y, Richard S, Parent A. The organization of the striatal output system: a single-cell juxtacellular labeling study in the rat. *Neurosci Res*. 2000;38(1):49–62.

Yelnik J. Functional anatomy of the basal ganglia. *Mov Disord*. 2002;17(Suppl 3):S15–S21.

Yelnik J, François C, Percheron G, Tandé D. Morphological taxonomy of the neurons of the primate striatum. *J Comp Neurol*. 1991;313(2):273–294.

Yelnik J, François C, Percheron G, Tandé D. A spatial and quantitative study of the striatopallidal connection in the monkey. *Neuroreport*. 1996;7(5):985–988.

Yelnik J, Percheron G. Subthalamic neurons in primates: a quantitative and comparative analysis. *Neuroscience*. 1979;4(11):1717–1743.

Yelnik J, Percheron G, François C. A Golgi analysis of the primate globus pallidus. II. Quantitative morphology and spatial orientation of dendritic arborizations. *J Comp Neurol*. 1984; 227(2):200–213.

Yger P, Spampinato GL, Esposito E, et al. A spike sorting toolbox for up to thousands of electrodes validated with ground truth recordings in vitro and in vivo. *Elife*. 2018;7:e34518.

Yi G, Grill WM. Average firing rate rather than temporal pattern determines metabolic cost of activity in thalamocortical relay neurons. *Sci Rep*. 2019;9(1):6940.

Yizhar O, Fenno LE, Prigge M, et al. Neocortical excitation/inhibition balance in information processing and social dysfunction. *Nature*. 2011;477(7363):171–178.

Yoshida M, Rabin A, Anderson M. Two types of monsynaptic inhibition of pallidal neurons produced by stimulation of the diencephalon and substantia nigra. *Brain Res*. 1971;30(1):235–239.

Yu C, Cassar IR, Sambangi J, Grill WM. Frequency-specific optogenetic deep brain stimulation of subthalamic nucleus improves parkinsonian motor behaviors. *J Neurosci*. 2020;40(22):4323–4334.

Yust-Katz S, Tesler D, Treves TA, Melamed E, Djaldetti R. Handedness as a predictor of side of onset of Parkinson's disease. *Parkinsonism Relat Disord*. 2008;14(8):633–635.

Zahm DS, Brog JS. On the significance of subterritories in the "accumbens" part of the rat ventral striatum. *Neuroscience*. 1992;50(4):751–767.

Zaidel A, Spivak A, Grieb B, Bergman H, Israel Z. Subthalamic span of beta oscillations predicts deep brain stimulation efficacy for patients with Parkinson's disease. *Brain*. 2010;133(Pt 7):2007–2021.

Zhang YF, Reynolds JNJ, Cragg SJ. Pauses in cholinergic interneuron activity are driven by excitatory input and delayed rectification, with dopamine modulation. *Neuron*. 2018;98(5):918–925.e3.

Zheng T, Wilson CJ. Corticostriatal combinatorics: the implications of corticostriatal axonal arborizations. *J Neurophysiol*. 2002;87(2):1007–1017.

Zhuangzi. *The Complete Works of Zhuangzi*. Translated by Burton Watson. Columbia University Press; 2013.

Zigmond MJ, Stricker EM. Deficits in feeding behavior after intraventricular injection of 6-hydroxydopamine in rats. *Science*. 1972;177(4055):1211–1214.

INDEX

Note: Figures and tables are indicated by "f" and "t" respectively, following page numbers.